Energy, Environment, and Sustainability

This books series publishes cutting edge monographs and professional books focused on all aspects of energy and environmental sustainability, especially as it relates to energy concerns. The Series is published in partnership with the International Society for Energy, Environment, and Sustainability. The books in these series are editor or authored by top researchers and professional across the globe. The series aims at publishing state-of-the-art research and development in areas including, but not limited to:

- Renewable Energy
- Alternative Fuels
- Engines and Locomotives
- Combustion and Propulsion
- Fossil Fuels
- Carbon Capture
- Control and Automation for Energy
- Environmental Pollution
- Waste Management
- Transportation Sustainability

More information about this series at http://www.springer.com/series/15901

Himanshu Tyagi · Avinash Kumar Agarwal
Prodyut R. Chakraborty · Satvasheel Powar
Editors

Applications of Solar Energy

Springer

Editors
Himanshu Tyagi
Department of Mechanical Engineering
Indian Institute of Technology Ropar
Rupnagar, Punjab
India

Prodyut R. Chakraborty
Department of Mechanical Engineering
Indian Institute of Technology Jodhpur
Jodhpur, Rajasthan
India

Avinash Kumar Agarwal
Department of Mechanical Engineering
Indian Institute of Technology Kanpur
Kanpur, Uttar Pradesh
India

Satvasheel Powar
School of Engineering
Indian Institute of Technology Mandi
Mandi, Himachal Pradesh
India

ISSN 2522-8366 ISSN 2522-8374 (electronic)
Energy, Environment, and Sustainability
ISBN 978-981-10-7205-5 ISBN 978-981-10-7206-2 (eBook)
https://doi.org/10.1007/978-981-10-7206-2

Library of Congress Control Number: 2017957678

Printed on acid-free paper

This Springer imprint is published by Springer Nature
The registered company is Springer Nature Singapore Pte Ltd.
The registered company address is: 152 Beach Road, #21-01/04 Gateway East, Singapore 189721, Singapore

Foreword

The monograph titled 'Applications of Solar Energy,' edited by Dr. Himanshu Tyagi, Prof. Avinash K. Agarwal, Dr. Prodyut R. Chakraborty, and Dr. Satvasheel Powar, is a collection of articles written by experts in this field. These articles are in fact the review of research work carried out in the recent past by the respective experts and presented in the recently held *First ISEES International Conference on 'Sustainable Energy and Environmental Challenges' (SEEC-2017)* from February 26 to 28, 2017, at Center of Innovative and Applied Bioprocessing (CIAB), Mohali, Punjab, India. Usually, such monographs suffer from being incongruent collection of research works, but in the current volume, articles are chosen keeping observation on the continuity and consistency of the theme of the monograph. The monograph covering various topics on diverse methods of utilizing efficiently the solar energy will be particularly useful to research students starting their work in this field. Most of the articles deal with different experimental techniques of using solar energy through thermal processes. One chapter deals with photovoltaic technique of using solar energy with maximum possible efficiency. Some chapters include computer simulation of two- and three-dimensional model systems for power generation from solar energy. In India, we have a number of research groups in various institutions working in these fields.

The editorial team has put tremendous effort in bringing out this monograph, especially in the review process of all chapters emanating from the short presentations at the conference. It is heartening to see young researchers in this area spending their valuable time in bringing out such a useful document for the benefit of this community. I wish it all the success.

Prof. Sarit K. Das
Director
Indian Institute of Technology Ropar, Rupnagar
India

Preface

Energy demand has been rising remarkably due to increasing population and urbanization. Global economy and society are significantly dependent on the energy availability because it touches every facet of human life and its activities. Transportation and power generation are major examples of energy. Without the transportation by millions of personalized and mass transport vehicles and availability of 24×7 power, human civilization would not have reached contemporary living standards.

First International Conference on 'Sustainable Energy and Environmental Challenges' (SEEC-2017) was organized under the auspices of 'International Society for Energy and Environmental Sustainability' (ISEES) by the 'Center of Innovative and Applied Bioprocessing' (CIAB), Mohali, from February 26 to 28, 2017. ISEES was founded at IIT Kanpur in January 2014 with an aim of spreading knowledge in the fields of energy, environment, sustainability, and combustion. The Society's goal is to contribute to the development of clean, affordable, and secure energy resources and a sustainable environment for the society and to spread knowledge in the above-mentioned areas and spread awareness about the environmental challenges, which the world is facing today. ISEES is involved in various activities such as conducting workshops, seminars, conferences in the domains of its interests. The Society also recognizes the outstanding works done by the young scientists and engineers for their contributions in these fields by conferring them awards under various categories.

This conference provided a platform for discussions between eminent scientists and engineers from various countries including India, USA, South Korea, Norway, Malaysia, and Australia. In this conference, eminent speakers from all over the world presented their views related to different aspects of energy, combustion, emissions, and alternative energy resource for sustainable development and cleaner environment. The conference started with four mini-symposiums on very topical themes, which included (i) New Fuels and Advanced Engine Combustion, (ii) Sustainable Energy, (iii) Experimental and Numerical Combustion, and (iv) Environmental Remediation and Rail Road Transport. The conference had 14 technical sessions on topics related to energy and environmental sustainability and a panel discussion on 'Challenges, Opportunities and Directions of Technical Education and Research in the Area of Energy, Environment and Sustainability' to wrap up the three-day technical extravaganza. The conference included 2 plenary talks, 12 keynote talks, 42 invited talks from prominent scientists, 49 contributed talks, and 120 posters. A total of 234 participants and speakers attended this three-day conference, which hosted Dr. V.K. Saraswat, Member NITI Ayog, India, as a chief guest for the award ceremony of the ISEES. This conference laid out the road map for technology development, opportunities, and challenges in this technology domain. The technical sessions in the conference included Advances in IC Engines and Fuels; Conversion of Biomass to Biofuels; Combustion Processes; Renewable Energy: Prospects and Technologies; Waste to Wealth-Chemicals and Fuels; Energy Conversion Systems; Numerical Simulation of Combustion Processes; Alternate Fuels for IC Engines; Sprays and Heterogeneous Combustion of Coal/Biomass; Biomass Conversion to Fuels and Chemicals—Thermochemical Processes; Utilization of Biofuels; and Environmental Protection and Health. All these topics are very relevant for the country and the world in the present context. The society is grateful to Prof. Ashok Pandey for organizing and hosting this conference, which led to germination of this series of monographs, which included 16 books related to different aspects of energy, environment, and sustainability. This is the first time that such voluminous and high-quality outcome has been achieved by any society in India from one conference.

The editors would like to express their sincere gratitude to the authors for submitting their work in a timely manner and revising it appropriately at short notice. We would like to express our special thanks to Dr. Chandan Banerjee, Dr. Anirban Bhattacharya, Dr. Santosh B. Bopche, Dr. Laltu Chandra, Dr. Sudipta De, Dr. Ambesh Dixit, Dr. Varun Goel, Dr. D.B. Jani, Dr. Shyam Karagadde, Dr. Vikrant Khullar, Dr. Pradeep Kumar, Dr. Manish Mishra, Dr. Dibakar Rakshit, Dr. Sandip Saha, Dr. Harjit Singh, Dr. Sanjeev Soni, and Dr. S. Suresh, who reviewed various chapters of this monograph and provided their valuable suggestions for improving the manuscripts. We acknowledge the support received from various funding agencies and organizations for the successful conduct of the first ISEES conference SEEC-2017, where these monographs germinated. These include Department of Science and Technology, Government of India (special thanks to Dr. Sanjay Bajpai); TSI, India (special thanks to Dr. Deepak Sharma); Tesscorn, India (special thanks to Sh. Satyanarayana); AVL India; Horiba, India; Springer (special thanks to Swati Mehershi); CIAB (special thanks to Dr. Sangwan).

This monograph is intended for researchers working in the field of solar energy, and we hope that the book would be of great interest to the professionals and postgraduate students involved in the study of solar thermal collectors, energy storage systems (using phase change materials), and power generation using solar energy. The main objective of this monograph is to promote a better and more accurate understanding of the various applications of solar energy.

Rupnagar, India	Himanshu Tyagi
Kanpur, India	Avinash Kumar Agarwal
Jodhpur, India	Prodyut R. Chakraborty
Mandi, India	Satvasheel Powar

Contents

Editors and Contributors

About the Editors

Dr. Himanshu Tyagi is currently working as Associate Professor in the School of Mechanical, Materials and Energy Engineering, Indian Institute of Technology Ropar. He has previously worked at the Steam Turbine Design Division of Siemens (Germany and India) and at the Thermal and Fluids Core Competency Group of Intel Corp (USA). He received his Ph.D. from Arizona State University, USA, in the field of heat transfer and specifically looked for the radiative and ignition properties of nanofluids. He and his co-workers proposed the concept of direct absorption solar collectors using nanofluids which won the Best Paper Award at the ASME Energy Sustainability Conference at Long Beach, CA. He also studied the ignition properties of nanoparticle-laden diesel droplets. The publication based on this work has appeared in Nano Letters and has been highlighted in Nature and Chemical and Engineering News. He obtained his master's degree from University of Windsor, Canada, and his bachelor's from IIT Delhi, both in the field of Mechanical Engineering. At present, he is working to develop nanotechnology-based clean and sustainable energy sources with a team of several Ph.D., postgraduate, and undergraduate students. Among other awards, he has received Summer Undergraduate Research Award (SURA) from IIT Delhi, International Graduate Student Scholarship from University of Windsor Canada, Indo-US Science and Technology Forum (IUSSTF) grant awarded for organizing an Indo-US Workshop on 'Recent Advances in Micro/Nanoscale Heat Transfer and Applications in Clean Energy Technologies' at IIT Ropar during December 2013.

Professor Avinash Kumar Agarwal joined IIT Kanpur in 2001 and is currently a Poonam and Prabhu Goyal Endowed Chair Professor. He was at ERC, University of Wisconsin, Madison, USA, as a Postdoctoral Fellow (1999–2001). His areas of interest are IC engines, combustion, alternative fuels, hydrogen, conventional fuels, lubricating oil tribology, optical diagnostics, laser ignition, HCCI, emission and particulate control, and large bore engines. He has published more than 160 peer-reviewed international journals and conference papers. He is Associate Editor of ASME Journal of Energy Resources Technology and International Journal of Vehicle Systems Modelling and Testing. He has edited 'Handbook of Combustion' (5 volumes; 3168 pages), published by Wiley VCH, Germany. He is a Fellow of SAE (2012), Fellow of ASME (2013), and a Fellow of INAE (2015). He is the recipient of several prestigious awards such as NASI-Reliance Industries Platinum Jubilee Award-2012; INAE Silver Jubilee Young Engineer Award-2012; Dr. C.V. Raman Young Teachers Award-2011; SAE International's Ralph R. Teetor Educational Award-2008; INSA Young Scientist Award-2007; UICT Young Scientist Award-2007; INAE Young Engineer Award-2005. He is the recipient of Prestigious Shanti Swarup Bhatnagar Award-2016 in Engineering Sciences. He is the first combustion/IC engine researcher to get this honor.

Dr. Prodyut R. Chakraborty is an Assistant Professor in the Mechanical Engineering, IIT Jodhpur, since February 2013. He received his bachelor's degree in Mechanical Engineering from the North Bengal University in 2000 and his M.Sc. in Engineering in 2004 and Ph.D. in 2011 both from the Department of Mechanical Engineering, Indian Institute of Science, Bangalore. Prior to his joining at IIT Jodhpur, he served two years at the Department of Material Physics in Space in German Aerospace Center (DLR), Cologne, as a postdoctoral research fellow with DLR fellowship. He also worked as a Research Analyst at the Applied CFD Lab, G.E. Global Research Center, Bangalore, during the years 2004 and 2005. His primary area of research is numerical modeling of alloy solidification, latent heat-based energy storage systems for high-temperature applications, thermal management and thermal comfort, and sorption cooling.

Dr. Satvasheel Powar is an Assistant Professor at the School of Engineering, IIT Mandi, since June 2015. He received bachelor's degree in Production Engineering from Shivaji University in 2003 and a M.Sc. (Mechanical Engineering) from the Dalarna University, Sweden, in 2005. He then served Greatcell Solar S.A., Switzerland, and G24i, UK, for about four years in total with a work focus on scalable process development. He received his Ph.D. in Chemistry/Materials Engineering from the Monash University, Australia, in 2013. Before his joining at IIT Mandi, he served two and half years at the Nanyang Technological University, Singapore, as a postdoctoral research fellow. His primary areas of research are new-generation solar photovoltaic and solar thermal utilization.

Contributors

Avinash Kumar Agarwal Department of Mechanical Engineering, Indian Institute of Technology Kanpur, Kanpur, Uttar Pradesh, India

Ankit Agarwal Numerical Experiment Laboratory (Radiation & Fluid Flow Physics), Indian Institute of Technology Mandi, Suran, Himachal Pradesh, India

Vishal Bhalla Department of Mechanical Engineering, Indian Institute of Technology Ropar, Rupnagar, Punjab, India

Santosh B. Bopche Department of Mechanical Engineering, National Institute of Technology Hamirpur, Hamirpur, HP, India

Prodyut Ranjan Chakraborty Department of Mechanical Engineering, Indian Institute of Technology Jodhpur, Karwad, Jodhpur, Rajasthan, India

Laltu Chandra Department of Mechanical Engineering, Indian Institute of Technology Jodhpur, Jodhpur, India; Centre for Solar Energy, Indian Institute of Technology Jodhpur, Jodhpur, India

Sudipta De Department of Mechanical Engineering, Jadavpur University, Kolkata, India

Ambesh Dixit Department of Physics, Indian Institute of Technology Jodhpur, Jodhpur, India; Centre for Solar Energy, Indian Institute of Technology Jodhpur, Jodhpur, India

Kunal Ghosh School of Computing and Electrical Engineering, Indian Institute of Technology Mandi, Mandi, Himachal Pradesh, India

Varun Goel Mechanical Engineering Department, National Institute of Technology, Hamirpur, India

D. B. Jani Gujarat Technological University, G.T.U., Ahmedabad, Gujarat, India

Priyanka Kajal School of Engineering, Indian Institute of Technology Mandi, Mandi, Himachal Pradesh, India

Mehmet Karakilcik Faculty of Sciences and Letters, Department of Physics, University of Cukurova, Adana, Turkey

Vikrant Khullar Department of Mechanical Engineering, Thapar University, Patiala, Punjab, India

Pradeep Kumar Numerical Experiment Laboratory (Radiation & Fluid Flow Physics), Indian Institute of Technology Mandi, Suran, Himachal Pradesh, India

Rajesh Kumar Department of Physics, Indian Institute of Technology Jodhpur, Jodhpur, India; Centre for Solar Energy, Indian Institute of Technology Jodhpur, Jodhpur, India

Suneet Kumar Department of Mechanical Engineering, National Institute of Technology Hamirpur, Hamirpur, HP, India

Balkrishna Mehta Department of Mechanical Engineering, Indian Institute of Technology Guwahati, Guwahati, Assam, India

Manish Mishra Department of Mechanical & Industrial Engineering, I.I.T., Roorkee, Uttarakhand, India

Aniket D. Monde Department of Mechanical Engineering, Indian Institute of Technology Jodhpur, Jodhpur, India

Dhiraj V. Patil School of Engineering, Indian Institute of Technology Mandi, Mandi, Himachal Pradesh, India

Satvasheel Powar School of Engineering, Indian Institute of Technology Mandi, Mandi, Himachal Pradesh, India

Dibakar Rakshit CES, IIT Delhi, New Delhi, India

Avishek Ray Department of Mechanical Engineering, Jadavpur University, Kolkata, India

P. K. Sahoo Department of Mechanical & Industrial Engineering, I.I.T., Roorkee, Uttarakhand, India

Prashant Saini School of Engineering, Indian Institute of Technology Mandi, Mandi, Himachal Pradesh, India

Srikanth Salyan Department of Mechanical Engineering, National Institute of Technology, Tiruchirappalli, Tamil Nadu, India

Satish Indian Institute of Technology Jodhpur, Jodhpur, India

Abhishek Saxena Mechanical Engineering Department, Moradabad Institute of Technology, Moradabad, India

Lakshya Sharma CES, IIT Delhi, New Delhi, India

Amit Shrivastava Department of Mechanical Engineering, Indian Institute of Technology Jodhpur, Jodhpur, India

Gurveer Singh Department of Mechanical Engineering, Indian Institute of Technology Jodhpur, Jodhpur, India; Centre for Solar Energy, Indian Institute of Technology Jodhpur, Jodhpur, India

Harjit Singh College of Engineering, Design and Physical Sciences, Brunel University London, Uxbridge, Middlesex, UK

S. Suresh Department of Mechanical Engineering, National Institute of Technology, Tiruchirappalli, Tamil Nadu, India

Himanshu Tyagi Department of Mechanical Engineering, Indian Institute of Technology Ropar, Rupnagar, Punjab, India

Ashok Verma Indian Institute of Technology Jodhpur, Jodhpur, India

Part I
General

Introduction to Applications of Solar Energy

Himanshu Tyagi, Avinash Kumar Agarwal, Prodyut Ranjan Chakraborty and Satvasheel Powar

Abstract We know that on the global scale till now the major sources of energy are fossil fuels like coal, oil, and natural gases. Because of following three reasons, it is pertinent to use the naturally available renewable energy in the form of solar energy to replace fossil fuel energy: (a) With the advent of science and technology and with ever increasing global population, the total energy consumption is increasing day by day, (b) the fossil fuel reserves world over are fast depleting, and (c) the fossil fuels emit green house gas during combustion, contributing significantly to global warming. Not surprising, during the last two, three decades, hectic research has been going on in almost all developed/developing countries to develop techniques to make maximum use of solar energy through various processes like thermal processes, photovoltaic processes. In India many research groups are actively engaged in studying utilization of solar energy in an efficient way using variety of techniques. The present monograph titled "*Applications of Solar Energy*" is the result of collection of work done by some such research groups.

H. Tyagi (✉)
Department of Mechanical Engineering, Indian Institute of Technology Ropar, Rupnagar 140001, Punjab, India
e-mail: himanshu.tyagi@iitrpr.ac.in

A. K. Agarwal
Department of Mechanical Engineering, Indian Institute of Technology Kanpur, Kanpur 208016, Uttar Pradesh, India
e-mail: akag@iitk.ac.in

P. R. Chakraborty
Department of Mechanical Engineering, Indian Institute of Technology Jodhpur, Karwad, Jodhpur 342037, Rajasthan, India
e-mail: pchakraborty@iitj.ac.in

S. Powar
School of Engineering, Indian Institute of Technology Mandi, Mandi 175005, Himachal Pradesh, India
e-mail: satvasheel@iitmandi.ac.in

H. Tyagi et al. (eds.), *Applications of Solar Energy*, Energy, Environment, and Sustainability, https://doi.org/10.1007/978-981-10-7206-2_1

Keywords Solar energy · Volumetric receiver · Direct absorption Concentration ratio · Energy storage · Phase change material Cooling · Solar cells

Energy is the key element of modern development. Fossil fuels are presently the major source of energy consumed worldwide. However, this is not sustainable. Fast depletion of limited resources and climate change problem due to emission from these fuels are of major concern. Increasing overall efficiency of energy conversion and use is one option. Moreover, exploring new systems with better utilization of renewable resources may be a possible future sustainable option. Decentralized polygeneration to meet energy and other utility demand with rational use of local resources is emerging as a suitable option. To match the varying demand with intermittent renewable resources, storage could be a good solution. However, suitable low cost and reliable storage are not yet fully developed. Hybridization of different intermittent renewable resources is a possible option. Countries with enough solar insolation like India may develop suitable hybrid polygeneration with solar inputs. Depending on local need, both solar thermal and photovoltaic utilities may be included in this polygeneration. Efficient system integration with possible optimization of different available local resources based on local needs is the key to successful polygeneration design. Multi-objective optimization of hybridization of different available resources is a useful technique for meeting local needs.

Heat transfer pays a very important role in solar thermal collectors. In case of solar central receiver systems, the major portion of incident solar radiation is reflected by the concentrating reflectors or heliostats toward the receiver by specular reflection and minimum portion is received through diffused radiation. In order to know this diffused component of radiation energy transfer, the absorptivity of reflectors as well as shape factors between the reflectors and the central receiver is must. This requires the estimation of radiative heat/energy exchange between the hot cavity receiver and reflecting collector surface along with the diffused configuration factor computation encompassing in solar reflector-collector systems of various orientations and geometries. The configuration factors are evaluated using contour integral technique, which is one of the accurate and faster tools for configuration factor estimation.

The finite element method has been proved to be the effective numerical tool for finding approximate solutions of two or even up to three-dimensional governing differential equations. It has an ability to handle irregular geometries and any number of complicated boundary conditions with ease. The cavity receiver of a solar thermal system operates at a very high temperature. It may also cause larger energy losses from the cavity by radiation as well as convection as dominant modes of heat transfer. The collection efficiency depends on the useful heat gained by the working fluids, which is influenced by the receiver energy losses. The energy losses from the cavity receiver depend on the cavity wall surface temperature. The temperature variation depends on the boundary conditions as well as geometrical orientation of the receiver. The knowledge of temperature distribution is one of the important factors needed for evolving an ideal design of a cavity receiver.

Solar collectors that can directly absorb radiation represent an emerging realm of solar thermal systems wherein the collection as well its subsequent conversion to the useful thermal energy happens within the working fluid. Nanofluids (stable dispersions of nanoparticles in the basefluid) have been found to be promising working fluids for realizing such direct solar to thermal energy conversion owing to their enhanced (and the ease of tuning) thermo-physical and optical properties. Seeding trace amounts of carefully chosen nanoparticles into the basefluid has been shown to significantly enhance the solar weighted absorptivity of the basefluid—hence rendering them suitable for solar thermal applications. A critical analysis of the fundamental limits of performance that can be achieved in the incumbent solar thermal systems reveals that solar selectivity could only be beneficial up to a certain temperature and solar concentration ratios, beyond which the efficiency cannot be further improved. Subsequently, the candidature of direct absorption solar thermal systems need to be assessed to ascertain if these could be deployed under conditions which are not so amenable for the conventional surface absorption-based solar thermal technologies. For this experimental studies show that even for low solar concentration ratios (conditions which are more suitable for the conventional surface-based absorbers), the two classes of solar thermal technologies can have comparable thermal efficiencies. It is envisaged that the benefits of the direct absorption-based solar thermal systems over the conventional ones shall be more pronounced for high flux conditions, i.e., high solar concentration ratios.

Solar thermal systems are one of the renewable energy systems used in the residential buildings for the heating purpose, and with these systems, the usage of non renewable energy resources decreases. To improve the performance of solar collectors, engineers and scientists are regularly working on it. Direct absorption-based solar thermal collectors (DASTC) are kind of solar collectors in which the fluid can be heated directly (without any absorption surface). Numerical models are prepared for direct absorption-based solar collector which can be used for residential purposes. The absorbed energy fraction, effect of the height, length of collector, and mass flow rate on the collector efficiency are some of the important factors. The collector efficiency increases with the increase of mass flow rate when the height of the fluid in the collector is same, and the efficiency of the collector deceases with the increase of channel length. Further, it is beneficial to use an optimum volume fraction of the nanoparticles in DASTC because at an optimum volume fraction the collector achieved maximum efficiency.

The concentrated solar thermal technologies (CST) are versatile in view of their multifaceted applications, such as, process heat, cooling, and electricity generation. These are of line and point focusing types with the later having much higher flux concentration (in Suns). This allows achieving a temperature in excess of 1200 K using, for instance, the open volumetric air receiver (OVAR). Such a high temperature is useful for applications, like the one which is developed at IIT Jodhpur, namely the solar convective furnace for heat treatment of aluminum. This requires a temperature of up to 750 K in the first phase of development. Thus, a suitable solar selective coating withstanding such a high temperature and having a thermal conductivity close to the base material for operating in an open atmosphere is desirable. Because of its atmospheric exposure, air- and dust-induced degradation is inevitable, which may lead to its failure. These challenges are to be addressed for adapting such high temperature CST technologies in arid deserts of India, the Middle East and Africa. This necessitates the study of various factors such as—design of OVAR including various subcomponents, flow-stability and the effect of heat flux distribution on an absorber pore and the developed coating and its characterization for OVAR.

Solar energy is a natural source of energy that is not depleted by its use. It is a promising option for replacing conventional energy resources partially or totally, but it is transient, intermittent, and unpredictable in nature. Because of this sporadic nature of solar energy across a given interval of hours, days and season, various practical problems arise. Variable DNI causes power plants to shut down for few hours of the day or to run at part load most of the time. This creates a demand for an effective subsystem which is capable of storing energy when available solar energy overshoots the demand during the interval of radiant sunshine, and to make it accessible during night or season. A similar problem arises for waste heat recovery systems where accessibility of waste heat and usage period are not the same and thus creates a need for thermal energy storage (TES) for energy conservation. TES has tempted a lot of researchers to improve its high energy storage capacity and efficiency. If solar energy system is not running with TES, a considerable section of energy demand has to depend on conventional resources which in result reduce the annual solar fraction. TES helps to reduce dependency over conventional resource by minimizing energy waste. TES is mainly described by the parameters like capacity, power, efficiency, storage period, charge and discharge time, and cost. There are different storage mechanisms by which energy can be stored: sensible, latent, and chemical reactions. In sensible type storage, energy is stored by increasing the temperature of solid or liquid storage media (e.g., sand-rock minerals, concrete, oils, and liquid sodium). These materials have excellent thermal conductivity and are cheaper, but due to low heat capacity, it increases system size. In latent type storage, energy is stored/released during phase change; thus, it has higher storage capacity than sensible, but suffers from the issue of low thermal conductivity. As the solid–liquid phase change process of pure or eutectic substances is isothermal in nature, it is beneficial for the application having limitations with working temperature. In chemical type TES, heat is absorbed/released due to breakdown or formation of chemical bonds. The technology is not much developed and has limited application due to possibility of degradation overtime and chemical instability. TES can also be classified as active and passive depending upon the solid or liquid energy storage medium. Active TES is further classified as direct active and indirect active depending on whether the storage fluid and the heat transfer fluid (HTF) are same or some other HTF is required to extract heat from solar field.

Various solar thermal energy harvesting techniques have been used which employ solar radiation incident on the optimal area with the help of concentrators.

The solar air heater is one of the solar thermal harvesting techniques, which has gained tremendous popularity in recent times. It is important to study the design and operations of various solar air heaters. Further, due to the intermittent nature of the solar energy, energy storage becomes an integral part of the solar energy harvesting and the solar energy may be stored using latent or sensible heat thermal energy storage. The integration of a solar air heater along with low cost or efficient, sensible heat thermal energy storage system is a very desirable feature.

Energy storage materials play a very important role in the design of TES. Energy storage materials store energy in form of sensible heat, latent heat, and thermo-chemical energy storage. Compared to the various forms of energy storage, latent heat-based energy storage system can store a lot of energy at isothermal temperature during melting and can release the stored heat during solidification. Phase change material (PCM) includes organic, inorganic, and eutectic. One of the primary dis-advantages of PCMs is their very low thermal conductivity. This challenge could be overcome by the addition of high thermal conductive additives to form a composite. Thermo-physical property determination of these composites is very important to determine the feasibility of using such composites as energy storage materials. Characterization techniques like differential scanning calorimetry (DSC), Fourier transform infrared (FTIR), thermogravimetric analysis (TGA), and laser flash apparatus (LFA) are very effective and useful methods to determine the potential use of PCM in TES applications. This involves the characteristics of pentaerythritol (PE) and D-Mannitol (DM), and the effect of adding high conductive graphene nanoplatelets (GnP) to form a phase change composites.

Solar thermal resources can be effectively utilized to meet the refrigeration and air-conditioning demands for both household and industrial purposes. Considerable fraction of total available electricity is consumed by the conventional vapor compression refrigeration systems (VCRS) during the summer season in countries with tropical climate. The leakage of VCRS refrigerants in the atmosphere has also been identified as one of the major contributors toward ozone layer depletion and hence global warming. The utilization of solar thermal energy for obtaining refrigeration and air-conditioning is the key to address these issues concerning high electricity demand as well as the environmental pollution. Solar thermal energy, being one of the leading resources of green energy, can reduce the carbon footprint considerably, when used for sorption cooling process. The advantages of using sorption cooling systems powered by solar thermal energy over VCRS are twofold when we consider environmental issues. Sorption-based solar thermal cooling reduces the electricity demand for cooling to a large extent, which in turn reduces usage of fossil fuels to produce this electricity, and thus leads to low carbon footprint. Also, the refrigerants used for sorption cooling are less prone to cause ozone layer depletion. Although sorption-based refrigeration systems driven by solar thermal energy are mature technologies, wide acceptability of such cooling system is yet to be achieved. Two major limitations of sorption-based solar thermal cooling are relatively low coefficient of performance (COP) and large volume requirement. Other than these two limitations, the intermittent nature of solar thermal resource and heat exchanger and control mechanism design complicacies also pose considerable challenge. Sorption cooling technology can be broadly classified based on absorption and adsorption processes. Absorption is a volumetric phenomenon where a substance of one state gets absorbed in another substance in a different state with or without having chemical reaction, such as liquid being absorbed by solid or gas being absorbed by liquid. On the other hand, adsorption is a surface phenomenon due to physical bonding forces such as van der Waals forces between a solid surface and adjacent fluid or due to chemical bonding between the two.

Solar assisted solid desiccant–vapor compression-based hybrid space cooling system for building thermal comfort is made by integrating desiccant-based dehumidification system as well as by use of solar heating system along with conventional vapor compression refrigeration (VCR) system. This is because in humid climates, excessive humidity during summer leads to inefficiencies of conventional cooling devices. Humidity increment in ambient air as well as ventilation requirement suddenly rises the latent load of the space to be cooled. Conventional VCR systems are not effective in handling both temperature and relative humidity of cooling air independently. The application of solid desiccant-based hybrid cooling systems significantly ameliorates the humidity control irrespective of temperature of supply air. The application of renewable free energy like as use of solar energy for regenerating the desiccant used in desiccant wheel helps us to alleviate the major requirement of electric energy needed by conventional VCR air-conditioning system for hot sunny days. So, it ameliorates overall energy efficiency and reduces energy costs. Performance of solid desiccant–vapor compression-based hybrid building space comfort cooling system can be used effectively during hot and humid period from April to September. Various important parameters which influence the performance of the system are—temperatures of supply air, room air, and regeneration air, etc. Regeneration temperature is one of the most important parameters having the key role in changing the performance of desiccant-based cooling system.

Solar energy is increasing, finding potential use for food processing (e.g., solar drying). Solar food processing method can be applied as direct absorption, air heater, and a combination of direct and indirect drying by solar radiation. Therefore, this process is one of the most accessible and hence the most widespread processing technology. Traditional solar drying involves keeping products in the direct sunlight. Solar drying and cooking processes take place at different temperature and time scale, and it depends on the nature of the food or substance. The amount of solar energy that reaches to the system and design parameters determines the performance of food processing and cooking systems. The drying and cooking time duration depend on the temperature of heated air and environment. The temperature distributions, mass, and ingredient of food have an important role in the performance of dryers and cooker boxes.

Offices which are mostly operated during day-lit hours are fascinated toward incorporation of solar daylighting systems so as to get benefitted of energy savings along with intangible benefits like good health, well-being, and productivity of their occupants. But discomfort glare is usually ignored in front of rigorous energy load optimization practices and researches, or even if considered, is not properly quantified, which leads to a false evaluation of the performance of daylighting in the context of lighting energy savings. Blinds when operated with optimum visual comfort prevention measures then only can create a functional day-lit environment. This necessitates the modeling and analysis of some typical office buildings located in India. This can be carried out using drafting tool Rhinoceros 3D 5.0 and simulation tool DIVA 4.0 in order to mathematically analyze the utilization of natural lighting for the office. The analysis includes properly arranged viewpoints similar to that experienced by any occupant, being placed in simulation environment so as to visualize and evaluate realistic glare scenario. The study further deals with the observation and analysis of the illuminance patterns inside the test space and glare values for different blind positions modeled.

Solar energy can also be used for direct generation of electric power using various devices, such as solar updraft towers and solar cells. The last two chapters of this monograph deal with these types of devices, respectively. The penultimate chapter deals with the study of a solar updraft tower. Here, a full-scale three-dimensional analysis of solar updraft tower power plant in Manzanares, Spain has been performed using commercially available CFD tool ANSYS Fluent. The two equations k-ε turbulent model with standard wall function has been utilized for the fluid flow. The soil has been modeled with the consideration of the fact that temperature at 20 m depth remains constant throughout the year. The surface-to-surface radiation model is also included in the heat transfer model. The simulation has been performed for the steady state without/with radiation, the transient state with/without thermal storage on 8th of June. And water has been taken as thermal storage. It has been found that results improve considerably by including radiation effect, closely matches with the results published from the plant. There is a reduction in the maximum velocity with the thermal storage; however, sufficient energy is available in thermal storage to overcome intermittency of insolation. The strong dependency of the plant on insolation can be reduced with the thermal storage.

The last chapter of this monograph discusses about the direct production of electric power using solar cells. Perovskite solar cells (PSCs) are in focus of the solar cell development research for the last few years due to their high efficiency, cost-effective fabrication, and bandgap tunability. Perovskite solar cell efficiency sharply increased from its initial reported efficiency of 3.8% in 2009 to 22.1% in 2016. This makes PSCs as the technology with the fastest growth rate in terms of the efficiency. Different device architectures have also been developed in an attempt to improve the PSC efficiency. At lab scale, a spin coating process is employed to deposit different layers of PSCs. Though spin coating process helps to achieve high efficiency, for large-scale production viability, researchers are developing different deposition techniques. A broad range of manufacturing techniques for Perovskite-based solar cells has been tested and reported comprising drop casting, spray coating, ultrasonic spray coating, slot die coating, electrodeposition, CVD, thermal vapor deposition, vacuum deposition, screen printing, ink jet printing, etc., with different device architectures. This chapter summarizes different PSC structures along with the corresponding manufacturing techniques.

This research monograph presents both fundamental science and applied innovations on several key and emerging technologies involving applications of solar energy. Specific topics covered in the manuscript include:

- Distributed Polygeneration Using Solar Energy: A Future Sustainable Energy System for India
- Effect of Reflector Absorptivity on Radiative Heat Exchange In Case of Solar Receiver Collection Systems
- Numerical Investigation of the Temperature Distribution of a Solar Cavity-Receiver-wall Using Finite Element Method
- Direct Absorption Solar Thermal Technologies
- Solar Thermal Energy: Use of Volumetric Absorption in Domestic Applications
- Thermal and Materials Perspective on the Design of Open Volumetric air Receiver for Process Heat Applications
- Solar Thermal Energy Storage
- Review on Integration of Solar Air Heaters with Thermal Energy Storage
- Solar Thermal Energy Storage using Graphene Nanoplatelets added Phase Change Materials
- Water–Lithium Bromide Absorption Chillers for Solar Cooling
- Solar Assisted Solid Desiccant—Vapor Compression Hybrid Air-Conditioning System
- Solar Food Processing and Cooking Methodologies
- Visual Comfort based Roller Shade Control & Estimating Potential Energy Savings
- Solar Updraft Tower—A Potential for Future Renewable Power Generation: A Computational Analysis
- Manufacturing Techniques of Perovskite Solar Cells.

The topics are organized in six different sections: (i) general, (ii) heat transfer aspects of solar thermal collectors, (iii) volumetric solar thermal collectors, (iv) thermal storage of solar energy, (v) various applications of solar energy: cooling, cooking, efficient buildings, and (vi) power generation using solar energy.

Distributed Polygeneration Using Solar Energy: A Future Sustainable Energy System for India

Avishek Ray and Sudipta De

Abstract Energy is the key element of modern development. Fossil fuels are presently the major source of energy consumed worldwide. However, this is not sustainable. Fast depletion of limited resources and climate change problem due to emission from these fuels are of major concern. Increasing overall efficiency of energy conversion and use is one option. Moreover, exploring new systems with better utilization of renewable resources may be a possible future sustainable option. Decentralized polygeneration to meet energy and other utility demand with rational use of local resources is emerging as a suitable option. To match the varying demand with intermittent renewable resources, storage could be a good solution. However, suitable low-cost and reliable storage is not yet fully developed. Hybridization of different intermittent renewable resources is a possible option. Countries with enough solar insolation like India may develop suitable hybrid polygeneration with solar inputs. Depending on local need, both solar thermal and photovoltaic utilities may be included in this polygeneration. Efficient system integration with possible optimization of different available local resources based on local needs is the key to successful polygeneration design. In this chapter, a comprehensive review of state-of-the-art solar-based polygeneration will be presented. Multi-objective optimization of hybridization of different available resources to meet local needs will be included in this chapter.

Keywords Polygeneration · Intermittent · Solar · Hybridization

1 Introduction

Energy is the critical input for the advancement of human civilization. There are wide variations in the energy consumption of the developing and the developed countries and also between the rich people and the poor people of the same country.

A. Ray · S. De (✉)
Department of Mechanical Engineering, Jadavpur University, Kolkata 700032, India
e-mail: de_sudipta@rediffmail.com

H. Tyagi et al. (eds.), *Applications of Solar Energy*, Energy, Environment, and Sustainability, https://doi.org/10.1007/978-981-10-7206-2_2

The per capita energy consumption is an indicator of the human development index (HDI) [1]. So with the advancement of science and technology, the total energy consumption in the world is increasing. It is predicted that there will be a 56% rise in the global energy consumption by 2040 [2]. Even in developing countries like India, with the increase in gross domestic product (GDP) the total energy consumption has almost doubled since 2010 [3]. Till now, the major sources of energy are fossil fuels like coal, oil. These fossil fuel reserves are fast depleting. The average annual growth rate of coal and oil consumption is 0.2% and 1.8%, respectively [4]. Moreover, they emit greenhouse gas (GHG) during combustion. In 2014, almost 42% of the global electricity was generated by the coal-fired thermal power plants [5]. At the same time, the coal-fired thermal power plants are responsible for 28% global GHG emission which is a significant contributor to global warming. Even in best practices, the electricity generation from coal emits 0.33 kg of CO_2/kWh [6]. So to meet the increasing load, the share of electricity from renewable sources must be increased as a sustainable option. The two strategies identified to reduce the carbon emission from the energy sector are either increasing the energy efficiency or increasing the share of renewable in the energy sector. Solar-based polygeneration system addresses both of these two issues. Solar energy has a good potential almost in all parts of India. In India, 15% of the total installed capacity is renewable. India ranks fourth in the world in terms of the solar energy installation [7]. The solar energy has the major problem of intermittency in nature. So solar energy is hybridized with some other forms of renewable energy like wind, biomass to mitigate this issue. Polygeneration can be looked after as a superior method of sustainable resource utilization than hybridization. In polygeneration, other utilities may also be generated along with electricity as per the availability of resources to maximally accommodate the variation of the load curve, say, when the electrical load is low but the available resource is high, then some other utilities like potable water, biofuels are generated. The efficient system integration increases the overall efficiency of the system. This reduces the levelized cost of electricity and has environmental benefits. Thus, the distributed polygeneration is a good option for developing sustainable energy systems [8]. It has been seen that in the solar-based polygeneration systems, the energy and the exergy efficiencies have increased marginally, but polygeneration shows a sustainable way of production of multiple commodities from the same system [9]. Ozlu et al. [10] have shown that in a solar-based polygeneration system producing electricity, heat, freshwater, and hydrogen the exergy and the energy efficiencies are found to be 36% and 44%, respectively. The polygeneration system has reduced the annual carbon emission by 476 tons than the conventional systems generating the same utility outputs. Khalid et al. [11] designed a multi-generation system producing electricity, space heating, cooling, and hot water. The energy and the exergy efficiencies of the system are to be 91 and 34% which are higher than the energy and the exergy efficiencies of the conventional systems.

In this chapter, the potential use of solar-based polygeneration as a distributed energy system is discussed. The possible method of design of reliable power supply with intermittent solar resources is discussed here. These systems need suitable control strategies to operate efficiently and with maximum reliability.

2 Advantages and Disadvantages of Solar Energy

The solar energy has many advantages [12]. Nowadays, the electricity consumption of the whole world is increasing. So, more power has to be generated by the conventional fossil fuel-based power plants. The fossil fuel-based power plants are polluting. The global emission factor for the coal-based thermal power plants is 532 g CO_2/kWh [13]. The solar energy is non-pollutant in nature. But solar energy has some disadvantages. It is intermittently available. So there is always a mismatch between the demand and supply. The maximum load occurs during the evening when solar energy is not available. Thus, there is a need of storage systems. Hybridization is another option to solve this problem. The problems of electrochemical storage are discussed later.

From the economic point of view, also the solar energy has many advantages and disadvantages. The maintenance cost of solar energy-driven devices and systems are very low. But the initial cost of installation of the solar energy systems is very high. Hence, for the developing countries like India, formulation of proper policy is needed to promote the growth of solar energy in India. The solar energy policy will simultaneously help to meet the global emission standards along with the reduction in the dependence on fossil fuels for energy.

2.1 Essence of Formulating Solar Energy Policy

The Intergovernmental Panel on Climate Change (IPCC) opined that the GHG emission must be reduced by 50% from that of 1990 standard by 2050 in order to keep global warming below 2 °C So the global policy makers met several times to discuss this issue [14, 15]. The different climatic policies in this regard are given in Table 1. Almost all the policies were introduced by United Nations Framework Convention on Climate Change (UNFCC).

Thus, in all these summits, it was unanimously decided by all policy makers to reduce the GHG emission. The two possible ways to reduce the GHG emission identified in the energy sector are either increasing energy efficiency or increasing the share of renewable in the power mix.

The solar polygeneration addresses both of these issues simultaneously. But from the above discussions, it is observed that solar energy has both advantages and disadvantages. Moreover, the disadvantages are not related to technology only, but there are also socio-economic problems. Hence, to accelerate the growth of solar energy, many countries of the world have adopted solar policies as shown in Table 2 [4].

Table 1 Some important climate change mitigation policies

Serial no	Protocol name	Place	Year	Main feature
1	Kyoto protocol	Kyoto, Japan	1997	The Kyoto Protocol was introduced by United Nations Framework Conventions on Climate Change to restrict emission. Emission targets were set to the participating nations
2	Cancun agreements	Mexico	2010	It was decided that the developed nations will annually mobilize 100 billion USD to developing nations as climate protection fund
3	Durban platform for enhanced action	South Africa	2011	Opinion of legal framework formulation was considered to all members of the United Nations. In this agreement, an attempt was made bring the emission targets under a legal framework
4	Paris agreement	Paris, France	2016	The climate change is considered as an urgent threat. The specific areas of fund and technology transfer of the developing nations are identified. The protocols of the technology transfer was vested on the individual nations

Table 2 Solar policies in different countries of the world

Name of country	Salient features
USA	• Incentives for solar energy projects under Public Utilities Regulatory Policy Act • Tax benefit was announced for both investment and selling • There was the provision for net metering
Canada	• Feed in tariff was applicable to the parties installing the solar energy devices • Government subsidy
European countries (Germany, Spain, France)	• Feed in tariff • Government subsidy • Incentives are given to both parties, buyer and seller, of electricity from renewable sources • Definite target was set for renewable power generation and carbon emission
China	• Framing of specific renewable energy law. The concept of renewable power purchase obligation was introduced • Purchasing of photovoltaic power was made mandatory to power companies • Increase in research fund for the development of these technologies • Government fund allocation was for the decentralized projects • PV industry chain was established

2.1.1 Initiatives by the Indian Government for Promotion of the Solar Energy

In India, Electricity Act 2003 was passed in the parliament which laid emphasis on rapid electrification of the unelectrified villages [16]. The supply of power from the large coal-based power plants is not feasible always due to economic reasons as well as adverse terrain conditions. Moreover, the number of consumers in many places is very low. So it is not profitable to use grid power to electrify these villages. In such places, the solar power is more relevant. After this, the Government of India has also launched many programmes like Rajiv Gandhi Grameen Vidyutikaran Yojana and Deendayal Upadhyay Yojana to expedite the rural electrification in India.

After this, the Jawaharlal Nehru National Solar Power Mission was launched which has the target of installing 20,000 MW by 2022 and achieved the grid parity by the same year [17]. Thus, in India, solar power development is mainly driven by international norms to reduce carbon footprint and rapid rural electrification. After the national policy, the various states of India have also adopted solar energy policies based on the resource availability and other socio-economic factors.

3 Availability of Solar Resources in Different Parts of India

Solar radiation is available in situ at the place of its utilization, i.e. it need not to be transported from distance places like coal, oil to the place of utilization. In India, the solar energy is used in many sectors such as water heating, cooking, cooling, desalination, drying, and steam generation. The route of utilization mainly depends on the need of the local people. To select the suitable technology, the resource estimation is very vital. In India, the statewise availability of solar resources is carried out by Ministry of New and Renewable Energy as shown in Fig. 1 [18].

The cooking is done by solar cookers. Solar dryers are common in rural areas. The desalination is mainly useful in the coastal areas. This is done by the solar still working on the principle of thermal distillation of water and its subsequent condensation at low temperature. The solar still generates potable water from the saline water. Solar cooling is done by the vapour absorption cooling system. Solar water heating is done by evacuated or the flat plate collectors using the thermosyphon system. The availability of the solar resource determines the particular type of technology for the use of solar energy.

Thus, it is shown in Fig. 1 that the maximum solar power potential lies in the western and the northern states of India. Polygeneration is an effective outcome of efficient system integration. In solar-based polygeneration, one or more other outputs are there with electricity. Other utility outputs may be decided according to local needs.

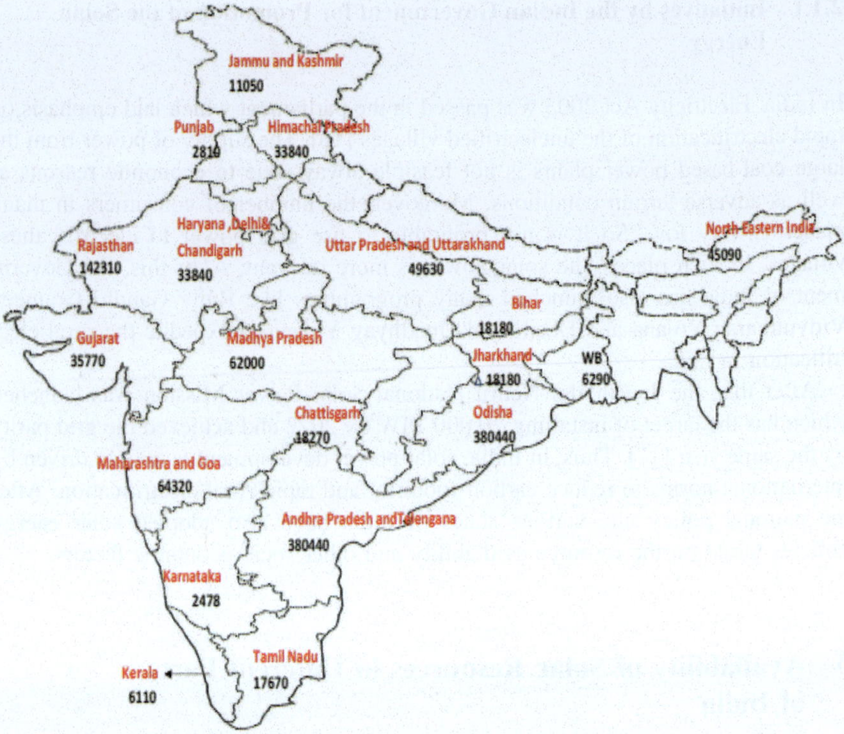

Fig. 1 Solar power potential in different states of India in MW

4 Routes of Solar Energy Utilization

The solar energy can be used in various routes as shown in Fig. 2. The solar thermal conversion implies the capturing of the solar heat and using it in many processes. Solar photovoltaic and thermophotovoltaic are the means for conversion of solar energy directly to electricity. The solar thermal energy is captured by the various types of solar collectors. The solar photovoltaic effect is achieved by the semi-conductor devices which can directly produce electricity when the photons are incident on it. Thermophotovoltaic devices directly convert heat to electricity.

4.1 Different Types of Solar Collectors

There are various types of solar collectors based on the principle of operation. The types of solar collectors are shown in Fig. 3. The various types of collectors have different efficiencies and areas of applications.

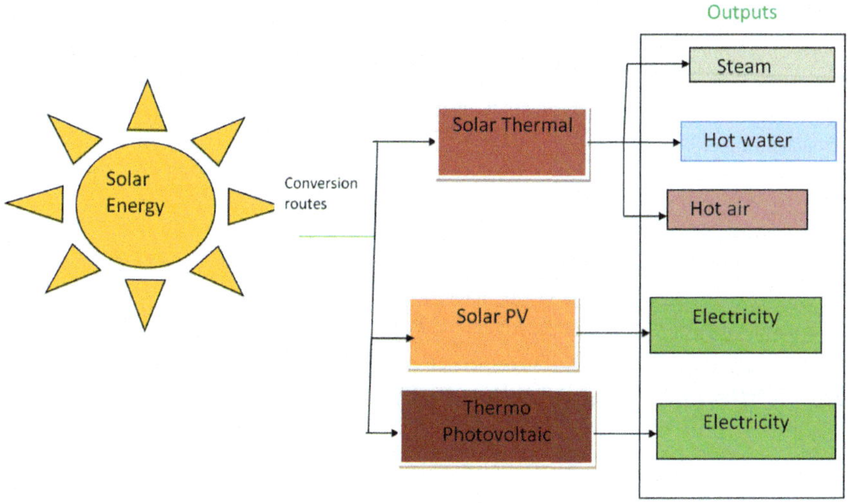

Fig. 2 Routes of solar energy conversion

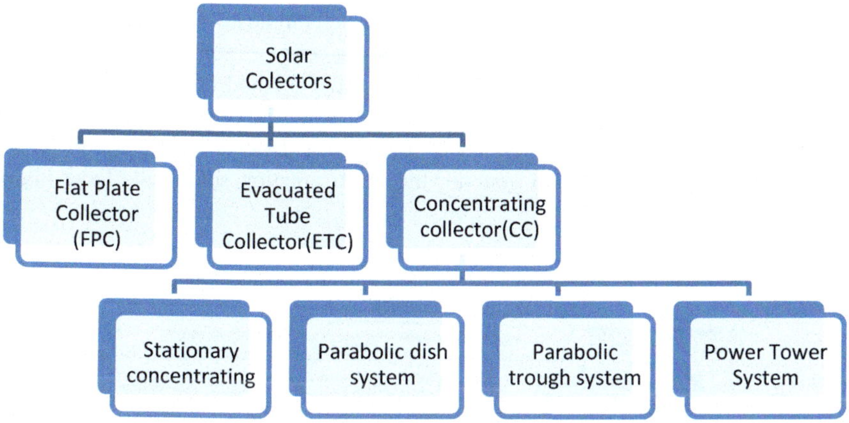

Fig. 3 Various types of solar collectors

The various application areas and efficiency ranges of these collectors are shown in Table 3 [19]. However, the most commonly used collectors are the evacuated tube collector (ETC) and the flat plate collector (FPC). The selection of collectors is made depending on the resource availability and economic issues.

Table 3 Application areas of different types of solar collectors

Type of collector	Temperature range	Application areas
Flat plate collector	80–100 °C	Low-temperature heat is output. Used for water heating, space heating
Evaluated tube collector	120–150 °C	Low-temperature heat is output. Used for water heating, space heating, industrial process heat, solar cooling, and desalination
Parabolic trough	150–300 °C	Industrial process heat
Parabolic dish	Greater than 1500 °C, but it is rarely used	Electricity generation and industrial process heat

4.2 *Different Types of Solar Cells*

With the advancement in the field of material science, different types of solar cells have come into existence. One has the advantage over the other, but has certain disadvantages also; for example, the organic solar cells need less input energy than the crystalline solar cells, but the efficiency is too low. The various types of solar cells based on the junction characteristics with highest laboratory scale efficiencies are given in Fig. 4 [20]. The numbers in the brackets indicate efficiencies. Till now, the most widely used solar cells are silicon solar cells.

The efficiency of the solar cells depends on the method of fabrication and the material used, say, the efficiency value of the thin film and the crystalline solar cells varies. Moreover, the energy trapped by the device depends on the nature of the semiconductor junction. So in general, the multi-junction solar cells have higher

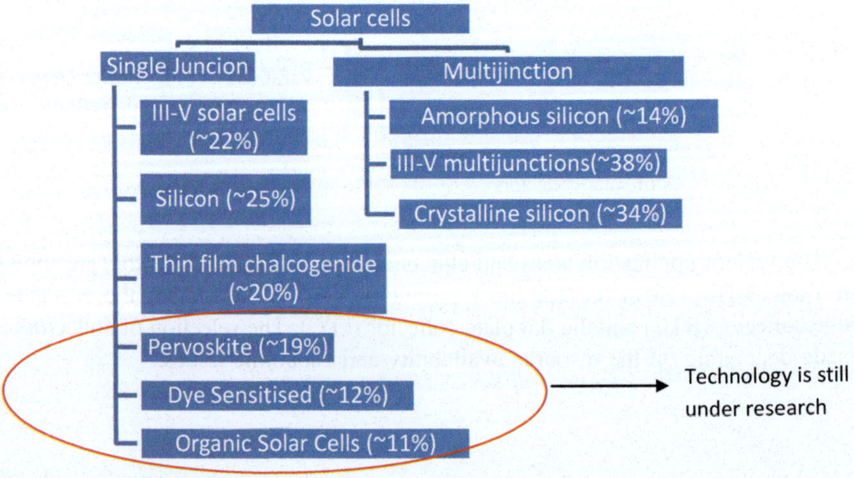

Fig. 4 Different types of solar cells

efficiencies than the single junction solar cells. But their fabrication is costly and hence less used commercially. These are III-V cells, GaAs and the InP cells. The thin film chalcogenide include the CIGS (CuInGaSe2) and CdTe cells [21]. The organic solar cells are made up of organic materials like polyaniline. Thus, with the suitability of application, each type of solar cell is used.

The main application areas of the solar photovoltaic (SPV) are home lighting, water pumping, desalination, and thermophotovoltaic application [21].

5 Intermittency of Renewable and Need for Polygeneration

Though the renewable energy sources are clean, these are intermittent and dilute too. The renewable sources of energy are not available as per the demand of the consumers, i.e. following the load curve. Hence, storage of energy is necessary for efficient operation. When the resource is available but the demand is low, the energy may be stored or utilized for some other purposes. Thus, to patch up the mismatch between resource availability and instantaneous power demand, polygeneration may be viewed as an effective way

5.1 Definition of Polygeneration

Polygeneration is the efficiently integrated process of generating multiple outputs from a single system [22].

In addition to electricity, potable water is also a generally required utility. The economic values of these utilities are added, and the system becomes economically more profitable and socially acceptable. Thus, the polygeneration system is beneficial from the economic, environmental, and the local resource utilization point of view [23]. The solar resources are available only at particular hours of the day. But consumption pattern does not vary in the same way as the insolation varies; i.e., solar radiation is unavailable at night, and the maximum electricity load occurs during evening [24]. To cope with such mismatch of demand and supply storage, hybridization and polygeneration are different options to give a reliable power supply [20]. Polygeneration proves to be environmentally and economically beneficial than the other two options, i.e. hybridization and storage [24, 25]. The basic schematic of a polygeneration system is shown in Fig. 5. Polygeneration can be viewed upon as a means of efficient resource utilization. Here, the possible available resources are utilized to cater the needs of the local people.

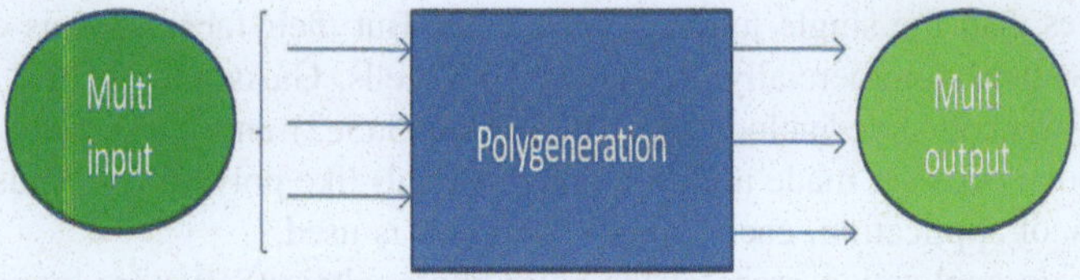

Fig. 5 Schematic of a polygeneration system

5.2 *Problems of Electrochemical Storage*

The need of storage is shown in Fig. 6a, b. The oldest form of electricity storage is done by using batteries [26]. The battery storage systems have the high efficiency but have many environmental impacts. Moreover, most of the batteries are to be replaced after five years. On the other hand, most of the renewable energy systems should have a life of above twenty years to be economically feasible.

During the period when the renewable energy (RE) generated is higher than the instantaneous load, power is fed to the storage device and vice versa. Thus, storage may be one of the options to meet fluctuations of the demand and supply. But it has several limitations regarding capacity, environment, and economic aspects.

5.3 *Hybridization of Solar Energy Devices for Polygeneration*

Solar energy is to be hybridized with some other renewable sources to get an uninterrupted power supply. The possible hybridization options of solar energy utilization are shown in Fig. 7.

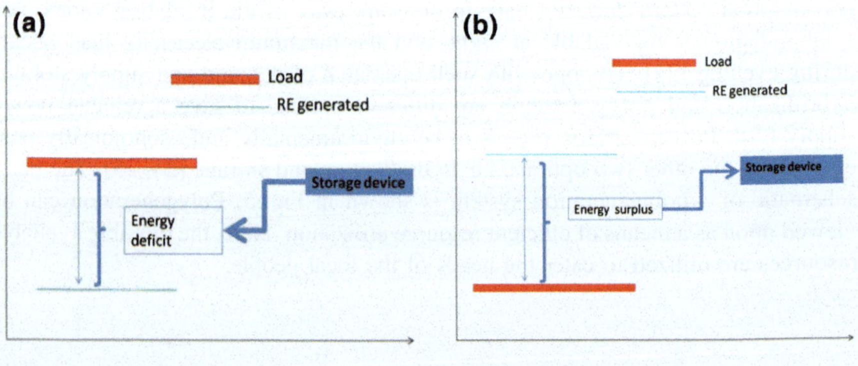

Fig. 6 **a** Energy from storage device (SD). **b** Energy to storage device (SD)

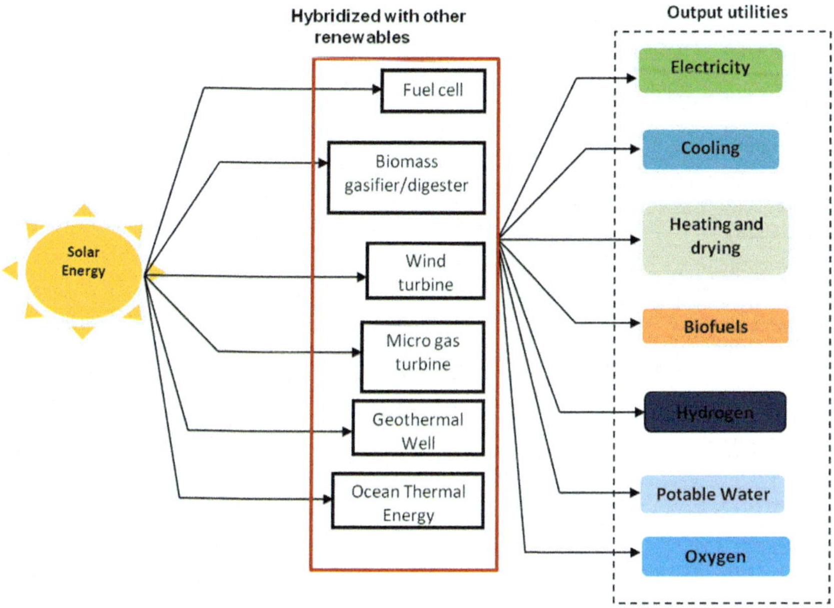

Fig. 7 Possible hybridization options and output utilities

In a polygeneration with solar and biomass inputs, solar–biomass hybrid system may be the option. When the solar radiation is high, then a part of the total syngas generated by the biomass gasifier system may be used to produce ethanol. During night, with no solar radiation, the entire syngas is used to generate electricity. Thus, both the utilities like ethanol and electricity are obtained [24]. Generally, solar energy alone (thermal or photovoltaic) cannot be used for polygeneration. Some literature of solar–hybrid polygeneration is referred to in Table 4.

Biomass is abundantly available in villages of India. So, solar energy is hybridized with biomass. In polygeneration, solar thermal integrates better than solar photovoltaic applications. The generic output utilities of solar-based polygeneration are electricity, potable water, heating, cooling, etc. However, this depends on location and the type of devices used for fabricating the system. The costs of generation of the same utility outputs are less than those generated by the stand-alone units. This is the purpose of developing integrated polygenerations.

5.4 Operation and Control of Polygeneration System

The basic schematic diagram of operation and control is shown in Fig. 8. The optimization of polygeneration is a multi-criteria problem. In most of the cases, the optimization is important for both the demand-side and the supply-side

Table 4 Solar hybrid polygeneration system

Author name (Year)	Routes		Type of hybridization	Input	Output
	Photovoltaic	Thermal			
Kribbus and Mittleman (2007) [27]	√	√	Heat engine	Solar energy	Electricity and heat
Calise (2011) [28]		√	Solid oxide fuel cell	Solar energy	Electricity, cooling, heating
Calise et al. (2012) [29]	√	√	Fuel cell	Solar energy	Electricity, cooling, heating
Calise et al. (2012) [30]		√	Reciprocating engine fed by vegetable oil	Solar energy and vegetable oils	Electricity, heating, and cooling
Rivalro et al. (2012) [31]		√	Wind turbine, wind turbine-based microgrid	Solar energy and wind	Electricity, heating and cooling
Ozturk and Dincer (2013) [32]		√		Solar energy	Electricity, heating, cooling, hydrogen, and oxygen
Al-Ali and Dincer (2014) [33]		√	Geothermal well	Solar and geothermal energy	Electrical power, cooling, space heating, hot water, and heat for industrial use
Aichmayer (2014) [34]		√	Microgas turbine	Solar-assisted microgas turbine	Electricity, hot water, and cooling
Suleman et al. (2014) [35]		√	Geothermal well	Solar and geothermal energy	Electricity, drying, and cooling
Buonomano et al. (2016) [36]	√	√	The entire cogeneration system is coupled with a gas turbine cogeneration	Solar energy, natural gas	Electricity, cooling, domestic hot water
Bai et al. (2015) [37]		√	Biomass gasifier	Solar energy and biomass	Electricity and methanol
Sahoo et al. (2015) [38]		√	Biomass gasifier	Solar energy and biomass	Electricity, cooling, heating and water
Khalid (2015) [39]		√	Biomass gasifier	Solar and biomass	Electricity, cooling, hot water, heated air

(continued)

Table 4 (continued)

Author name (Year)	Routes		Type of hybridization	Input	Output
	Photovoltaic	Thermal			
Ahmadi et al. (2015) [40]	√	√	Ocean thermal energy	Solar and ocean thermal energy	Electricity, freshwater, cooling, and hydrogen
Khan and Martin (2015) [41]	√		Biogas from a digester	Solar energy and biogas	Electricity, biogas, drinking water
Calise et al. (2017) [42]		√	Geothermal energy	Solar and geothermal energy	Electricity, desalinated water, heating, cooling
Islam and Dincer (2017) [43]		√	Geothermal energy	Solar and geothermal energy	Electricity, heating, and cooling
Mata-Torres et al. (2017) [44]		√	None	Solar energy	Electricity and desalinated water
Sahoo et al. (2017) [23]		√	Biomass gasifier	Solar energy and biomass	Electricity, cooling, and potable water
Dincer et al. (2017) [11]		√	Biomass gasifier	Solar energy and biomass	Electricity, cooling, hot water, and space heating

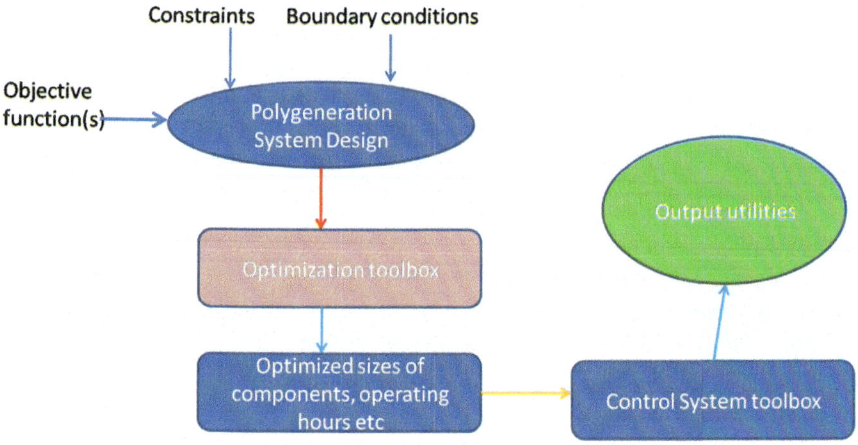

Fig. 8 Scheme of optimization and control of polygeneration

management of the polygeneration systems. Various optimization algorithms such as genetic algorithm, mixed-integer linear programming and nonlinear programming, genetic algorithm are used for the design of the polygeneration systems [45, 46]. Another approach for the optimization is using the expectancy and the probabilistic models taking into consideration. This is very important when the prospective planning of an energy supply of a particular area is needed along with the system design.

Control of solar-based polygeneration system is very important from the demand-side management (DSM) point of view. In most of the cases, the solar energy is hybridized with some other forms of energy to get the reliable power supply. In this context for most of the cases, the model predictive control and the optimal control strategies are used [47, 48]. This control system is achieved by the applications of programmable logic controller (PLC) [49].

6 Conclusions

With the increasing energy demand and threat of climate change due to existing fossil fuel-based energy systems, development of efficient renewable energy systems is a present imperative need. With abundant availability of solar radiation, solar energy systems are considered to be effective for Indian energy need in a sustainable way. However, non-availability at night and unpredictability of availability of solar power over the year, demands for "hybridization" of solar energy systems and such systems emerge to provide reliable continuous power supply. Polygeneration including solar power hybridized with renewable resources like biomass and several other utility outputs like potable water, ethanol, heating, cooling can be a future sustainable option, specifically for Indian application. Polygeneration is a process to build a new energy system through efficient process integration. The process integration increases the efficiency of the system. The better process integration will result in the better output of the polygeneration system. A review of available technologies along with state-of-the-art development as available in the literature is reported in this chapter.

References

1. Gae A (2008) Access to energy and human development. http://hdr.undp.org/en/content/access-energy-and-human-development. Accessed 20 May 2017
2. US energy information administration. https://www.eia.gov/todayinenergy/detail.php?id=12251. Accessed 20 May 2017
3. India energy outlook. https://www.iea.org/publications/freepublications/publication/IndiaEnergyOutlook_WEO2015.pdf. Accessed 11 May 2017

4. Solangi KH, Islam MR, Saidur R, Rahim NA, Fayaz H (2011) A review on global solar energy policy. Renew Sustain Energy Rev 15:2149–63. https://doi.org/10.1016/j.rser.2011.01.007
5. Electricity production from coal sources. http://data.worldbank.org/indicator/EG.ELC.COAL.ZS. Accessed 20 May 2017
6. Carbon dioxide uncontrolled emission factors. https://www.eia.gov/electricity/annual/html/epa_a_03.html. Accessed 20 May 2017
7. http://pib.nic.in/newsite/PrintRelease.aspx?relid=155612. Accessed 21 May 2017
8. Alanne K, Saari A (2006) Distributed energy generation and sustainable development. Renew Sustain Energy Rev 10:539–58
9. Hoggerward J, Dincer I, Naterer GF (2017) Solar energy based integrated system for power generation, refrigeration and desalination. Appl Therm Eng 121:1059–1069
10. Ozlu S, Dincer I (2016) Performance assessment of a new solar energy based multigeneration system. Energy 112:164–178
11. Khalid F, Dincer I, Rosen M (2017) Thermo-economic analysis of a solar-biomass-integrated multigeneration system for a community. Appl Therm Eng. https://doi.org/101016/j.applthermaleng.2017.03.040
12. Tsoutsos T, Frantzeskaki N, Gekas V (2005) Environmental impacts from the solar energy technologies. Energy Policy 33:289–96
13. Rydh CJ, Svärd B (2003) Impact on global metal flows arising from the use of portable rechargeable batteries. Sci Total Environ 302:167–84
14. European environment agency. http://eu/data-and-maps/daviz/dominant-remediation-technolog. Accessed 28 April 2017
15. Paris agreement. http://www.ec.europa.eu/clima/policies/international/negotiations/paris/index. Accessed 01 May 2017
16. Electricity act. http://www.cercind.gov.in/Act-with-amendment.pdf. Accessed 18 May 2017
17. MNRE. http://www.mnre.gov.in/solar-mission/jnnsm/introduction-2/. Accessed 18 May 2017
18. MNRE. http://www.data.gov.in/catalog/state-wise-renewable-energy-potential. Accessed 10 May 2017
19. Rabl A (1985) Survey of solar equipments and applications. Active solar collectors and their applications, p 3–11
20. Green M, Emery K, Hishikawa Y, Warta W, Dunlop E, Levi D (2017) Solar cell efficiency tables(version 49). Prog photovolt Res Appl 25:3–13
21. Pandey AK, Tyagi VV, Selvaraj JA, Rahim NA, Tyagi SK (2016) Recent advances in solar photovoltaic systems for emerging trends and advanced applications. Renew Sustain Energy Rev 53:859–84. https://doi.org/10.1016/j.rser.2015.09.043
22. Serra LM, Lozano MA, Ramos J, Ensinas AV., Nebra SA (2009) Polygeneration and efficient use of natural resources. Energy 34:575–86. https://doi.org/10.1016/j.energy.2008.08.013
23. Sahoo U, Kumar R, Pant PC, Chaudhary R (2017) Development of an innovative polygeneration process in hybrid solar-biomass system for combined power, cooling and desalination. Appl Therm Eng 120:560–7. https://doi.org/10.1016/j.applthermaleng.2017.04.034
24. Ray A, Jana K, De S (2017) Polygeneration for an off-grid Indian village: optimization by economic and reliability analysis. Appl Therm Eng 116:182–96. https://doi.org/10.1016/j.applthermaleng.2016.11.020
25. Jana K, De S (2017) Environmental impact of biomass based polygeneration—A case study through life cycle assessment. Bioresour Technol 227:256–65. https://doi.org/10.1016/j.biortech.2016.12.067
26. Aneke M, Wang M (2016) Energy storage technologies and real life applications—a state of art review. Appl Energy 179:350–77
27. Kribbus A, Mittleman G (2008) Potential of polygeneration with solar thermal and photovoltaic systems. Int J Sol Energy Eng 130:1101-1-01101–5
28. Calise F (2011) Design of polygeneration system with solar collectors and solid oxide fuel cell: dynamic simulation and economic assessment. Int J Hydrogen Energy 36:6128–50

29. Calise F, Polombo A, Vanoli L (2012) Design and dynamic simulation of a novel polygeneration system fed by vegetable oil and solar energy. Energy Convers Manag 60:32–7
30. Calise F, D'Accadia MD, Vanoli L (2012) Design and dynamic simulation of a novel solar trigeneration system based on hybrid photovoltaic/thermal collectors (PVT). Energy Convers Manag 60:214–25. https://doi.org/10.1016/j.enconman.2012.01.025
31. Rivalro M, Greco A, Massardo A (2013) Thermo-economic optimization of the impact of renewable generators on polygeneration smart grids including hot thermal storage. Energy Convers Manag 65:75–83
32. Ozturk M, Dincer I (2013) Thermodynamic analysis of a solar-based multi-generation system with hydrogen production. Appl Therm Eng 51:1235–44. https://doi.org/10.1016/j.applthermaleng.2012.11.042
33. Al-Ali M, Dincer I (2014) Energetic and exergetic studies of a solar multigenerational geothermal system. Appl Therm Eng 71:16–23
34. Aichmayer J, Spelling J, Lambert B (2013) Small scale hybrid solar plants for polygeneration in rural areas. Energy Procedia 57:1536–45
35. Suleman F, Dincer I (2014) Development of an integrated renewable energy system for multigeneration. Energy 78:196–204
36. Rady M, Amin A, Ahmed M (2015) Conceptual design of a small scale multi-generation concentrated solar plant for a medical centre in Egypt. Energy Procedia 83:289–98
37. Bai Z, Liu Q, Lei J, Jin H (2015) A polygeneration system for methanol production and power generation with solar-biomass thermal gasification. Energy Convers Manag 102:190–201
38. Sahoo U, Kumar R, Pant P, Chaudhury R (2015) Scope and sustainability of hybrid solar-biomass plant with cooling, desalination in polygeneration process in India. Renew Sustain Energy Rev 51:304–16
39. Khalid F, Dincer I, Rosen M (2015) Energy and exergy analyses of a solar biomass integrated cycle for multigeneration. Sol Energy 112:209–99
40. Ahmadi P, Dincer I, Rosen M (2015) Performance assessment of novel solar and ocean thermal energy conversion based multi generation system for coastal areas. J Sol Energy Eng 137:11013-1-011013–12
41. Khan E, Martin A (2015) Optimization of hybrid renewable energy polygneration system with membrane distillation for rural households in Bangladesh. Energy 93:1116–27
42. Calise F, Macaluso A, Piacentino A, Vanoli L (2017) A novel hybrid polygeneration system supplying energy and desalinated water by renewable sources in pantelleria island. Energy. https://doi.org/10.1016/j.energy.2017.03.165
43. Islam S, Dincer I (2017) Development, analysis and performance assessment of a combined solar and geothermal energy based integrated system for multigeneration. Sol Energy 328–43
44. Mata-Torres C, Rodrigo A, Escobar R, Cardemil J, Simsek Y, Matute J (2017) Solar polygeneration for electricity production and desalination. Renew Energy 101:387–98
45. Rubiyo-Mata C, Marcuello-Uche J, Martinez-Gracia A, Bayod-Rujula A (2011) Design optimization of polygeneration plant fuelled by natural gas and renewable energy sources. Appl Energy 88:449–57
46. Bracco S, Delfino F, Pamparo F, Robba M, Rossi M (2015) A dynamic optimization-based architecture for polygeneration microgrids. Energy Convers Manag 96:511–20
47. Menon R, Paolone M, Marechal F (2013) Study of optimal design of polygeneration in optimal control strategies. Energy 55:134–41
48. Menon R, Paolone M, Marechal F (2016) Intra-day electro-thermal model predictive control for polygeneration microgrid. Energy 104:308–19
49. Delfino F, Rossi M, Miniciardi R, Robba M (2015) An optimization based architecture for local systems managed by PLC devices. IEEE Int Symp Syst Eng

Part II
Heat Transfer Aspects of Solar Thermal Collectors

Effect of Reflector Absorptivity on Radiative Heat Exchange in Case of Solar Receiver Collection Systems

Santosh B. Bopche

Abstract In case of solar central receiver systems, the major portion of incident solar radiation is reflected by the concentrating reflectors or heliostats towards the receiver by specular reflection and minimum portion is received through diffused radiation. In order to know this diffused component of radiation energy transfer, the absorptivity of reflectors as well as shape factors between the reflectors and the central receiver is must. The present chapter is dealing with the estimation of radiative heat/energy exchange between the hot cavity receiver and reflecting collector surface along with the diffused configuration factor computation encompassing in solar reflector-collector systems of various orientations and geometries. The configuration factors have been evaluated using contour integral technique, which has been reported as one of the accurate and faster tools for configuration factor estimation. The integral cum analytical expressions are also demonstrated for configuration factor, and the methodology adopted for exchange of radiative heat between the hot cavity receiver and the reflecting surface of parabolic dish collector is also discussed.

Keywords Radiation · Configuration factor · Cavity receiver
Heat transfer

Nomenclature

A	Surface area, m^2
C	Specific heat of the material, J/kg-K
D	Diameter, m^2
dA	Differential elemental area of hemispherical cavity receiver surface, m^2
E_b	Emissive power of black surfaces, W/m^2

S. B. Bopche (✉)
Department of Mechanical Engineering, National Institute of Technology Hamirpur, Hamirpur 177005, HP, India
e-mail: santoshb.bopche@yahoo.co.in; santoshbopche@nith.ac.in

H. Tyagi et al. (eds.), *Applications of Solar Energy*, Energy, Environment, and Sustainability, https://doi.org/10.1007/978-981-10-7206-2_3

H	External irradiation (W/m^2) or Y dimension of the geometry or height of plane, m
h	Convective heat transfer coefficient inside the cavity, W/m^{2}K
L	Z dimension of the geometry or width of plane, m
N	Number of elements in the enclosure
q	Heat flux, W/m^2
r	Radial dimension, m
r	Radius of cavity opening, m
R	Radial dimension of the geometry or radius of a cavity receiver, m
S	Spacing between the interacting surfaces, m
s	Vector pointing from emitting surface contour to the point on receiving surface contour, m
T	Axial offset distance of surfaces, m or temperature, °C
V	Volume, m^3
X	Constant or collector dimension, m
Y	Vertical dimension of a geometry, m
x, y, z	Axes of coordinate system

Non-dimensional parameters

F	Configuration factor value
S/r	Spacing ratio
T/H	Offset ratio
L/r	Width ratio
r/H	Radius ratio or geometry dimension ratio along '*y*' direction
RR	Radius ratio (R_1/R_2)
r/R	Opening ratio/aperture ratio (*r/R*)

Symbols

ρ	Density of material, kg/m^3
ψ	Pitch angle of the parabolic collector
δ	Rim angle of the collector at that particular location
θ	Azimuthal location or angle included between surface normal and line joining two infinitesimal surfaces or angular positions over the contour, *rad*
φ	Azimuthal location or tilt angle of receiver, *rad*
ψ	Angle position or tilt angle of reflector surface, *rad*
ϵ	Emissivity values of respective surfaces

Subscripts

i, 1	Radiation-emitting surface/source surface
j, 2	Radiation-absorbing surface/sink surface
air	Ambient condition

c	Collector
f	Fluid
i	Inside
o	Opening or outer
in	Initial condition
r	Receiver
$r\text{-}o$	Receiver opening
$r\text{-}surface$	Inside surface area of a receiver
$r\text{-}wall$	Receiver wall
col	Parabolic collector related

1 Introduction

It is known that cavity receivers are most extensively used high operating temperature type point-focusing systems for Brayton or Stirling cycle based systems. It is due to its competence to meet the temperature (1000–1500 K) and pressure (3–35 bar) which are the requirements of a typical gas turbine with sound efficiency [1]. Since the receiver operates above 1000 K, thermal radiation plays a dominant role in losing heat from the opening and exchange with the outer interacting or partially absorbing structures. Such heat losses through convection as well as radiation affect considerably the efficiency and cost-effective performance of the parabolic dish-type solar thermal power system.

Wu et al. [2] have discussed convective heat escape/loss mechanisms of cavity-type receiver dependent on various parameters, e.g. atmospheric wind conditions, air-temperature enclosed inside a cavity and the geometries/shapes of the receiver. In case of the absence of wind effect, the convection heat transfers/loss from the aperture is observed to be 5–15% of radiative heat losses for smaller receivers. It means radiation heat loss contributes around 75–85% of combined convective cum radiative heat/energy transfers/losses from the receiver.

The studies examining the effect of cavity receiver tilt angle on the losses have claimed that the convective heat losses show a reducing trend with increase in tilt angle from 0° (horizontal) to 90° (vertically downward position of a receiver) and the contribution of radiation heat loss rises with tilt angle [3]. Reddy and Kumar [4] have seen that the convective heat transfer is dominated by radiative energy loss/exchange for tilting angles above 45°.

The distribution of coefficient of heat transfer around the parabolic trough collector (reflector) and around the circular receiver has been examined by Hachicha et al. [5], in the presence of a wind flow. The zero degree pitch position of the collector is like the open end of the collector facing the wind blow whereas 90° is with collector open end facing vertically up and 180° is with collector open end facing down.

Uhlig et al. [6] have presented the ways to improve the performance of receiver that is by increasing the receiver size without changing the aperture size, by reducing the aperture size, by providing additional concentrator like compound parabolic collector, by facing down the receiver opening, by obstructing an air inflow or outflow from the cavity enclosure and by providing glazing at the aperture of receiver.

Various shapes have been suggested for minimizing these losses. Shuai et al. [7] have discussed about six different types of solar receiver cavity shapes, e.g. cylindrical, dome end, heteroconical, elliptical, spherical and conical. Kumar and Reddy [8] have compared three receiver types: (1) cavity-type receiver, (2) semi-cavity-type receiver and (3) modified cavity-type receiver. The schematics of cross-sectional views are as shown in Fig. 1.

Cavity receiver exhibits maximum heat loss among the three. It is attributed to the formation of static or stagnant fluid zone inside the cavity receiver. The minimum convective heat loss is seen in case of modified type of cavity receiver. The information about the energy loss through radiation mode is not reported [8]. The studies examining collective convection as well as radiation heat loss from the modified-type cavity receiver due to inclination, depth of insulation, working temperature and cavity surface emission property are also available [9]. It is seen that the tilt angle influences dominantly the free convection transfers or losses from the cavity and emissivity controls the losses due to radiation. The authors have also proposed a correlation for obtaining total energy loss from the cavity receiver. The forced convection energy transfer or loss is 8–10 times the free convection for wind speeds of about 5 m/s, whereas for lower wind speeds, the natural convection plays a dominant role in losing energy from receiver cavity opening [9]. The non-uniform cavity wall temperature dictates variation in radiative heat exchange between the differential element and the circular opening of a modified cavity. The information of configuration factor is must for finding radiative heat transfer contribution.

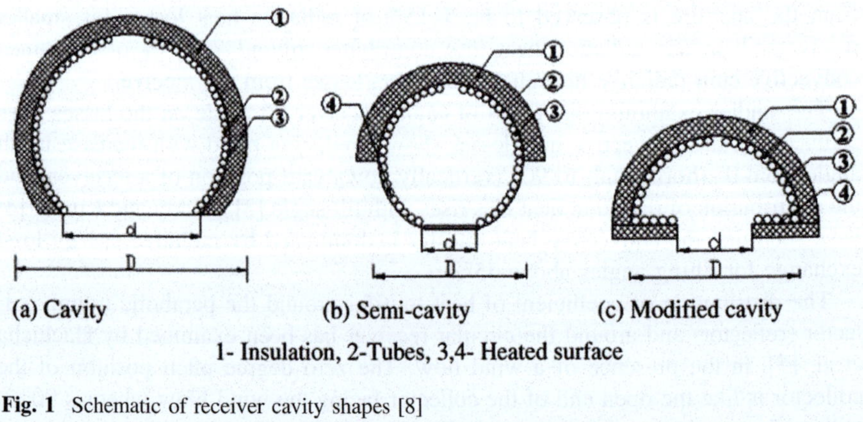

Fig. 1 Schematic of receiver cavity shapes [8]

A unit sphere/hemisphere method is the one in which with the evaluation of area of sink surface after two times projection on the base of a hemisphere, the configuration factor can be easily obtained, as shown in Fig. 2. The unit sphere method is a method useful for experimental determination of view factor between one infinitesimal and one finite area, through optical projection of finite area twice on the plane of infinitesimal area. The expression reported by Modest for the estimation of view factor is $F_{dA_1-A_2} = \frac{A_2''}{\pi R^2}$, where A_2'' is the area of second projection of finite area A_2 and R is the radius of hemisphere [10].

These projected areas have been scaled through AutoCAD software. The configuration factor outcomes are demonstrated for cases of two spheres, cylinder and a rectangle, two equal parallel rectangles and for an element on a sphere. Using CAD software and simulation schemes, the configuration factor values have been obtained by Cabeza-Lainez et al. [11] for configuration or view factor between a sphere, circle and a differential element situated at arbitrary locations.

The deterioration of surface reflectivity properties may increase radiative energy absorption by the surfaces. The effect of reflectivity of surfaces on the configuration factor values or radiative heat losses from inside of the cylindrical tube with internal specular reflections has been discussed by Dixon [12]. It is seen that the interacting surfaces with lower emissivity values offer larger radiative heat losses. It is attributed to multiple numbers of reflections before absorption between the poorly emitting surfaces.

The numerical solution of configuration factor integral equations can be obtained using mathematical formulations. The results of various quadrature formulas like trapezoidal, Simpson and Gaussian have been tested for configuration factor determination and compared for geometries with plain as well as curved boundaries by Rao and Sastri [13]. The Gaussian quadrature technique has been proven to be the accurate and computationally lesser time demanding method for configuration factor evaluation. Using direct integration technique, configuration factor values have been obtained between circumferential ring elements (inner and outer) and the annular end (inner/outer) of a cylindrical enclosure by Diaconu et al. [14]. The obtained solution is applicable to the non-isothermal surfaces of an annulus.

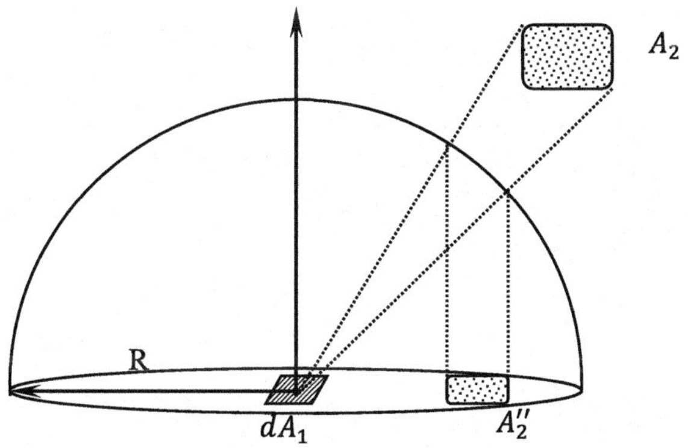

Fig. 2 Schematic of unit sphere method

Similar to present study, Maor and Appelbaum [15] have discussed the configuration factor evaluation for a photovoltaic collector system positioned on inclined ground surfaces of any angular orientation. The cross-string technique is used to validate the obtained results. This chapter aims at providing analytical–numerical solution for configuration factors among the geometries, e.g. the circular opening of receiver and the collector surfaces along with the radiative heat exchange for two values of reflector reflectivity (0.82 and 0.9). The configuration factor situations discussed are applicable for radiative heat exchange evaluation among the components of solar central cavity receiver collector systems.

2 Methodology

The present paper focuses on the estimation of radiation heat exchange between the cavity receiver opening and the parabolic dish collector's reflecting surfaces. The receiver collector system is considered as an enclosure and the solar radiation entering in, is considered as an external irradiation. The integral equation has been solved in order to obtain net heat flux at the respective surfaces. The configuration factor details needed for the estimation of radiation exchange between the cavity opening and the parabolic collector's reflecting surface are estimated using contour integral technique, for various orientations of circular receiver opening and the rectangular finite surfaces of the solar reflecting surface/collector.

2.1 *Methodology for Configuration Factor Evaluation*

The expression for configuration factor between the diffusely reflecting and emitting surfaces is expressed as

$$A_1 F_{A_1-A_2} = \int_{A_1}^{A_1} \int_{A_2}^{A_2} \frac{\cos\theta_1 \cos\theta_2}{\pi S^2} dA_1 dA_2 \tag{1}$$

The angles θ_1 and θ_2 are the included angles between surface normal and line joining the two differential areas. This equation includes two area integrals, the solution of which is very difficult to yield. By applying Stokes' theorem twice, double area integral equation can be expressed in a simpler form containing double contour integrals, as depicted in Eq. (2).

$$A_1 F_{A_1-A_2} = \frac{1}{2\pi} \oint\oint \ln S\, ds_1 ds_2 \tag{2}$$

where S is distance between two points on contours of surfaces 1 and 2.

$$|S| = \sqrt{\left(x_i - x_j\right)^2 + \left(y_i - y_j\right)^2 + \left(z_i - z_j\right)^2} \tag{3}$$

The contours of two surfaces are described by the two vectors 's_1' and 's_2'. The Eq. (2) is expressed in Cartesian coordinates as in Eq. (4).

$$A_1 F_{A_1 - A_2} = \frac{1}{2\pi} \oint\oint \ln S(dx_1 dx_2 + dy_1 dy_2 + dz_1 dz_2) \tag{4}$$

Equation (4) is used to obtain the necessary configuration factor values of the geometries involved in the analysis. The methodology adopted for radiative heat exchange is discussed in the subsequent Sect. 2.2.

2.2 Methodology for Radiative Heat Transfer

The parameters considered for studying the paraboloid dish collector system with hemispherical cavity receiver are, reflectivity of parabolic dish collector's reflecting surface (ρ), geometrical concentration ratio, C as

$$\left\{ C = \frac{Area\ of\ collector\ aperture}{Area\ of\ receiver\ opening} \right\},$$

Inside diameter of hemispherical receiver (D_R),
Diameter of paraboloid collector opening (D_c),
Thickness of a modified cavity made of stainless steel 306 (r_2-r_1),
Mass flow of coolant (air) and diameter of the coolant tube, d_t.

The heat flux reflected back from the parabolic collector surface is $q_c = \rho\, q_{solar} (\mathrm{W/m^2})$.

The heat flux concentrated at the circular opening of a receiver is $q_r = C \rho\, q_{solar} (\mathrm{W/m^2})$.

The heat flux at the internal surface of a hemispherical cavity is $q_{r-surface} = \frac{\rho\, C\, q_{solar} A_r}{A_{R-surface}} (W)$.

Assuming, hemispherical modified cavity receiver with coolant tubes being integral with the cavity wall.

The length of coolant tube depends on the number of turns which depends on diameter of receiver. It is assumed that the cavity receiver is subjected to a uniform radiation heat flux internally and assumed as perfectly absorptive. The outer boundary is assumed adiabatic. The coolant (air) flows through the tube, for which the heat transfer coefficient is obtained using Dittus-Boelter correlation, as given in Eq. (5)

$$Nu = \frac{hd_t}{K} = 0.023\, Re^{0.8} Pr^{0.3} \tag{5}$$

The energy balance at the cavity wall results into following equation, (neglecting radiative energy transfer from within the cavity receiver). Lumped mass analysis is carried out as follows:

$$\begin{aligned}&\textit{Energy gain} - \textit{Energy carried by tube flow}\\&\quad - \textit{Convective energy loss inside the cavity}\\&\quad = \textit{Rate of change on Internal energy of cavity wall}\end{aligned}$$

$$q_r A_{R-s} - [h_{air} A_s (T - T_{air})]_{tube-fluid} - [h_i A_s (T - T_{fi})]_{Inside\ Cavity} = \rho\, C\, V \frac{dT}{dt} \tag{6}$$

Separating the variables and rearranging Eq. (6) terms, we obtain Eq. (7) which can get cavity wall temperature variation with respect to time and at steady state. In present chapter, the outcome of radiative heat transfer is obtained for steady state.

$$\begin{aligned}\frac{(T - T_{air}) + (T - T_{fi})}{(T_{in} - T_{air}) + (T_{in} - T_{fi})} &= \frac{q_r A_s}{[h_{air} A_{s-tube}(T_{initial} - T_{air})] + [h_i A_s (T_{initial} - T_{fi})]}\\&\quad + \left[1 - \frac{q_r A_s}{[h_{air} A_{s-tube}(T_{initial} - T_{air})] + [h_i A_s (T_{initial} - T_{fi})]}\right]\\&\quad exp\left(\frac{-(h_{air} A_{s-tube} + h_i A_s)t}{\rho\, C\, V}\right)\end{aligned} \tag{7}$$

In a similar fashion, the temperatures of paraboloid dish collector's reflecting surfaces are obtained using Eq. (7). The convective energy/heat transfer coefficient for convective energy/heat loss from the parabolic collector, in the presence of wind, is obtained using Nusselt number data provided by Hachicha et al. [5].

The surface area needed for the analysis for temperature distribution of parabolic collector reflecting surfaces is calculated using the following expression (8).

$$A_{c-s} = \frac{\pi D_c}{12Y^2}\left[\left(\frac{D_c^2}{4} + 4Y^2\right)^{1.5} - \frac{D_c^3}{8}\right] or = \frac{\pi X}{6Y^2}\left[(X^2 + 4Y^2)^{1.5} - X^3\right] \tag{8}$$

The collector surface is divided into three segments as shown in Figs. 3 and 4. The intensity of solar radiation reduces with the tilting of reflecting surface. Assuming, intensity of solar radiation flux incident on the collector reflector is normal to Segment I. For subsequent sections, its intensity reduces with the tilt angle of the parabolic contour.

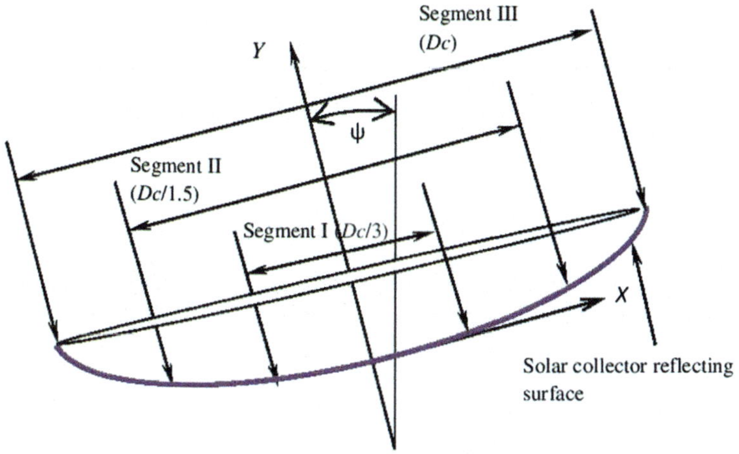

Fig. 3 Schematic indicating pitch angle, *X* and *Y* dimensions of paraboloid dish collector

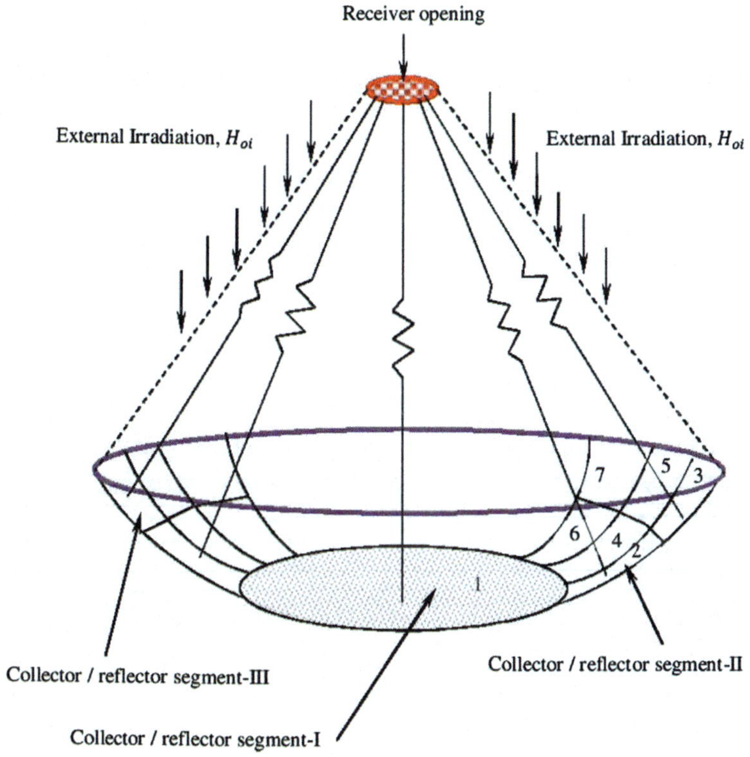

Fig. 4 Development of a radiation enclosure and network diagram

The intensity at the centre of the respective segments,
i.e. intensity at Segment I = q_r = 1000 W/m^2 (solar intensity)

The intensity at the geometrical centre of segment II = $q_r cos\,(\delta)$

$$= q_r \cos\left[tan^{-1}\left[\frac{X}{(f-y)}\right]\right].$$

where δ is the rim angle at the geometrical centres of the respective segments.

The radiation enclosure formed is as shown in Fig. 4. The receiver opening/aperture is assumed as a circular disc maintained as the cavity wall temperature. The assumption is well justified because the radiation emerging by the cavity wall shall pass through the receiver circular aperture or opening. The collector's reflecting surface is divided into three circular or concentric circular segments as I, II and III. Segment I is undivided since it is capturing solar radiation directly, whereas the energy intercepted by segments II and III depends on the cosine of the inclination angle.

The Segment 1 of parabolic dish is assumed as a circular disc, represented by element 1. The Segment II is divided circumferentially into 20 elements of equal average widths, represented by elements 2–40. Similarly, the Segment III is divided peripherally into 20 elements of same average height represented by elements 3–41. For obtaining radiation heat exchange, the configuration factor values are obtained between the circular aperture of receiver and each of the elements of parabolic dish collector. The details of configuration factor are disclosed in Sect. 3.

The range of parameters and realistic details taken for the study are discussed as follows.

2.3 Data Selection

The intensity of solar radiation is taken as $q_{solar} = 1000\,\mathrm{W/m^2}$.

The air flows through the tube which is integral with the cavity wall.

The air side heat transfer area is obtained as A_s = 0.217 (m^2).

The air properties are taken at bulk mean temperature of T_{air} = 200 °C and assumed constant.

The Dittus-Boelter correlation is used for air side heat transfer coefficient as

$$Nu = \frac{hd_t}{K} = 0.023\,Re^{0.8}Pr^{0.3} \quad (9)$$

For a mass flow rate 0.05 kg/s, the tube side heat transfer coefficient for half-inch size is obtained as $h_{air} = 1396.8\,\mathrm{W/m^2K}$. The wall temperature is increased by reducing this heat transfer coefficient (Table 1).

Thermal conductivity of stainless steel (SS 316) cavity material at 500 °C is $K_{ss316} = 21.5\,\mathrm{W/mK}$ [16].

Table 1 Data selection

Sr. no.	Input parameters	Values taken
1.	Reflectivity of collector (reflector), ρ	0.82 (Aluminium) [10] 0.90 (Mirror polished surface)
3.	Inside diameter of hemispherical receiver, D_R	0.3 (m)
4.	Diameter of circular aperture of receiver, $D_{r,\ o}$	0.25 (m)
5.	Diameter of parabolic dish collector opening, D_c	2 (m)
6.	Thickness of a modified cavity made of stainless steel 306, $(r^2\text{-}r^1)$	0.003 (m)
7.	Mass flow of coolant (air)	0.05 kg/s
8.	Diameter of coolant tube, d_t	0.0125 (m)
9.	Receiver wall temperature, in °C followed by 'respective coolant side heat transfer coefficient values, in W/m^{2}K', as depicted in brackets […]	182.2 [1396.8], 200 [380], 250 [124.5], 300 [73.5], 500 [25.6], 650 [14.8], 800 [8.6], 1000 [3.8]

The Biot number based on cavity wall thickness is $Bi = \frac{h_{air}(r_2 - r_1)}{K_{SS316}} = 0.04\ (Max.)$ which is less than 0.1.

For the estimation of cavity wall temperature through lumped mass analysis, the details are provided as follows:

The convective heat transfer coefficient inside the cavity, h_i, is 5 W/m^{2}K for cavity wall at 500 °C, which is kept constant while analytically estimating the radiative heat exchange between the cavity wall and the parabolic reflector surfaces [6].The analysis is continued for the temperature of fluid trapped inside the cavity is taken as T_{fi} = 150 °C.

Density of stainless steel (Grade 316) = 8070 kg/m^3 [17],

Specific heat of stainless Steel (Grade 316), C = 530 J/kg/K [17],

Continuing the above analysis, by using the analytical expression (7), the steady-state temperatures of Segment I, Segment II and Segment III are obtained as 28.5, 38.33 and 50.33 °C for an ambient temperature of 25 °C.

3 Results and Discussion on Configuration Factor Evaluation

The cases studied for configuration factor are (i) two circular parallel discs separated by a distance, S, (ii) between a circular disc and a rectangular plane and (iii) between a circular disc and a rectangular plane tilted by angle, 'ψ' and offset by distance 'T'. The problem formulation using contour integral technique and results are discussed.

3.1 Configuration Factor Between Two Circular Parallel Discs

The situation is applicable for the circular aperture of a receiver and circular Segment I of the parabolic dish collector. The centre coordinates of Segment I (*X*, *Y*) ≡ (0, 0) and centres of both the circular discs are aligned. Separation distance between the collector Segment I and the receiver aperture is taken as focus, *F*, of a parabolic dish. The problem is formulated as follows.

The problem formulated is as follows. Two circular discs are positioned in y-z plane (Fig. 5).

Formulation:

$x_i = 0, x_j = h, dx_i = dx_j = 0, y_i = R_1 sin\,\theta_1, y_j = R_2 sin\,\theta_2, dy_i = R_1 cos\,\theta_1 d\theta_1,\ dy_j = R_2 cos\,\theta_2$ $d\theta_2, z_i = R_1 cos\,\theta_1, z_j = R_2 cos\,\theta_2, dz_i = -R_1 sin\,\theta_1 d\theta_1, dz_j = -R_2 sin\,\theta_2 d\theta_2$

$$
\begin{aligned}
&R = \sqrt{h^2 + [R_2 sin\theta_2 - R_1 sin\theta_1]^2 + (R_2\cos\theta_2 - R_1 cos\theta_2)^2} \\
&F_x = 0; \\
&F_y = \frac{1}{2\pi A_i}\int_{\theta_1=0}^{2\pi}\int_{\theta_2=0}^{-2\pi} \ln\,(R)\,R_1 cos\theta_1 R_2 cos\theta_2 d\theta_1 d\theta_2 \\
&F_z = \frac{1}{2\pi A_i}\int_{\theta_1=0}^{2\pi}\int_{\theta_2=0}^{-2\pi} \ln\,(R)\,R_1 \sin\theta_1\, R_2 sin\theta_2 d\theta_1 d\theta_2
\end{aligned}
\tag{10}
$$

$$
\begin{aligned}
F_{i-j} = F_x + F_y + F_z = &\frac{1}{2\pi A_i}\int_{\theta_1=0}^{2\pi}\int_{\theta_2=0}^{-2\pi} \ln\,(R)\,R_1 cos\theta_1\, R_2 cos\theta_2 d\theta_1 d\theta_2 \\
&+ \frac{1}{2\pi A_i}\int_{\theta_1=0}^{2\pi}\int_{\theta_2=0}^{-2\pi} \ln\,(R)\,R_1 \sin\theta_1\, R_2 sin\theta_2 d\theta_1 d\theta_2
\end{aligned}
$$

A uniformly spaced computational grid has been chosen for obtaining numerical solution from double contour integral equations as shown in Eq. (10). The results obtained are compared with the analytical solution presented by Modest [10]. It is the analytical outcome of contour integral technique. The results are compared as shown in Fig. 6.

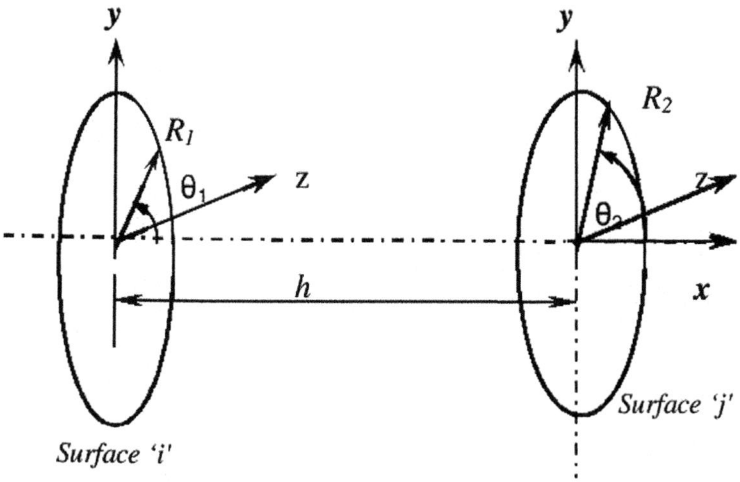

Fig. 5 Two circular parallel discs

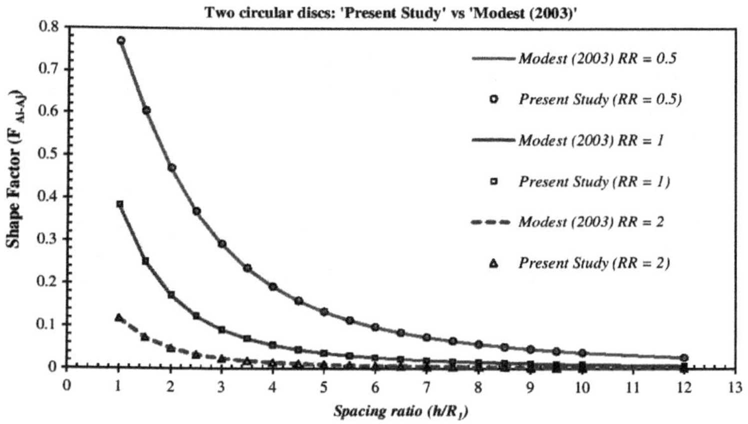

Fig. 6 Schematic showing comparison between analytical solution by [10] and present numerical study

Similarly, the problem is formulated for a circular disc and a parallel rectangular plate as discussed ahead.

3.2 *Configuration Factor Between a Circular Disc and a Parallel Rectangular Plate*

The schematic with the geometrical nomenclatures is as shown in Fig. 7.

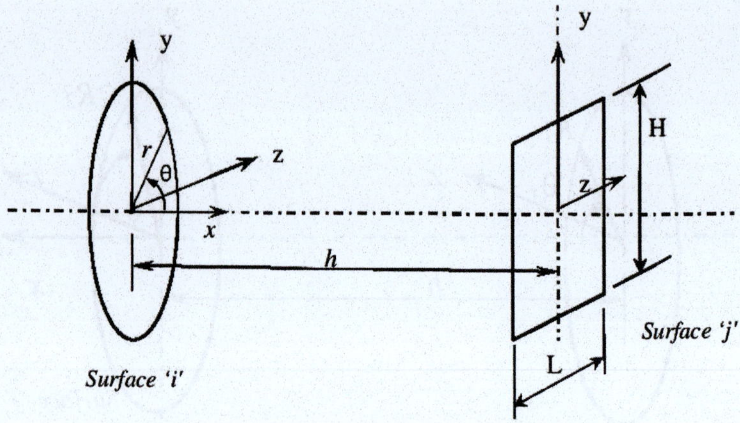

Fig. 7 Geometrical formulation of circular disc and a rectangular plane

Coordinates of respective surfaces are $x_i = 0, x_j = h, dx_i = dx_j = 0, dy_i = r\ \sin\ \theta,$ $y_j = H, dy_i = r\ \cos\ \theta\, d\theta, dy_j = dH, z_i = r\ \cos\ \theta, z_j = L, dz_i = -r\ \sin\ \theta\, d\theta, dz_j = dL$

$$R = \sqrt{h^2 + [r\,sin\theta - H]^2 + (r\,cos\theta - L)^2}$$

$$F_y = \int_{\theta=0}^{2\pi} \int_{\frac{-H}{2}}^{\frac{H}{2}} \ln\,(R)\, r\,cos\,\theta\, d\theta\, dH + \int_{\theta=0}^{2\pi} \int_{\frac{H}{2}}^{\frac{-H}{2}} \ln\,(R)\, r\,cos\,\theta\, d\theta\, dH \qquad (11)$$

$$F_z = \int_{\theta=0}^{2\pi} \int_{\frac{L}{2}}^{\frac{-L}{2}} -\ln\,(R)\, r\,sin\,\theta\, d\theta\, dL + \int_{\theta=0}^{2\pi} \int_{\frac{-L}{2}}^{\frac{L}{2}} -\ln\,(R)\, r\,sin\,\theta\, d\theta\, dL \qquad (12)$$

The results have been plotted for the spacing ratio (*S/r*) ranging from 1 to 10 and width ratio, (*L/r*) 1–5 (Fig. 8).

The results obtained using this case are compared with the inside sphere method [10] by approximating rectangular plane as a circular disc (by equating the areas) and inscribing it inside the only sphere. The results are shown in Figs. 9, 10, 11, 12 and 13 for various normalized width ratio (*L/r*) and spacing ratio (*S/r*) values. The error seen between the results may be attributed to the approximation of rectangular area by a circular one.

The results deviate extensively at closer spacing values for almost all the width ratio (*L/r*) values considered in this study. It is a common behaviour seen from Figs. 9, 10, 11, 12 and 13. The inside sphere method yields diverging outcome for closer spacing values.

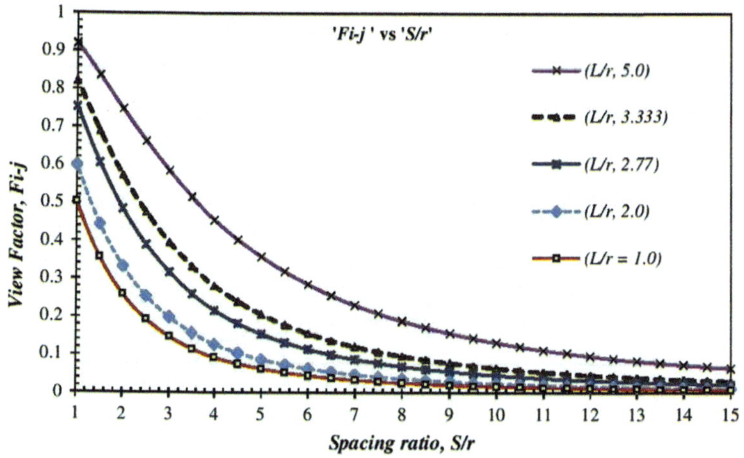

Fig. 8 Variation of configuration factor with spacing ratio (*S/R*) and width ratio (*L/R*)

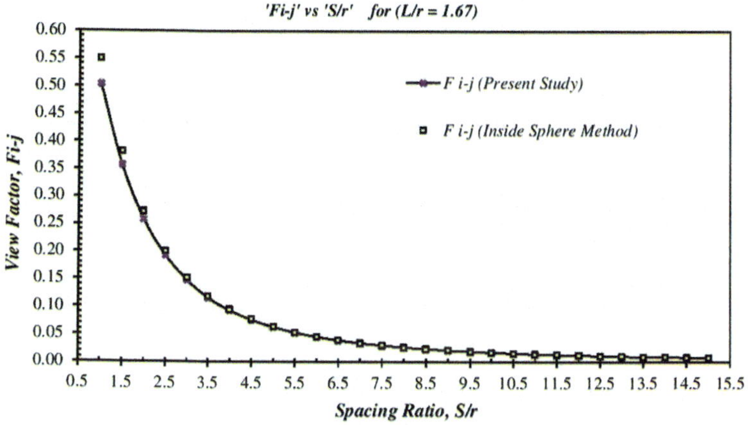

Fig. 9 Comparison of present results with inside sphere method [10] for $L/r = 1.67$

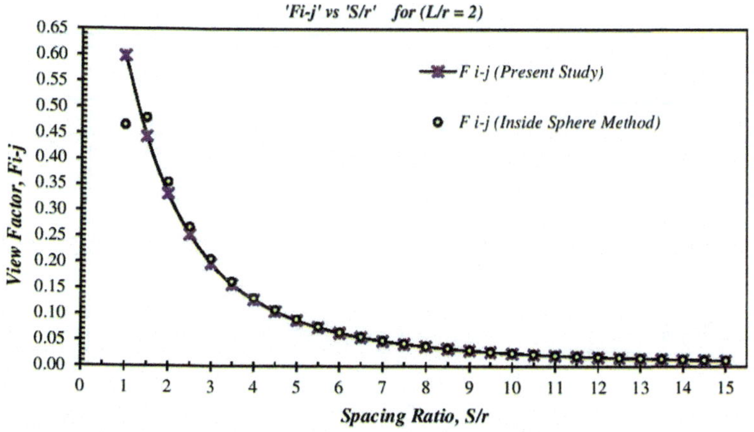

Fig. 10 Comparison of present results with inside sphere method [10] for $L/r = 2.0$

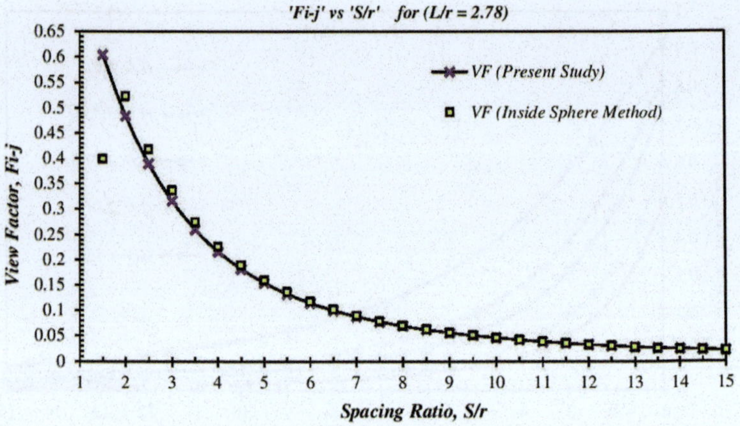

Fig. 11 Comparison of present results with inside sphere method [10] for $L/r = 2.78$

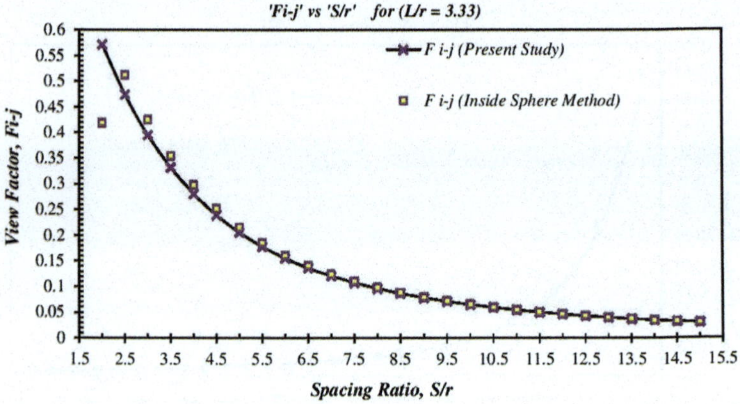

Fig. 12 Comparison of present results with inside sphere method [10] for $L/r = 3.33$

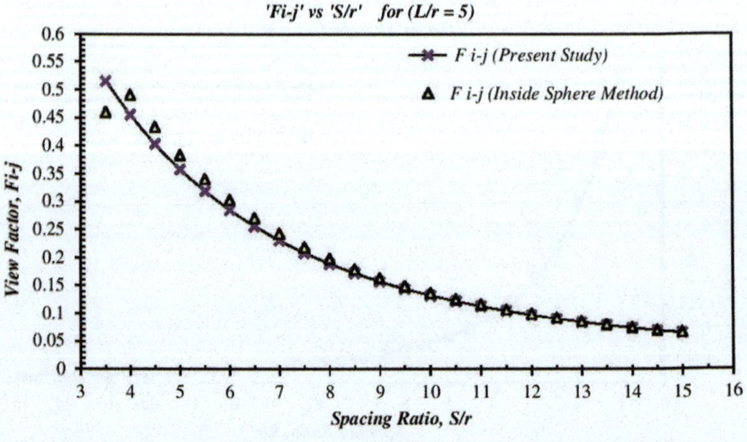

Fig. 13 Comparison of present results with inside sphere method [10] for $L/r = 5$

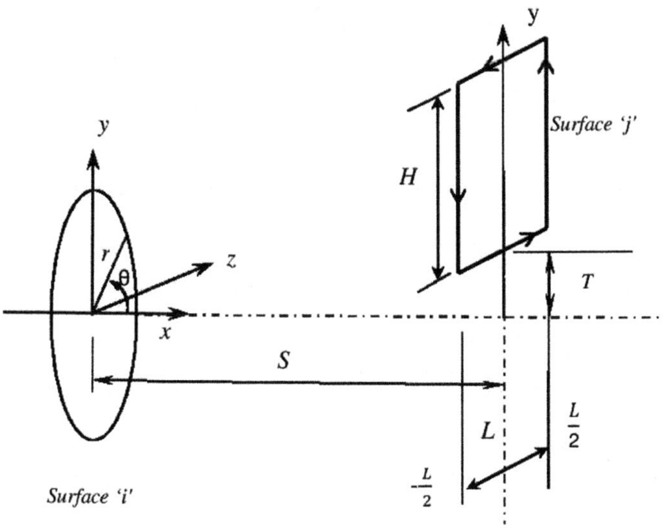

Fig. 14 Problem formulation (axis z is perpendicular to the plane of paper)

3.3 *Configuration Factor Between a Circular Disc and an Offset Rectangular Plate*

In this case, the rectangular finite element of a paraboloidal dish collector is assumed at distance, *T*, away from the receiver central axis. The problem can be formulated as follows (Fig 14).

Formulation:

$x_i = 0, x_j = S, dx_i = dx_j = 0, y_i = r\ \sin\theta, y_j = T + H, dy_i = r\ \cos\theta\, d\theta, dy_j = dH, z_i = r$
$\cos\theta, z_j = L, dz_i = -r\ \sin\ \theta\, d\theta, dz_j = dL$

$$R = \sqrt{S^2 + [r\,sin\theta - (T + H)]^2 + (r\,cos\theta - L)^2}$$

$$F_y = \int_{\theta=0}^{2\pi} \int_{0}^{H} \ln\,(R)\,r\,cos\theta\,d\theta\,dH + \int_{\theta=0}^{2\pi} \int_{H}^{0} \ln\,(R)\,r\,cos\theta\,d\theta\,dH \qquad (13)$$

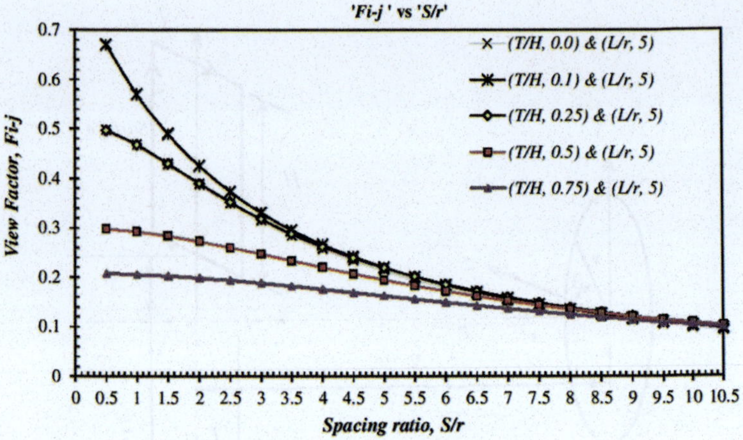

Fig. 15 Variation of configuration factor for different 'offset distance ratio' values

$$F_z = \int_{\theta=0}^{2\pi} \int_{\frac{L}{2}}^{\frac{-L}{2}} -\ln(R) r \sin\theta \, d\theta \, dL + \int_{\theta=0}^{2\pi} \int_{\frac{-L}{2}}^{\frac{L}{2}} -\ln(R) \, r \sin\theta \, d\theta \, dL \tag{14}$$

The configuration factor values have been plotted for various values of 'offset distance ratio' (*T/H*) and 'width ratio' (*L/r*), keeping either of the parameters constant, as shown in Figs. 15 and 16. The common trend is reduction of configuration factor values with increase in spacing between the interacting surfaces. The configuration factor first increases and then reduces with rise in rectangular plane offset distance from the central receiver axis (i.e. *T/H*), as shown in Figs. 15 and 16. The configuration factor improves with increase in width (*L*) of the rectangular plane, as shown in Fig. 17.

While obtaining results for variation of configuration factor with width ratios of 5, 3.33 and 2, the width was kept constant as 50 units and radius values of circular disc were taken as 10, 15 and 25 units, respectively. The size (radius) of circular disc as a radiation-emitting surface influences the configuration factor results at closer spacing ratios (0.5 and 1.0) between the interacting surfaces. It is shown in Fig. 17.

The tilting of rectangular plate produces a tilt angle at the centre of the parabolic dish collector, represented by ψ, as shown in Fig. 18.

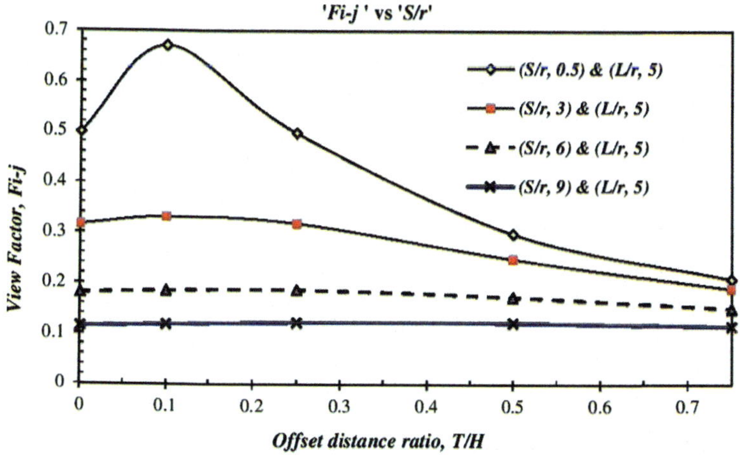

Fig. 16 Variation of configuration factor with offset distance ratio, T/H

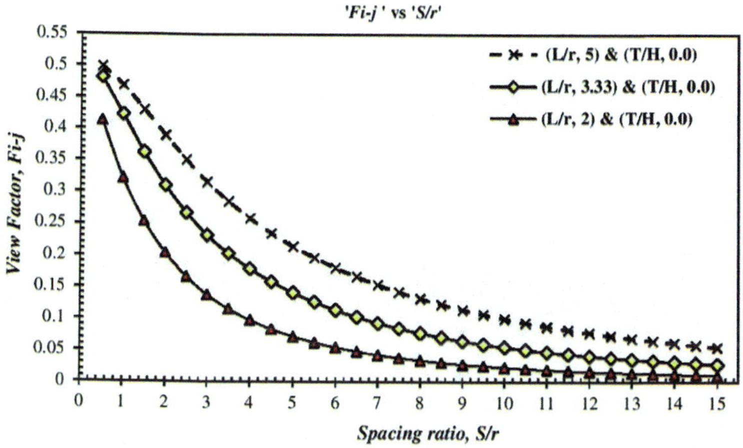

Fig. 17 'Configuration factor' versus 'width ratio'

Fig. 18 Schematic of receiver with circular opening and rectangular plane of reflector

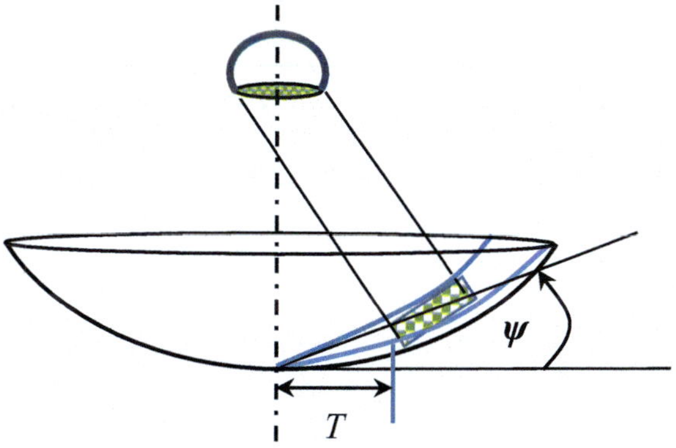

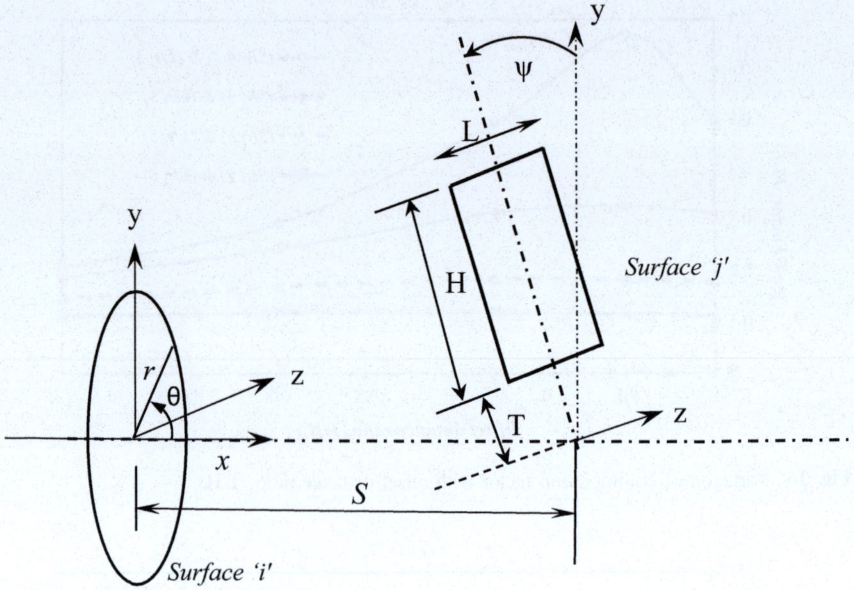

Fig. 19 Problem formulation for tilted rectangular plane

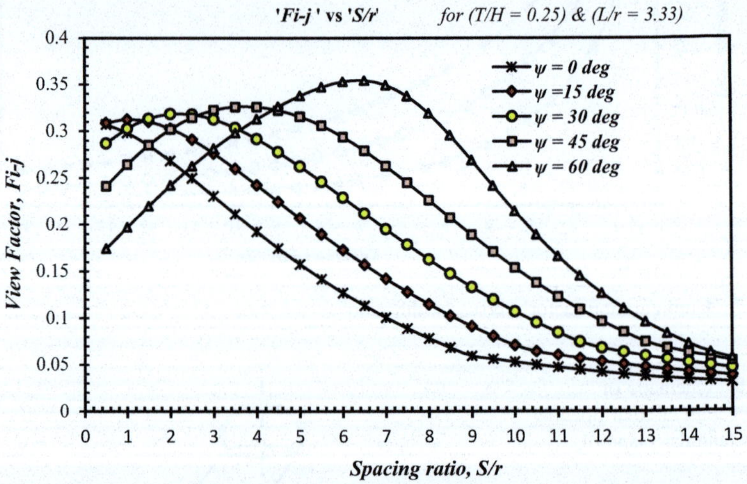

Fig. 20 Configuration factor values for various tilt angles of rectangular plane

3.4 Configuration Factor Between a Circular Disc and Offset and Tilted Rectangular Plate

The present case is applicable for shape factor or radiative exchange between cavity receiver opening and reflector surface of parabolic dish collector as shown in Fig. 18.

In this case, the rectangular plane is tilted by an angle of ψ. The rectangular plane can be assumed with straight edges for smaller sizes. The problem is formulated as follows (Figs. 19 and 20).

Formulation:

$x_i = 0, x_j = S - (T+H)\,sin\,\psi, dx_i = 0, dx_j = -r\,sin\,\psi\,dH, y_i = r\,\sin\theta, y_j = (T+H)$ $cos\,\psi, dy_i = r\,\cos\theta\,d\theta, dy_j = cos\psi\,dH, z_i = r\,\cos\theta, z_j = L, dz_i = -r\,\sin\theta\,d\theta, dz_j = dL$

$$R = \sqrt{(S-(T+H)\sin\psi)^2 + [r\,sin\,\theta - (T+H)\cos\psi]^2 + (r\,cos\,\theta - L)^2}$$

$$F_y = \int_{\theta=0}^{2\pi}\int_{0}^{H} \ln(R)\,r\,cos\,\theta\,\cos\psi\,d\theta\,dH + \int_{\theta=0}^{2\pi}\int_{H}^{0} \ln(R)\,r\,cos\,\theta\,\cos\psi\,d\theta\,dH \quad (14)$$

$$F_z = \int_{\theta=0}^{2\pi}\int_{\frac{L}{2}}^{\frac{-L}{2}} -\ln(R)\,r\,sin\,\theta\,d\theta\,dL + \int_{\theta=0}^{2\pi}\int_{\frac{-L}{2}}^{\frac{L}{2}} -\ln(R)\,r\,sin\,\theta\,d\theta\,dL \quad (15)$$

After tilting rectangular surface as a radiation intercepting/receiving surface, interesting trend of configuration factor values was found. Tilting of rectangular plane surface may either improve or decrease the value of shape factor with increase in spacing between the circular and rectangular plane surfaces. For surfaces with no tilting of rectangular plane, shape factor values exhibit reducing trend with spacing ratio, as usual. For tilted rectangular surface geometry, shape factor values first increase and then decrease after attaining a maxima. Tilting by 15° has reported an optimum spacing ratio of 1. The optimum spacing ratio values for tilt angles of 30°, 45° and 60° were found to be 2.0, 3.5 and 6.5, respectively. For higher spacing ratios, shape factor values reduce asymptotically to zero.

4 Radiation Analysis

The integral Eq. (16) used in present chapter for the estimation of radiation heat transfer among the grey and diffused surfaces is referred [10].

$$\frac{q_i}{\epsilon_i} - \sum_{j=1}^{N} \left(\frac{1}{\epsilon_j} - 1\right) F_{i-j} q_j + H_{oi} = \sum_{j=0}^{N} F_{i-j}(E_{bi} - E_{bj}), i = 0, 1, 2, \ldots\ldots N \quad (16)$$

where

q_i	is the net radiative heat transfer rate at surface ‘*i*’, W/m^2
q_j	is the net radiative heat transfer rate at surface ‘*j*’, W/m^2
ϵ_i *or* ϵ_j	are the emissivity values of surface ‘*i*’ or surface ‘*j*’,
E_{bi} *or* E_{bj}	Emissive power of black surface ‘*i*’ or surface ‘*j*’, W/m^2
H_{oi}	External irradiation reaching at surface ‘*i*’, W/m^2
N	is the number of elements in the enclosure (total equal to 41)

It is known that for a parabolic concave surface, the configuration factor with itself is nonzero for both normal incident radiation and diffuse emission from the reflector surface. The energy exchange and therefore the configuration factor among the surface elements of the collector both are neglected since the temperature variation over the collector surface is seen negligible.

The boundary, through which the external solar irradiation is entering, is assumed adiabatic for the energy leaving out of the enclosure.

Let the receiver opening be considered as surface ‘0’ and parabolic collector surface elements as ‘1–40’. The Surface ‘i’ is considered here as heat source surface and ‘j’ as a sink surface. The segments II and III are divided into 20 elements each. The circular disc-shaped Segment I of collector is represented by differential element 1, the Segment II is divided by elemental surfaces 2, 4, 6, ...40 and elements 3, 5, 7, ...41 are encompassed in Segment III of the collector.

The configuration factor situations discussed are applicable for the estimation of configuration factors between a circular opening of a receiver and elements 2, 4, 6, ...40 of Segment II and elements 3, 5, 7, ...41 of Segment III. The additional details geometrical details required for the configuration factor evaluation are taken as follows.

Centre coordinates of Segment II are $X = \frac{Dc}{6} + \frac{\left(\frac{Dc}{1.5} - \frac{Dc}{3}\right)}{4}$ and $Y = \frac{X^2}{4f}$;

Rim angle at the middle of Segment II is $\delta_2 = tan^{-1}\left[\frac{X}{(f-y)}\right]$

The spacing between the surfaces ‘*i*’ and ‘*j*’ is obtained as $S = \frac{(F-Y)}{(\cos\,\delta_2)^2}$,

r = radius of receiver opening is taken as 0.0125 m and offset distance as ‘*T*’ equal to *Dc*/3.

$$\text{Height of rectangular plane, } H = \left(\frac{D_c}{1.5} - \frac{D_c}{3}\right)/2\cos(\delta_2)$$

Width of the strip, L (circumference at centre of strip/no. of strips, 2 PI * X/N), Tilt angle of rectangular plate, $\psi \sim$ Rim angle at the middle of the strip, X, Y. The details needed for Segment III are also obtained in a similar fashion.

The outcomes of configuration factors obtained for a present analytical model are as follows. These are for parabolic collector diameter (D_c) of 2 m and for diameter of receiver opening ($D_{r\text{-}o}$) of 0.25 m.

Configuration factor between receiver opening and the collector Segment I is equal to 0.09875.

Configuration factors between the cavity opening and the collector segments II and III are equal to 0.03715 and 0.052, respectively. Similarly, various values of configuration factors are obtained for different concentration ratio values ranging from 60 to 2194 for getting receiver wall temperature in the range of 182.8–1000°C. The distribution of external irradiation entering into the collector receiver system over the respective segments I, II and III is obtained as follows.

Table 2 Radiative heat exchange rates for collector emissivity of 0.18

Net radiative heat flux rates at the respective surfaces of the enclosure for emissivity of collector, 0.18

Receiver wall temperature, $T_{r\text{-}w}$ (°C)	182.8	200	250	300	500	650	800	1000
$q_{r\text{-}o}$ (W/m^2)	2073	2971.3	6198.1	10496	42965	90997	169120	338500
$q_{collector}$ (W/m^2)	−4490.4	−3773.4	−1192.7	2238	28193	66569	128980	264550
% $q_{r\text{-}o}$	1.97	0.88	0.62	0.63	0.99	1.44	2.03	3.08
% q_{collr}	−4.27	−1.12	−0.12	0.13	0.65	1.05	1.55	2.40

Table 3 Radiative heat exchange rates for collector emissivity of 0.1

Net radiative heat flux rates at the respective surface temperatures of the enclosure for emissivity of collector, 0.1

Receiver wall temperature, $T_{r\text{-}wall}$ (°C)	182.8	200	250	300	500	650	800	1000
$q_{r\text{-}o}$ (W/m^2)	2599.6	3820.8	8207.7	14050	58193	123490	229690	459980
q_{col} (W/m^2)	−995.8	147	4256.4	9740	51067	112050	211570	427470
% $q_{r\text{-}o}$	2.47	1.14	0.82	0.84	1.34	1.95	2.76	4.19
% q_{collr}	−0.95	0.043	0.43	0.58	1.18	1.77	2.54	3.89

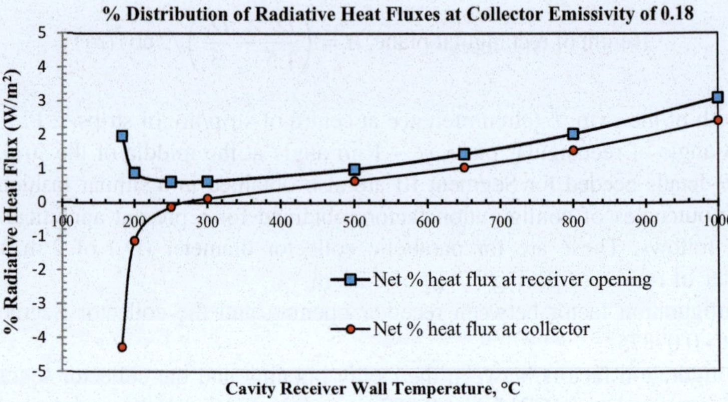

Fig. 21 Percent distribution of radiative heat flux at receiver opening and the parabolic collector for collector emissivity of 0.18

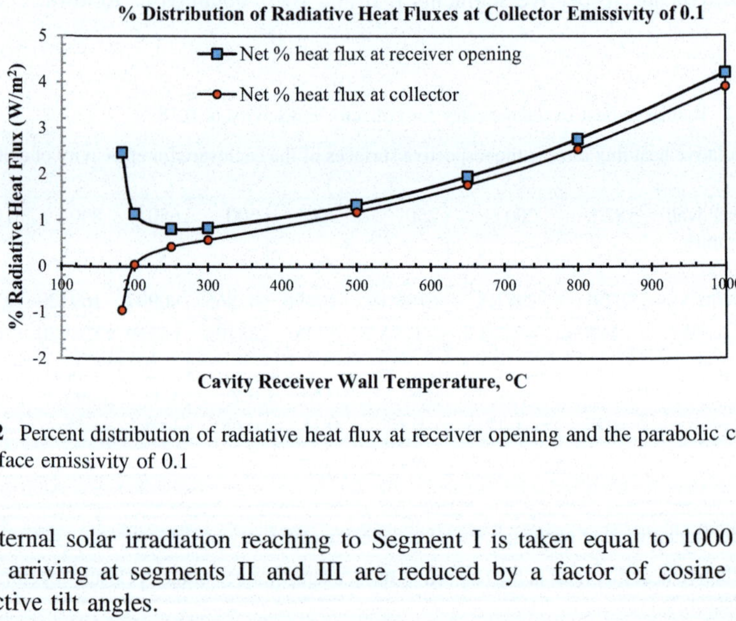

Fig. 22 Percent distribution of radiative heat flux at receiver opening and the parabolic collector for surface emissivity of 0.1

External solar irradiation reaching to Segment I is taken equal to 1000 W/m^2 while arriving at segments II and III are reduced by a factor of cosine of the respective tilt angles.

4.1 Results and Discussion on Radiative Heat Exchange

The algebraic equations obtained through network analysis for 41 unknown heat fluxes at the respective interactive surfaces (e.g. $i = 0, 1, 2, 3, \ldots\ldots.41$) are solved simultaneously. The equations are solved using MATLAB code based on Gauss elimination method. The outcome of radiative heat exchanges at the respective surfaces for various cavity receiver wall temperatures is presented as follows in Tables 2 and 3.

The average receiver wall temperature depends on the concentration ratio, *C*, of the collection system, which decides the concentrated input heat energy or flux at the receiver opening. The radiative heat flux values at the respective surfaces like receiver opening and at the parabolic collector elements are obtained in the presence of continuous external irradiation. The net radiative heat fluxes at the receiver opening and at the collector along with their percentages with respect to the input radiative heat flux at the receiver opening are presented in Tables 2 and 3 for various cavity wall temperatures at collector emissivity values of 0.18 and 0.1.

The variation of per cent distribution of radiative heat fluxes at the receiver opening and the parabolic dish collector for the receiver wall temperatures is as shown in Figs. 21 and 22. The radiative heat exchange shows an increasing trend with cavity wall temperature. In the presence of (solar) external irradiation of 1000 W/m^2, the net heat flux at receiver opening attains a minima at around 250 °C cavity wall temperature and shows increasing trend with receiver wall temperatures. The net heat flux at parabolic dish collector shows a continuously rising trend with the receiver wall temperature.

The values are shifting upwards for reduction in emissivity of parabolic dish collector surface.

5 Conclusions

The contour integral technique was employed to formulate and compute the diffused surface configuration factor values between two surfaces involved in solar parabolic dish-type collector cum cavity-type receiver systems. The numerical outcome of the three cases of configuration factor has been compared with the analytical solution presented by Modest [10]. The numerical results were found to be in good agreement with analytical solution. The radiative heat exchange is obtained using integral equation of radiative heat transfer [10], incorporating the obtained configuration factor values and wall temperatures. The present study concludes as follows.

(1) The proposed mathematical formulations produce correct shape factor results using contour integral technique with error even less than 1%.
(2) The integral equations reported in this chapter for configuration factor may be used for the calculation of diffuse radiation heat exchanges between the collector and receiver surfaces.
    The key objective of discussed work is to estimate the radiative heat transfer between the receiver and the cylindrical parabolic dish collector through which the radiation energy is concentrated in the cavity receiver.
(3) The net radiative heat flux (net heat flux transfers) at the receiver surface is 2% of the flux concentrated by the collector surface at the cavity-type receiver

opening at 182.8 °C. This percentage then reduces to 0.6% at around 250 °C and exhibits continuously rising trend with the cavity wall temperatures up to 1000 °C, for emissivity of reflector surface as 0.18. These values increase or shift up by around 0.5%–1% in case if reflector surface emissivity is maintained at 0.1.

It is believed that the present findings contribute considerably in the area of radiation energy exchange for providing accurate estimation of the energy transfers.

Acknowledgements The author acknowledges financial support of *Science and Engineering Research Board, Department of Science and Technology, Govt. of* INDIA, under ECR 2017 (Project Ref. No. *ECR/2017/000225*), for accomplishing configuration factor evaluation tasks presented in this chapter.

Conflicts of Interest None declared.

References

1. Wang W, Laumert B (2017) Effect of cavity surface material on the concentrated solar flux distribution for an impinging receiver. Sol Energy Mater Sol Cells 161:177–182
2. Wu SY, Xiao L, Cao Y, Li YR (2010) Convection heat loss from cavity receiver in parabolic dish solar thermal power system: a review. Sol Energy 84:1342–1355
3. Shavazi EA, Hughes GO, Pye JD (2015) Investigation of heat loss from a solar cavity receiver. Energy Procedia 69:269–278
4. Reddy KS, Kumar NS (2008) Combined laminar natural convection and surface radiation heat transfer in a modified cavity receiver of solar parabolic dish. Int J Thermal Sci 47:1647–1657
5. Hachicha AA, Rodriguez I, Castro J, Oliva A (2013) Numerical simulation of wind flow around a parabolic trough solar collector. Appl Energy 107:426–437
6. Uhlig R, Flesch R, Gobereit B, Giuliano S, Liedke P (2014) Strategies enhancing efficiency of cavity receivers. Energy Procedia 49:538–550
7. Shuai Y, Xia XL, Tan HP (2008) Radiation performance of dish solar concentrator/cavity receiver systems. Sol Energy 82:13–21
8. Kumar NS, Reddy KS (2008) Comparison of receivers for solar dish collector system. Energy Convers Manag 49:812–819
9. Reddy KS, Srihari Vikram T, Veershetty G (2015) Combined heat loss analysis of solar parabolic dish modified cavity receiver for superheated steam generation. Sol Energy 121:78–93
10. Modest MF (ed) (2003) Radiative heat transfer, 2nd Edn. Academic Press, New York, Chapter 4, pp 131–158
11. Cabeza-Lainez JM, Pulido-Arcas JA, Castila MV (2013) new configuration factor between a circle, a sphere and a differential area at arbitrary positions. J Quan Spectro Rad Transf 129:272–276
12. Dixon JC (1984) Heat transfer in specularly reflecting tubes. Int J Heat Fluid Flow 5(4):222–228
13. Rammohan Rao V, Sastri VMK (1996) Efficient evaluation of diffuse view factors for radiation. Int J Heat Mass Transf 39(6):1281–1286
14. Diaconu BM, Cruceru M, Gheorghian AT, Popescu LG (2012) View factors in a finite length axy-symmetric cylindrical annulus enclosure. J Quant Spectros Radiat Transf 113:2100–2112

15. Maor T, Appelbaum J (2012) View factors of photovoltaic collector systems. Sol Energy 86:1701–1708
16. National Physical Laboratory website. http://www.kayelaby.npl.co.uk/general_physics/2_3/2_3_7.html, 07/03/2017
17. http://www.azom.com/properties.aspx?ArticleID=863, 07/03/2017

Numerical Investigation of the Temperature Distribution of a Solar Cavity Receiver Wall Using Finite Element Method

Suneet Kumar and Santosh B. Bopche

Abstract The finite element method has been proved to be the effective numerical tool for finding approximate solutions of two or even up to three-dimensional governing differential equations. It has an ability to handle irregular geometries and any number of complicated boundary conditions with ease. The cavity receiver of a solar thermal system operates at a very high temperature. It may also cause larger energy losses from the cavity by radiation as well as convection as dominant modes of heat transfer. The collection efficiency depends on the useful heat gained by the working fluids, which is influenced by the receiver-energy losses. The energy losses from the cavity receiver depend on the cavity wall surface temperature. The temperature variation depends on the boundary conditions as well as geometrical orientation of the receiver. The knowledge of temperature distribution is one of the important factors needed for evolving an ideal design of a cavity receiver. The present chapter focuses on the design aspects as well as approximate estimation of wall temperature distribution of a cavity receiver of cylindrical shape. The step by step formulation of the problem using finite element method is also presented in the present chapter. The computational domain of a receiver wall is discretised into number of triangular elements and the simultaneous equations are solved using MATLAB.

Keywords Finite element method · Solar receiver · Cavity · Temperature
Heat transfer

S. Kumar · S. B. Bopche (✉)
Department of Mechanical Engineering, National Institute of Technology Hamirpur,
Hamirpur 177005, HP, India
e-mail: santoshb.bopche@yahoo.co.in

S. Kumar
e-mail: suneet2k6@gmail.com

H. Tyagi et al. (eds.), *Applications of Solar Energy*, Energy, Environment,
and Sustainability, https://doi.org/10.1007/978-981-10-7206-2_4

Nomenclature

A	Area, m^2
B, C	Constants for a given triangle of mesh
C_p	Coefficient of heat of air at constant pressure, J/kg K
g	Gravitational acceleration, m/s^2
h	Coefficient of heat transfer, W/m^2K
k	Coefficient of conduction conductivity, W m^{-1} K^{-1}
l, m, n	Surface normals or direction cosines
N	Shape function
p	Pressure
q_g	Heat generated kW/m^3
q	Heat flux W/m^2
S	Surfaces
T	Temperature, °C
T	Surrounding temperature, °C
T_i	Fluid temperature inside the cavity, °C
x, y, z	Cartesian coordinates

Subscripts

c	Collector opening
e	Element related
o	Receiver opening
s	Surface
i, j, k	Nodes of the triangular element
i	Internal details

1 Introduction

As is well known, the finite element method (FEM) is a numerical tool for handling and analysing technological problems having practical interests. Obtaining an analytical solution has certain limitations. So, for problems with complicated geometries and boundary conditions, finite element is an effective technique.

Lot of research has been reported on performance improvement of solar cavity receiver systems through various designs of a solar cavity receiver. Some contributions are discussed in this chapter for providing brief information of the subject dealing with. Jarvinen [1] has studied different shapes of solar cavity receiver such as domed surface, aperture and un-apertured surfaces and also examined the various heat losses from the solar cavity receiver. The material chosen for construction of solar cavity is ceramic and jet impingement technique is adopted for cooling the receiver wall. With ceramic as a cavity wall material higher thermal conversion efficiency up to 80–90% can be achieved. Ceramic materials like alumina, cordierite, silicon nitride, etc., can also be used as a material for construction of solar cavity. In cavity receivers, domes with low radius are used due to room limitations. Higher thermal conversion efficiency is achieved by using jet impingement technique in solar cavity receivers. Thermal efficiency of 80% is achieved in open-ended solar cavity and this value increases up to 87% by using entrance kind of reflector and provision of aperture baffle reduces the cavity entry space by about 30%. A solar-based Brayton cycle with such a receiver operates at 38% overall cycle efficiency and by using deeper solar cavity receivers, overall efficiency of the cycle can be increased little more.

The high temperature compressed air can be produced using solar cavity receivers which may be used for various applications such as electric power, fuels and chemicals or in the field of industrial process heating [2]. Various designs of heat exchangers are developed with different configurations and have been tested for some important parameters such as pressure drop, structural reliability and fabricability. The most favourable design recommended by the author is 'Ceramic receiver wall with fins'. As discussed earlier, a large temperature solar receiver with transparent glass (made of quartz) is cooled by impinging air jets originating from six to nine locations, located symmetrically around the hemispherical solar cavity receiver, as shown in Fig. 1. In second configuration, swirl is developed inside the solar window cavity by changing the angles of the nozzles [3].

The heat transfer distribution is seen differently in cases of swirled as well as non-swirled nozzle arrangement of 'jet air cooling system'. Adopting this jet impingement cooling technique the top of the case is not seen to be cooled to the desired level in symmetric jet case. The unsymmetric air jet cooling scheme has been seen to be the promising cooling technique to homogenise the heat transfer distribution and to produce the necessary heat fluxes at the top of the window cavity. It is also reported that the nozzle configuration can be altered for specified solar cavity receiver temperature.

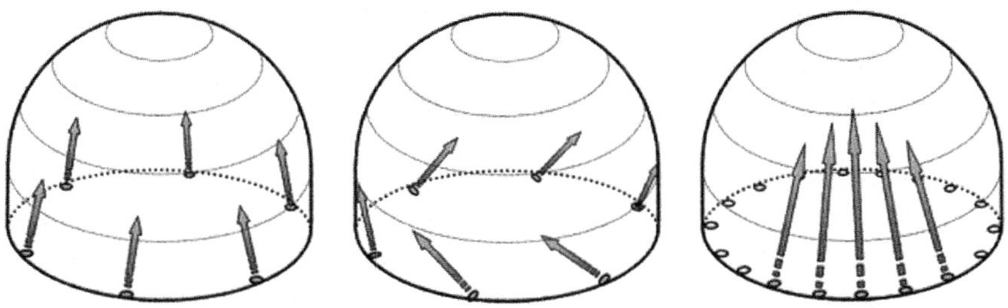

Symmetric (6 and 9 nozzles altenatively)

Symmetric,swireled (6 nozzles)

Asymmetric, modulated (3/4/5/6 nozzles,with mass flow modulation)

Fig. 1 Nozzle arrangement of impinging jets [3]

Sauceda et al. [4] have examined the solar cavity receiver of conical shape lighted by a parboiled dish concentrator. The main aim has been to check whether it accomplishes the demand of energy as well as quality of a solar thermal cooling system. The conical receiver under investigation has diameter of 19 cm and height 20 cm located at the focus point 80 cm from the vertex of perfectly reflecting parabolic dish of 190 cm diameter. The study is performed in three cases such as (a) smooth receiver, (b) smooth receiver having glass cover and (c) receiver having fins inside of conical surface. It is seen that the conical receiver provided with fins and without glass cover has highest efficiency. The performance of receiver with no glass cover is about 10% higher than that with smooth receiver.

Kribus et al. [5] have established operation of the receiver system with two consecutive stages under huge temperatures and larger fluxes of concentrated radiation. In this test series, the air temperature at exit reaches up to 1000 °C and receiver wall is found capable to handle a maximum temperature of 1200 °C. The optical loss in the optical subsystems is one of the main reasons of reducing maximum designed wall temperature in the receiver. It is also to be noted that there is no straight measurement of concentrated power on the central receiver type solar collection systems. Keeping on priority the equivalent radiation heat flux, the authors have recommended the downward facing type cavity receivers [6]. The uniformity of solar flux concentrated on the receiver wall is compared with five traditional geometries as cylindrical, dome, hetero-conical, elliptical, spherical, and conical. The results show that shape of the solar cavity receiver has considerable effect on distribution of overall heat flux. A receiver with spherical cavity and with quite better radiation distributive performance initiates the process of shape optimisation for optimised solar thermal performance.

Buck et al. [7] have conceptualised a dual type receiver which improves the applicability of the central receiver for the solar steam based thermal power plant. The results show various types of advantages of new idea such as increase in the thermal performance of the solar cavity receiver, decreased in receiver wall temperature and heat transfer losses. With the help of these modifications the annual output can be increased by about 27% as compared to air heating system based on solar system as well as cost is also reduced by reducing spread area of heliostats and receiver wall temperatures.

In case of solar cavity receivers, the convective as well as radiative heat transfer losses are the most important constituents of the thermal losses. Prakash et al. [8] have found that the output of solar cavity receiver is influenced badly by optical as well as thermal heat losses. They have carried out a numerical as well as experimental investigation for calculation of convective mode of energy losses from a down-facing cylindrical solar cavity type receiver for steady state condition. Experimental study is carried out at different tilt angles of the receiver cavity such as 0°, 30°, 45°, 60° and 90° for fluid temperatures at entry ranging from 50 °C to 75 °C. For numerical investigation, Fluent CFD software is used for examining the effect of cavity tilting and coolant inlet temperature (ranging from 50 °C to 300 °C) on the efficiency of the solar thermal receiver system.

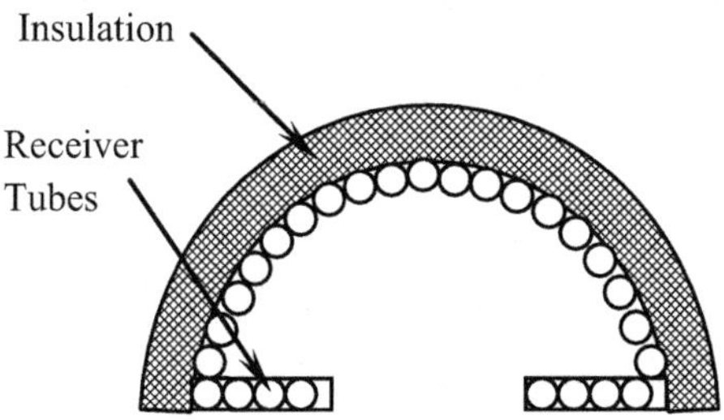

Fig. 2 Modified solar cavity receiver

Fewer studies are contributed on modified cavity receives in the literature. The schematic of modified cavity type receiver is as shown in Fig. 2. Before studying the three-dimensional model of the modified cavity receiver Reddy and Sendhil Kumar [9] have performed the investigations of a 2-D model of modified solar cavity receiver to look at the losses of natural convective type from the cavity opening. They have pursued the study for 2-D model of solar cavity for two different boundary conditions as; (1) with no insulation and (2) with insulation at the bottom of the cavity and for three-dimensional model with no insulation at the bottom of the aperture boundary condition. A separate study is performed for both the cases of 2-D as well as 3-D to develop the Nu (Nusselt Number) correlations. The natural convective heat losses are also compared between the two models. The results show that for both 2-D and 3-D model of the solar cavity receiver maximum convective energy loss occurs at cavity inclination angle of 0° and minimum heat escape occurs at cavity tilt angle of 90°.

The difference between the Nusselt numbers obtained through two-dimensional and three-dimensional analyses is larger for smaller tilting angles of a cavity and is smaller for larger tilting angles of a cavity. Two-dimensional solar cavity models have been suggested for accurate calculation of free convective losses for larger tilting angles of cavity and three-dimensional models can be used for exact calculation of convective losses for lower tilting angles. The computational efforts required for analysing 3-D computational model is five times larger than 2-D model for a grid of 15 times finer than that of 2-D model.

Prakash et al. [10] have performed the investigations to calculate the natural convective energy transfers from open cavity to outer surrounding. Three different types of solar cavity receiver shapes are studied such as hemispherical, cubical and spherical of equal heat transfer surface areas. For the said shapes of the cavity a three-dimensional numerical study is performed by using Fluent CFD software. For the hemispherical and cubical cavity shapes the effect of opening ratios; 1, 0.5 and 0.25 on natural convective heat losses have been examined and for spherical cavity opening ratio is taken as 0.5 and 0.25. The effect of inclination angles; 0°, 30°, 45°, 60° and 90° on convective heat losses has also been studied. It is seen that the magnitude of natural convective heat losses exhibits an increasing trend with aperture ratio. It is also found that when the opening ratio of receiver cavity is increased from 0.25 to 0.5 for almost all shapes and for different tilting angles of the cavity the natural convective heat transfer rate increases from 30% to 80% with the surrounding. With the increase of the cavity wall temperature, convective losses also show an increasing trend. When cavity tilting angle is decreased the heat loss rises and larger heat loss from the solar cavity is found at the tilting angle of 0° and lowest at the 90° inclination angle from the horizontal axis. When all the cavity shapes are studied for aperture ratio of 0.25 and 0.5, it is found that the hemispherical cavity has the largest heat losses as compared to the cubical and spherical cavities. Shavazi et al. [11] have also developed an experimental facility to guesstimate the heat loss from the solar cavity type receiver. A cylindrical cavity of aspect ratio of 1 and 2 and aperture opening ratio of 0.5 and 1 is chosen for the investigation. The tests have also been continued for constant heat input through heat cables and for various inclination angles. The variation of cavity wall temperature is noticed on account of heat losses. The convective as well as radiative energy transfers found varying with tilting angle of solar cavity.

Yuan and Ho [12] have calculated the energy loss from the solar cavity type receiver using CFD-FLUENT 13.0 and SOLIDWORKS 2011. The results have been compared with the experimental outcomes of a cubical solar cavity. On simulating a cubical cavity, it is found that convective heat transfer is under-predicted by 45% and radiative energy loss is correctly estimated for a constant wall temperature boundary condition. However, for constant wall flux boundary condition convective energy loss is found to be under-predicted only by 32%. For cylindrical cavity receiver, the convective heat loss is over predicted for larger inclination angles, i.e. above 60° but for other angles the results are in good agreement.

For the direct normal irradiance of 800 and 1000 W/m^2, the temperature difference in the absorber is reduced to 130 °C and 149 °C, respectively, which are lower than previous cavity designs. Neglecting the natural convective heat transfer loss from the aperture opening of the cavity and conduction loss from the cavity insulation material, the average air outlet temperature and thermal efficiency of the receiver are found to be 798.4 °C and 82.7%, respectively, for direct normal irradiance of 800 W/m^2. For all the cases the pressure drop is limited to 3%.

Cao [13] used a heat-pipe-based solar receiver system that improves the efficiency of the solar cavity receiver. It is found that this type of receiver has larger value of heat transfer coefficient, larger collection efficiency, higher heat transfer capability and lower cost. It is driven by gravity has simple structure and uses low temperature alloy materials. Similar such solar cavity receiver is developed by Bannister and Mayer [14] for white cliffs solar thermal power plant. Computer simulation is used to design the solar cavity receiver and the outcomes are compared and found agreeing well with the physically obtained results.

Affandi et al. [15] have analysed a parabolic dish collector, capable of converting 31.25% solar irradiance into the electricity. The reflecting materials, e.g. Aluminium and silver are used with intercept factor of 0.9–1. It is reported that the convective as well as radiative heat losses from the cavity receiver depend on the solar irradiation, material reflectivity and intercept factor. There is direct influence of receiver reflectivity on the solar radiation concentrated at the entry to the cavity receiver.

Wang and Laumert [16] have used a ray-tracing methodology in order to investigate the effect of material of the solar cavity on the distribution of solar radiative flux inside the receiver. Three metals: Inconel (Grade-600), Aust. SS (Stainless steel) and Kanthal-APM and two coating type materials (Pyromark 2500 and YSZ TBC coating) have been tested for better performance. It was found that emissivity of the top cavity surface majorly influences the solar flux distribution and optical efficiency as compared to the absorptivity of the bottom surface.

Pozivil et al. [17] have designed a pressurised type solar central receiver for 3 and 35 kW solar to thermal power unit based on Brayton cycle. An ultimate goal is to enhance the thermal performance of the receiver maximum up to around 80%. Zheng et al. [18] have performed the study in order to enhance the convective energy transfer rate in a solar cavity type receiver tube filled up by a porous material. It is concluded that better thermal performance can be accomplished by reducing thermal conductivity ratio between the porous materials and cooling medium even less than 1000.

Lopez et al. [19] have investigated the influence of the optical properties of the coated material inside the cavity surface on thermal performance of the solar cavity type receiver. The cavity surface is coated with material having high thermal absorptivity in order to increase the light-thermal heat conversion. Different types of coating materials have been tested numerically and the outcomes of which have been compared with that of the standard coating 'Pyromark-2500'. It is seen that thermal efficiency of the solar cavity receiver rises with the gain in absorptivity of coating material.

Reddy and Sendhil Kumar [20] have investigated the collective influence of free-convection and radiative energy losses in case of modified solar cavity type receiver. The effect of the working temperature, absorptivity of the cavity surface, orientation and geometry on the overall energy loss from the cavity has been studied. It's seen that convective energy transfer from the receiver is mainly affected by the tilt angle of the cavity and radiative heat loss specifically by the optical properties of the cavity surface. It is seen that the distribution of convective as well as radiative energy losses in case of modified cavity receiver having area ratio of about 8 and operating wall temperature of 400 °C is 52 and 71.34% of the overall energy loss for 0° tilt angle and 40.7 and 59.28% for 90° tilting angle, respectively. In addition to above studies reported Wang et al. [21] have examined the effect of multiple nozzles embedded around the solar cavity receiver. Solar cavity receiver material is taken as stainless steel, and solar cavity is combined with micro gas turbine and solar collector system. It is reported that these cavity materials operates at temperatures much lesser than their melting points even at direct normal irradiance of 800 and 1000 W/m^2.

This chapter discusses the estimation of cylindrical solar cavity wall temperature distribution using FEM. The external boundary surface is subjected to a forced convection boundary condition, i.e. of coolant. The receiver opening and the internal side of a solar cavity wall both are subjected to a concentrated solar radiation heat flux, radiation intensified by a 2.5 m diameter parabolic dish type solar collector. This chapter specifically investigates the distribution of receiver wall temperature for a cylindrical shaped cavity type receiver made of stainless steel designed for direct normal irradiance of 1000 W/m^2. The additional objectives of the present chapter are to formulate the problem of forced cooling of a cylindrical solar cavity by using finite element method and to obtain a steady state temperature distribution of the walls of a hollow cylindrical solar cavity receiver.

2 Methodology

The methodology adopted for fulfilling the above-mentioned objectives is discussed.

The concentrated solar radiation flux is focussed at the receiver opening. It is assumed that radiation flux is distributed uniformly inside the receiver cavity surface. The outside of the receiver wall is subjected to the forced convection boundary condition of air as a cooling medium. The design of the analytical model developed is provided as follows.

2.1 Problem Formulation Using FEM

The solid wall of the cavity receiver is discretised by uniform triangular elements which are formulated using finite element method. The governing differential equation with constant wall temperature, constant wall flux and convection or radiation boundary conditions is used for initialising the numerical simulation. The schematic is as shown in Fig. 3a.

The governing differential equation for heat transfer through a slab with uniform heat generation is

$$\frac{\partial}{\partial x}\left(k_x\frac{\partial T}{\partial x}\right)+\frac{\partial}{\partial y}\left(k_y\frac{\partial T}{\partial y}\right)+\frac{\partial}{\partial z}\left(k_z\frac{\partial T}{\partial z}\right)+q_g=0 \tag{1}$$

With the following boundary conditions

(1) Surface S_1 is at constant temperature,

$$\mathrm{T}=\mathrm{T_b} \tag{2}$$

(2) Surface S_2 is subjected to a constant and uniform heat flux (e.g. radiative)

$$k_x\frac{\partial T}{\partial x}l+k_y\frac{\partial T}{\partial y}m+k_z\frac{\partial T}{\partial z}n+q=0 \tag{3}$$

(3) Surface S_3 is subjected to a convective heat transfer boundary

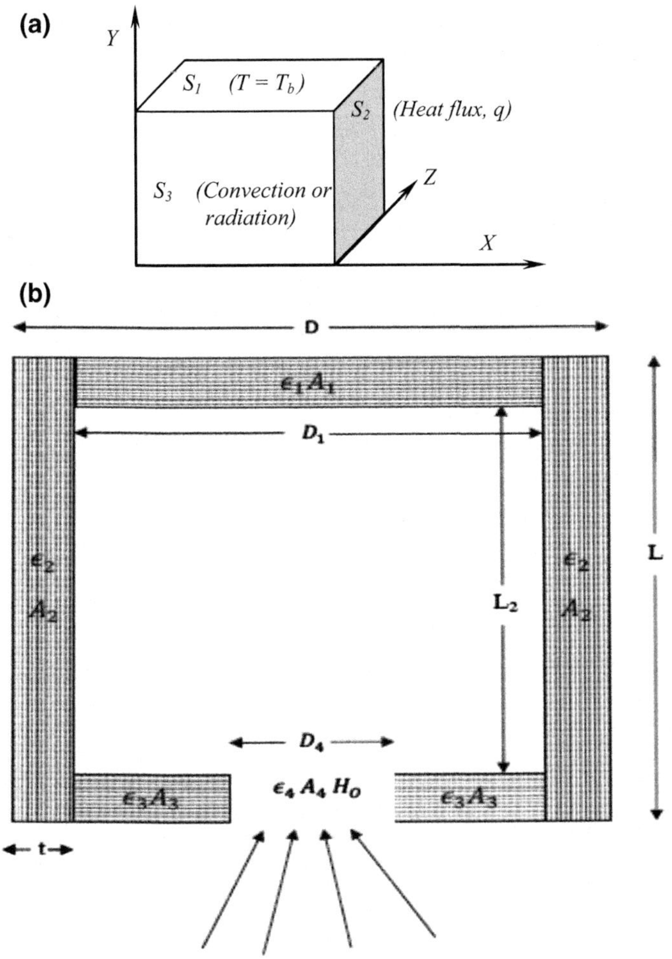

Fig. 3 **a** Schematic showing boundary conditions **b** computational domain

$$k_x \frac{\partial T}{\partial x} l + k_y \frac{\partial T}{\partial y} m + k_z \frac{\partial T}{\partial z} n + h\,(T - T_\infty) = 0 \tag{4}$$

Where l, m and n are surface normals or direction cosines along X, Y and Z axes, respectively.

The problem solution yield in mathematical form can be applied to a metallic wall which is subjected to different types of boundary conditions.

The variational integral I corresponding to the above differential equation with its boundary conditions is given as,

$$\mathrm{I(T)} = \frac{1}{2}\int\int\int\left[k_x\left(\frac{\partial T^e}{\partial x}\right)^2 + k_y\left(\frac{\partial T^e}{\partial y}\right)^2 + k_z\left(\frac{\partial T^e}{\partial z}\right)^2 - 2q_g T\right]dv$$
$$+ \int\int q\,T\,ds + \int\int 0.5\,h\,(T - T_\infty)^2 ds \quad (5)$$

$$\mathrm{I_e} = \int\int\int\frac{1}{2}\left[k_x\left(\frac{\partial T^e}{\partial x}\right)^2 + k_y\left(\frac{\partial T^e}{\partial y}\right)^2 + k_z\left(\frac{\partial T^e}{\partial z}\right)^2 - 2q_g T^e\right]dv$$
$$+ \int\int qT^e ds + \int\int 0.5\,h\,(T^e - T_\infty)^2 ds \quad (6)$$

where

$$T^e = [N]\{T\} = [N_1, N_2, N_3, N_4 \ldots\ldots\ldots\ldots N_r][T_1, T_2 \ldots\ldots\ldots\ldots T_r]^T$$

The gradient matrix is written as

$$\{g\} = \begin{Bmatrix} \frac{\partial T^e}{\partial x} \\ \frac{\partial T^e}{\partial y} \\ \frac{\partial T^e}{\partial z} \end{Bmatrix} = \begin{bmatrix} \frac{\partial N_1}{\partial x} & \cdots & \frac{\partial Nr}{\partial x} \\ \frac{\partial N_1}{\partial y} & \cdots & \frac{\partial N_r}{\partial y} \\ \frac{\partial N_1}{\partial z} & \cdots & \frac{\partial N_r}{\partial z} \end{bmatrix} \begin{bmatrix} T_1 \\ \cdots \\ T_r \end{bmatrix} = [B]\{T\} \quad (7)$$

Consider

$$\{g\}^T[D]\{g\} = \left\{\frac{\partial T^e}{\partial x}\frac{\partial T^e}{\partial y}\frac{\partial T^e}{\partial z}\right\}\begin{bmatrix} k_x & 0 & 0 \\ 0 & k_y & 0 \\ 0 & 0 & k_z \end{bmatrix}\begin{Bmatrix} \frac{\partial T^e}{\partial x} \\ \frac{\partial T^e}{\partial y} \\ \frac{\partial T^e}{\partial z} \end{Bmatrix}$$
$$= k_x\left(\frac{\partial T^e}{\partial x}\right)^2 + k_y\left(\frac{\partial T^e}{\partial y}\right)^2 + k_z\left(\frac{\partial T^e}{\partial z}\right)^2 \quad (8)$$

Substitute above equation in (6) we get

$$I_e = \frac{1}{2}\int\int\int\left[\{g\}^T[D]\{g\} - 2q_g T^e\right]dv + \int\int qT^e ds + \int\int 0.5h(T^e - T_\infty)^2 ds \quad (9)$$

Finally we have

$$\frac{\partial T^e}{\partial\{T\}} = \int\int\int [B]^T[D][B]\{T\}dv$$
$$- \int\int q_g[N]^T dv + \int\int h[N]^T[T]dsh - \int\int h\,[N]^T T_\infty ds = 0 \quad (10)$$

Above equation can be written in compact form

$$[\boldsymbol{K}]\{\boldsymbol{T}\} = \{\boldsymbol{f}\} \tag{11}$$

Where

$[K] = \int\int\int [B]^T[D][B]\,dv + \int\int hN^T[N]\,ds$, *Matrix* $\{T\}$ are the unknown nodal temperatures and

$\{f\} = \int\int\int q_g[N]^T dv - \int\int q[N]^T ds + \int\int hT_\infty [N]^T ds$ involves heat generation as well as boundary terms

The finite element formulation of the equations for situations with and without the convective heat transfer inside the solar cavity receiver is provide as follows,

Equations for without inside convection heat transfer is given in the form of

$$[\boldsymbol{K}]\{\boldsymbol{T}\} = \{\boldsymbol{f}\},$$

as

$$[K] = \int\int\int [B]^T[D][B]dv + \int\int hN^T[N]ds$$

(or)

$$[K] = \frac{tk}{4A}\begin{bmatrix} B_i^2 + C_i^2 & B_iB_j + C_iC_j & B_iB_k + C_iC_k \\ B_iB_j + C_iC_j & B_j^2 + C_j^2 & B_jB_k + C_jC_k \\ B_iB_k + C_iC_k & B_jB_k + C_jC_k & B_k^2 + C_k^2 \end{bmatrix} + \frac{h \times t \times l_{jk}}{6}\begin{bmatrix} 0 & 0 & 0 \\ 0 & 2 & 1 \\ 0 & 1 & 2 \end{bmatrix}. \tag{12}$$

and,

$$\{f\} = \int\int q\,[N]^T ds + \int\int hT_a\,[N]^T ds \tag{13}$$

$$[f] = \frac{q \times t \times l_{ij}}{2}\begin{bmatrix} 1 \\ 1 \\ 0 \end{bmatrix} + \frac{h \times T_a \times t \times l_{jk}}{2}\begin{bmatrix} 0 \\ 1 \\ 1 \end{bmatrix} \tag{14}$$

Equations for analysing situation with inside convection heat transfer is given as

$$[\boldsymbol{K}]\{\boldsymbol{T}\} = \{\boldsymbol{f}\} \tag{15}$$

Where

$$[K] = \iiint [B]^T[D][B]dv + \iint hN^T[N]\,ds + \iint h_iN^T[N]\,ds$$

$$\{f\} = \iint q\,[N]^T ds + \iint hT_a[N]^T ds + \iint h_iT_i\,[N]^T ds$$

$$[K] = \frac{tk}{4A}\begin{bmatrix} B_i^2 + C_i^2 & B_iB_j + C_iC_j & B_iB_k + C_iC_k \\ B_iB_j + C_iC_j & B_j^2 + C_j^2 & B_jB_k + C_jC_k \\ B_iB_k + C_iC_k & B_jB_k + C_jC_k & B_k^2 + C_k^2 \end{bmatrix} + \frac{h \times t \times l_{jk}}{6}\begin{bmatrix} 0 & 0 & 0 \\ 0 & 2 & 1 \\ 0 & 1 & 2 \end{bmatrix} + \frac{h_i \times t \times l_{jk}}{6}\begin{bmatrix} 0 & 0 & 0 \\ 0 & 2 & 1 \\ 0 & 1 & 2 \end{bmatrix}$$

$$[f] = \frac{q \times t \times l_{ij}}{2}\begin{bmatrix} 1 \\ 1 \\ 0 \end{bmatrix} + \frac{h \times T_a \times t \times l_{jk}}{2}\begin{bmatrix} 0 \\ 1 \\ 1 \end{bmatrix} + \frac{h_i \times T_i \times t \times l_{jk}}{2}\begin{bmatrix} 0 \\ 1 \\ 1 \end{bmatrix}$$

These set of matrices give simultaneous equations in terms of unknown temperatures at respective nodes of the computational domain. These simultaneous algebraic equations are solved using a computer code based on 'Gauss elimination technique' which is compiled in MATLAB. The obtained temperatures are plotted using MATLAB software in the form of colour contours.

2.2 Data Selection

As discussed the computational domain encompass radiative heat flux from inside, conduction through the wall and the heat from the wall is carried by the coolant (air) from outside. The details essential for computational analysis are discussed as follows.

The mass flow and temperature details of a 5 kW gas turbine power plant are considered. The combustion chamber is hereby replaced by the present cavity receiver.

1. Mass flow rate of air as a coolant = 0.1 kg/s [21]
2. Temperature of air, inlet to compressor = 300 K (Open cycle)
3. Turbine inlet pressure = 3 Bar
4. Compressor pressure ratio = 3 and using $\frac{T_2}{T_1} = \left(\frac{P_2}{P_1}\right)^{\frac{1.4-1}{1.4}}$

   The temperature at the outlet of the compressor is obtained as, $T_2 = 137$ °C which is entry temperature to the cavity receiver. It is approximated to 150 °C at entry to the receiver.

5. Required working fluid temperature at the outlet of receiver = 800–1000 °C (taking, 850 °C)
6. The specific heat capacity of air as a working fluid, C_p = 1.005 kJ/*kg-K*

Carrying out an energy balance for heat gained by the coolant and the heat transferred at the surface,

$$q = m \times C_p \times \Delta T = h \times A \times (T_{wall} - T_{mean}) \tag{16}$$

$$q = m \times C_p \times \Delta T = 0.1 \times 1.005 \times (850 - 150) = 70.35 \text{ kW}$$

Where, $T_{mean} = \frac{(850+150)}{2} = 500\,^{\circ}\text{C}$

The coolant side surface area of a receiver, $A_s = (3.14 \times 0.5 \times 0.6) + \left(\frac{3.14}{4 \times 0.5^2}\right) = 1.13825 \text{ m}^2$

$$\begin{aligned} &q = h \times A \times (T_{wall} - T_{mean}) \\ &70.35 \times 10^3 = h \times 1.13825 \times (1000 - 500) \\ &h = 123.61 \text{ W/m}^2\text{K} \end{aligned} \tag{17}$$

So, the coolant side heat transfer coefficient is taken as, $h = 123.61 \text{ W/m}^2\text{K}$.
Material of construction = Stainless steel [21]
Thermal conductivity, $\text{K} = 18 \text{ W/mK}$ for operating temperature of 500 °C.

Calculation of concentration ratio, C

The diameter of a parabolic collector opening is taken as 2 m while the aperture diameter of receiver aperture is as 25 cm. The concentration ratio and the intensity of concentrated heat flux at the receiver opening are obtained as follows;

$$\begin{aligned} &\text{Receiver aperture area}, A_o = \frac{3.14}{4 \times 0.25^2} = 0.049 \text{m}^2 \\ &\text{Solar collector area of opening}, A_c = \frac{3.14}{4 \times 2.5^2} = 4.906 \\ &\text{Concentration ratio (C. R.)} = \frac{\textit{Solar collector area}, A_c}{\textit{Aperture area}, A_o} = \frac{4.906}{0.049} = 100 \\ &\textit{Concentration ratio} = \frac{\textit{Local Intensity}}{\textit{Incident intensity}} \\ &100 = \frac{I_{local}}{800} \\ &I_{local} = 80 \text{ kW/m}^2 \end{aligned} \tag{18}$$

Considering emissivity of the receiver surface from inside is equal to 0.8.

$$\begin{aligned} I_{local} &= 80 \times 0.8 \text{ kW/m}^2 \\ &= 64000 \text{ W/m}^2 \end{aligned}$$

The heat flux at the entry to the receiver opening is taken as 64000 W/m^2.

The heat transfer coefficient inside (entrapped volume in the cavity) the receiver cavity is taken as $6 \text{ W/m}^2\text{K}$ [22].

The additional assumptions of the study include uniform cavity surface emissivity of 0.8, distribution of radiative heat energy is uniform inside the cavity and radiative energy losses from the receiver opening are neglected.

2.3 *Computational Domain and Discretisation*

The computational domain consists of a cylindrical cavity receiver of dimensions as shown in Fig. 3b.

External dimension (diameter) of the receiver, $D = 50\,\text{cm}$,
Height of the cavity of receiver, $L = 60\,\text{cm}$,
Internal depth of the receiver, $L_2 = 58.8\,\text{cm}$,
Thickness of cavity wall, $t = 6\,\text{mm}$,
Internal diameter of the receiver, $D_1 = 48.8\,\text{cm}$,
Diameter of the receiver opening, $D_4 = 25\,\text{cm}$,
Emissivity of the surfaces, $\epsilon = 0.8$,
Surface areas the cavity wall, A.

External solar irradiation concentrated at the receiver opening, $H_o\,\text{W/m}^2$.

The cylindrical receiver wall is discretised symmetrically using 18 triangular elements having overall 20 number of nodes. This is the coarser mesh (mesh 1 with 18 elements) of the present analysis. The outcome is compared with the finer meshes, e.g. mesh 2 with 72 elements, mesh 3 with 336 elements. The temperature distribution with the studied meshes is as shown in Fig. 4.

The temperature contours are seen to be smoothly approximating with increase in fineness of the computational mesh. Minor temperature variation is seen between the outcomes with 20 and 72 elements whereas considerable differences in outcome are noticed with 336 elements (i.e. finer grid).

2.4 *Grid Independence Test*

The outcome of the temperatures at the respective nodes are compared herewith the experimental results presented by [2] as well as analytically obtained average wall temperature. The lumped mass analysis is pursued for obtaining the average metallic wall temperatures of the cylindrical receiver. The outcomes compared are as shown in Fig. 5.

The nodal temperature values are found in close proximity with the analytically obtained solution. However, the present finite element results deviate from the experimentally obtained outcome, exhibiting 24% and 5.8% as maximum and minimum deviation values, respectively.

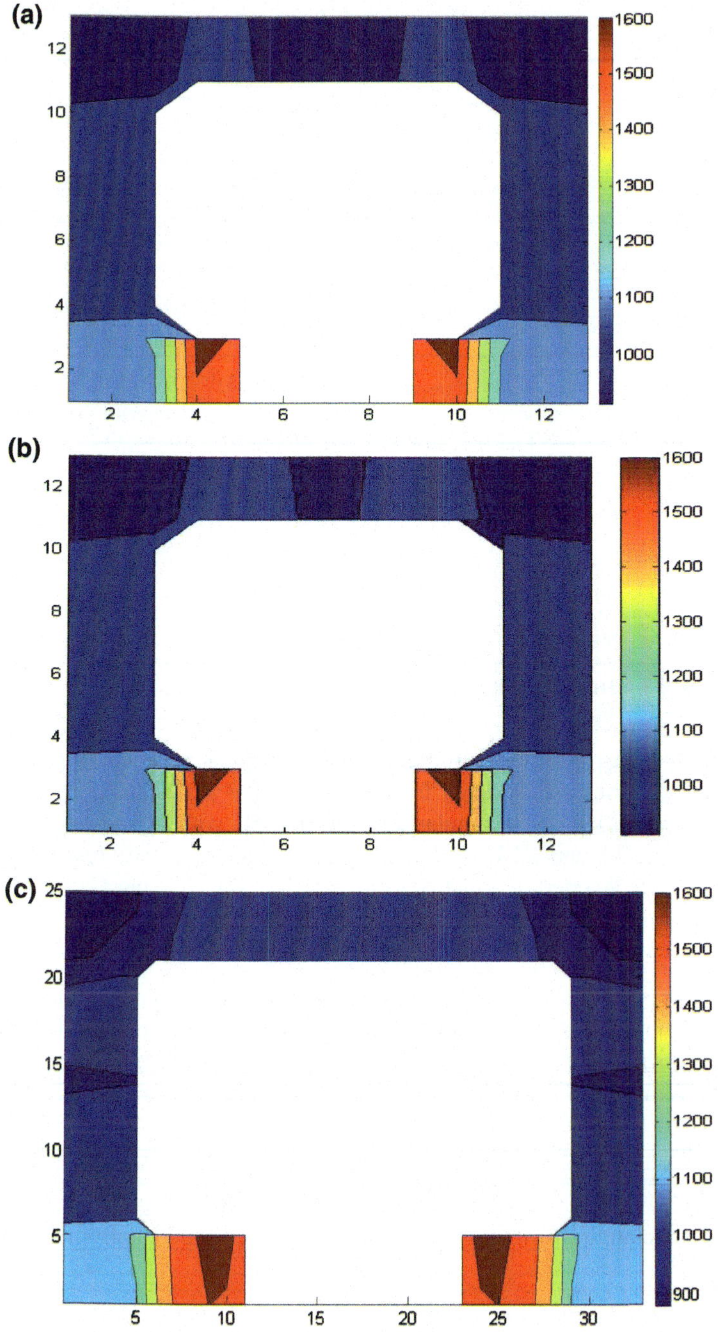

Fig. 4 **a** Temperature contours with 20 elements **b** temperature contours with 72 elements **c** temperature contours with 336 elemetns

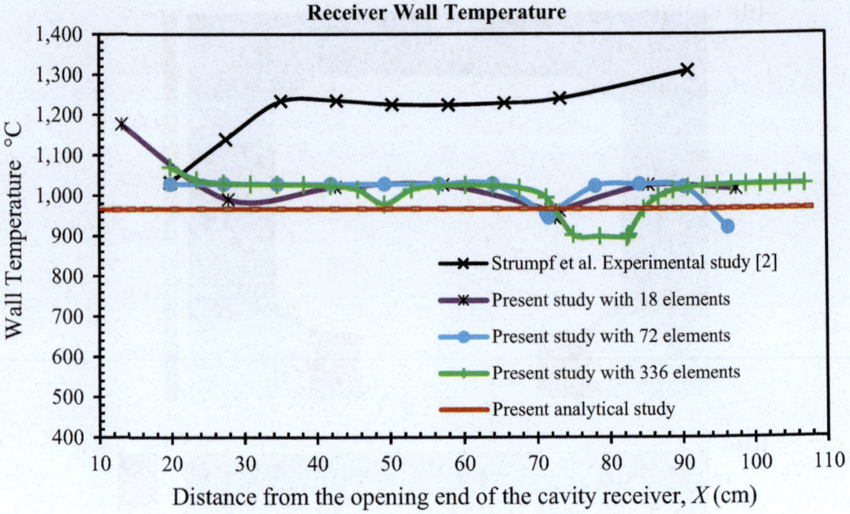

Fig. 5 Comparison of results

3 Results and Discussion

As discussed, the data selected is incorporated in the mathematical formulation. The algebraic equations are obtained for unknown nodal temperatures of the receiver wall. For mesh 1 with 18 triangular elements, mesh 2 with 72 and mesh 3 with 336 elements, 20, 54 and 215 algebraic equations are obtained, respectively. The equations are converted in a matrix form, which is solved using Gauss elimination technique through a MATLAB computer program. The temperature distribution of the receiver wall is obtained in the present chapter for following four cases.

Case 1: Temperature distribution with no internal convection heat transfer.
Case 2: Temperature distribution with internal fluid temperature of 100 °C (for $h_i = 6$ W/m^{2}K and $T_i = 100$ °C).
Case 3: Temperature distribution with internal fluid temperature of 400 °C (for $h_i = 6$ W/m^{2}K and $T_i = 400$ °C).
Case 4: Temperature distribution with internal fluid temperature of 700 °C (for $h_i = 6$ W/m^{2}K and $T_i = 700$ °C).

The temperature distribution obtained is discussed sequentially, as follows.

3.1 Temperature Distribution with no Internal Convection Heat Transfer

The present analysis is pursued with the heat flux at entry to the receiver opening as 80 kW, coolant side heat transfer coefficient equal to 123.61 W/m^{2}K and bulk mean temperature of 500 °C. The temperature distribution of a discretised computational domain with 336 equal sized triangular elements obtained is presented as shown in Fig. 6. The inside surface emissivity is taken as 0.8 and radiative heat losses from the bottom opening are neglected. The heat flux entering through the receiver cavity opening is assumed uniformly distributed among the inner walls of the cavity.

As discussed earlier, the concentrated solar radiation flux is entering into the receiver cavity though the opening. It gets uniformly distributed inside the cavity wall. The metallic surface subjected to this incident energy exhibits peak temperature whereas outer wall which is subjected to forced convection boundary condition is comparatively cooler. Across the thickness the temperature reduces gradually.

The bottom wall which is not subjected to effective cooling method results an unsafe temperature of around 1600 °C. In case, the bottom wall of receiver if subjected to a natural convection boundary (with heat transfer coefficient value of 6 W/m^{2}K [22]) outside of it, would have mitigated the temperature of the bottom wall to a safe level.

At the top right and left corner, slight increase in temperature is found. It may be attributed to the accumulation of trapped heat at the corners.

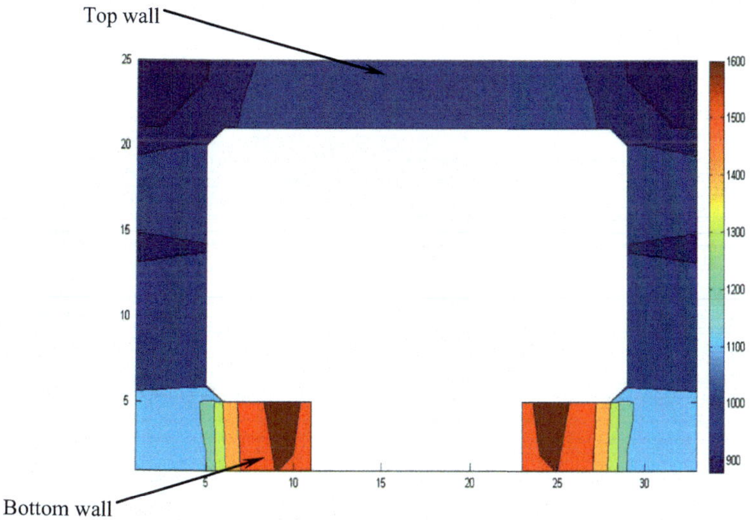

Fig. 6 Temperature distribution of the receiver cavity wall

3.2 Temperature Distribution with Internal Fluid Temperature of 100 °C (for $h_i = 6$ W/M^{2}K and $T_i = 100$ °C)

As discussed in previous case, it was assumed that there is neither heat loss by radiation nor by convection through the opening of the receiver cavity. In this case, internal convection is considered with internal convective heat transfer coefficient equal to 6 W/m^{2}K [25] and internally trapped fluid temperature of 100 °C (Fig. 7).

Slightly more uniformity in temperatures of side wall as well as top wall is observed by incorporating internal convective heat loss component in the analysis.

3.3 Temperature Distribution with Internal Fluid Temperature of 400 °C (for $h_i = 6$ W/M^{2}K and $T_i = 400$ °C)

The particulars of this case are: heat flux entering from the opening of a receiver = 80, 000 W/m^2, inside coefficient of heat transfer as 6 W/m^{2}K, temperature of trapped fluid of 400 °C inside the cavity, metallic wall temperature of conductivity 18 W/m K (for stainless steel), the coolant side heat transfer coefficient as 123.61 W/m^{2}K and bulk mean temperature of coolant (air) is equal to 500 °C. The temperature distribution obtained is as shown in Fig. 8.

The temperature of the cavity wall increases by about 31.4 °C maximum and 11.3 °C minimum by increasing the temperature of fluid trapped inside the cavity from 100 °C to 400 °C. Little rise of inside surface temperatures of the side walls and that of top wall has been noticed.

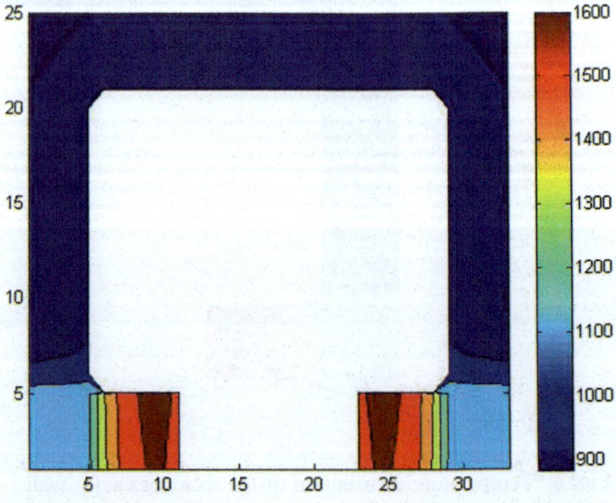

Fig. 7 Temperature distribution of the receiver cavity wall

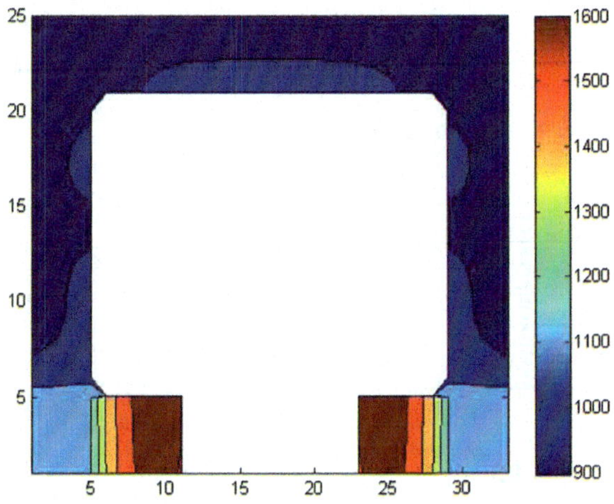

Fig. 8 Temperature distribution of the receiver wall

3.4 *Temperature Distribution with Internal Fluid Temperature of 700 °C (for* $\mathbf{h_i}$ *= 6 W/M*^{2}*K and* T_i *= 700 °C)*

It is seen that by increasing the internally trapped fluid temperature without altering the inside heat transfer coefficient value, the overall cavity wall shows a hike in all the nodal temperatures. As a result of increasing the temperature of fluid inside the receiver cavity from 400 to 700 °C, the average wall temperature shoots up by around 31.4 °C. The temperature contours are as shown in Fig. 9.

Quite considerable temperature variation across the wall is observed in this case as compared to the previous cases discussed in Sects. 3.2 and 3.3. It is attributed to poor heat transfer from within the cavity receiver. This case approaches to Case 1, i.e. with no internal convection.

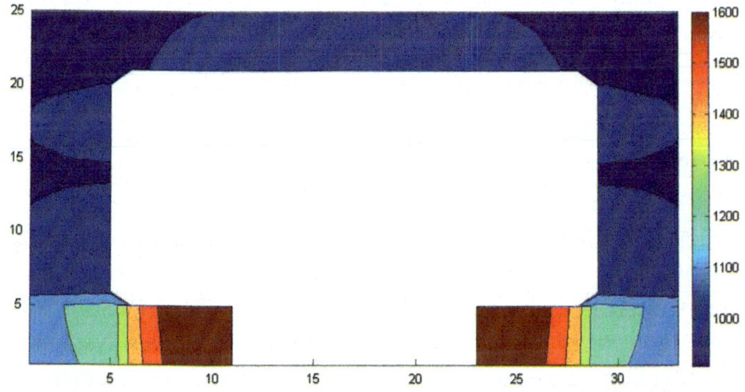

Fig. 9 Temperature distribution of the receiver wall

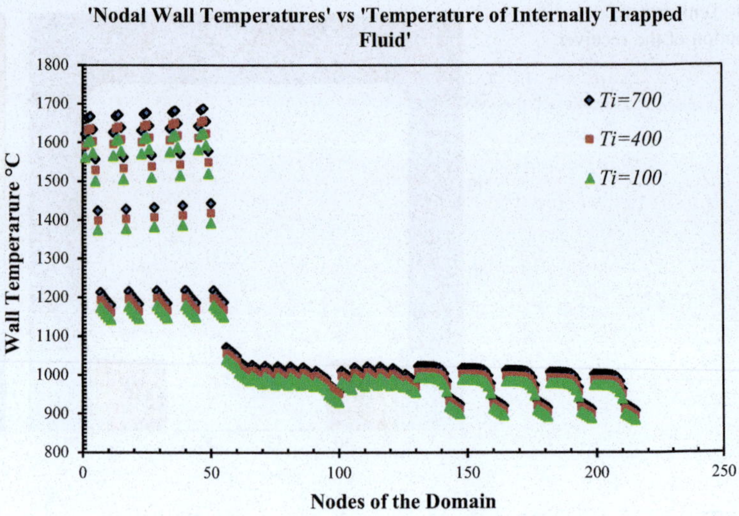

Fig. 10 Temperatures at the respective nodes of computational domain

So, it can be concluded that with decrease in inside solar cavity temperature, peak wall temperature of the receiver cavity wall can be reduced.

The shifting of nodal temperatures at the selected internal fluid temperatures is as shown in Fig. 10.

The bottom wall is represented by nodes numbering from 1 to 50, cylindrical side wall by nodal numbers from 51 to 130 and top wall from 131 to 215. The shooting up of bottom wall temperature to the unsafe value is attributed to the absence of effective boundary condition. Additionally, the consideration of uniform radiative energy flux distribution inside the receiver cavity may be the reason behind this temperature gain.

The work aiming at bringing the bottom wall temperature of the receiver to a safe limit will be accomplished in future. It is seen from the present chapter that finite element technique can be considered as an effective tool for accurate prediction of receiver wall temperatures and other important heat transfer information.

4 Conclusions

The finite element technique is used as a tool to foretell the temperature of the receiver cavity wall. The finite element method can also be used not only to calculate temperature but also few other important details can also be obtained that may assist design related activities of the solar thermal power systems. To initiate work in this direction, some cases have been discussed in this Chapter. The findings of present work are listed as follows

(1) The boundary conditions influence strongly the wall temperature distribution of a receiver.
(2) The temperature of internally trapped fluid inside the receiver cavity alter considerably the wall temperature of a receiver that may affect the performance of the solar system.
(3) The consideration of hypothetical boundary condition i.e., the cavity, subjected to uniformly distributed concentrated solar radiation from inside, have yielded unsafe values of cavity bottom wall temperatures, since it is not in contact with any active coolant. In real situation, the radiation distribution might not be uniform.

Acknowledgements The authors thank the National Institute of Technology Hamirpur for providing access of computational facilities.

Conflicts of Interest: None declared.

References

1. Jarvinen PO (1975) Solar-heated-air receivers. Sol Energy 19:139–147
2. Strumpf HJ, Kotchick DM, Coombs MG (1982) High-Temperature ceramic heat exchanger element for a solar thermal receiver. Sol Energy Eng 104:305
3. Roger M, Buck R, Muller Steinhagen H (2005) Numerical and experimental investigation of a multiple Air Jet Cooling system for application in a solar thermal receiver. ASME J Heat Transf 127:863
4. Sauceda D, Velazquez1 N, Beltran R, Quintero M (2006) Thermal analysis of a conical receiver in a parabolloid dish to be used as generator in an advanced solar thermal cooling system. ASME, p 0019
5. Kribus A, Doron P, Rubin R, Karni J, Reuven R, Duchan S, Taragan E (2000) A multistage solar receiver: the route to high temperature. Sol Energy 67:311
6. Shuai Y, Xia X-L, Tan H-P (2008) Radiation performance of dish solar concentrator/cavity receiver systems. Sol Energy 82:13–21
7. Buck R, Barth C, Eck M, Steinmann W-D (2006) Dual-receiver concept for solar towers. Sol Energy 80:1249–1254
8. Prakash M, Kedare SB, Nayak JK (2009) Investigations on heat losses from a solar cavity receiver. Sol Energy 83:157–170
9. Reddy KS, Sendhil Kumar N (2009) An improved model for natural convection heat loss from modified cavity receiver of solar dish concentrator. Sol Energy 83:1884–1892
10. Prakash M, Kedare SB, Nayak JK (2012) Numerical study of natural convection loss from open cavities. Int J Therm Sci 51:23–30
11. Abbasi-Shavazia E, Hughesb GO, Pye JD (2015) Investigation of heat loss from a solar cavity receiver. Energy Procedia 69:269–278
12. Yuan JK, Ho CK (2015) Numerical simulation of natural convection in solar cavity receivers. J Sol Energy Eng 137:031004-1–10
13. Cao Y (2013) Heat pipe solar receivers for concentrating solar power (CSP) plants. In: Proceedings of the ASME 2013 7th international conference on energy sustainability ES2013, pp 1–8

14. Bannister P, Mayer IF (1997) Developmental solar thermal receiver studies for the white cliffs solar thermal power plant. J Sol Energy Eng 119:61–67
15. Affandi R, Ruddin M, Ghani A, Ghan CK, Pheng LG (2015) The impact of the solar irradiation, collector and the receiver to the receiver losses in parabolic dish system. World Conf Technol Innov Entrep Soc Behav Sci 195:2382–2390
16. Wang W, Laumert B (2017) Effect of cavity surface material on the concentrated solar flux distribution for an impinging receiver. Sol Energy Mater Sol Cells 161:177–182
17. Pozivil P, Agab V, Zagorskiyb A, Steinfeld A (2014) A pressurized air receiver for solar-driven gas turbines. Energy Procedia 49:498–503
18. Zheng Z-J, Li M-J, He Y-L (2015) Optimization of porous insert configuration in a central receiver tube for heat transfer enhancement. In: The 7th international conference on applied energy—ICAE2015, Energy Procedia, vol 75, pp 502–507
19. Lopez-Herraiz M, Fernandez AB, Martinez N, Gallas M (2017) Effect of the optical properties of the coating of a concentrated solar power central receiver on its thermal efficiency. Sol Energy Mater Sol Cells 159:66–72
20. Reddy KS, Sendhil Kumar N (2008) Combined laminar natural convection and surface radiation heat transfer in a modified cavity receiver of solar parabolic dish. Int J Therm Sci 47:1647–1657
21. Wang Wujun, Wang Bo, Li Lifeng, Laumert Bjorn, Strand Torsten (2016) The effect of the cooling nozzle arrangement to the thermal performance of a solar impinging receiver. Sol Energy 131:222–234
22. Uhlig R, Fleschb R, Gobereita B, Giulianoa S, Liedke P (2014) Strategies enhancing efficiency of cavity receivers. Energy Procedia 49:538–550

Part III
Volumetric Solar Thermal Collectors

Direct Absorption Solar Thermal Technologies

Vikrant Khullar, Harjit Singh and Himanshu Tyagi

Abstract Solar collectors that can directly absorb radiation represent an emerging realm of solar thermal systems wherein the collection as well its subsequent conversion to the useful thermal energy happens within the working fluid. Nanofluids (stable dispersions of nanoparticles in the basefluid) have been found to be promising working fluids for realizing such direct solar to thermal energy conversion owing to their enhanced (and the ease of tuning) thermo-physical and optical properties. Seeding trace amounts of carefully chosen nanoparticles into the basefluid has been shown to significantly enhance the solar weighted absorptivity of the basefluid—hence rendering them is suitable for solar thermal applications. Firstly, a brief description relevant to the incumbent surface absorption-based solar thermal technologies has been presented. A critical analysis of the fundamental limits of performance that can be achieved in the incumbent solar thermal systems reveals that solar selectivity could only be beneficial up to a certain temperature and solar concentration ratios, beyond which we cannot further improve the efficiency. Subsequently, the candidature of direct absorption solar thermal systems has been assessed to ascertain if these could be deployed under conditions which are not so amenable for the conventional surface absorption-based solar thermal technologies. Finally, a representative experimental study is presented that points out that even for low solar concentration ratios (conditions which are more suitable for the conventional surface-based absorbers), the two classes of solar thermal technologies can have comparable thermal efficiencies. It is envisaged that the benefits of the direct absorption-based solar thermal systems over the conventional ones shall be more pronounced for high-flux conditions, i.e. high solar concentration ratios.

V. Khullar (✉)
Department of Mechanical Engineering, Thapar University, Patiala 147004, Punjab, India
e-mail: vikrant.khullar@thapar.edu

H. Singh
College of Engineering, Design and Physical Sciences, Brunel University London, Uxbridge, Middlesex UB8 3PH, UK

H. Tyagi
Department of Mechanical Engineering, Indian Institute of Technology Ropar, Nangal Road, Rupnagar 140001, Punjab, India

H. Tyagi et al. (eds.), *Applications of Solar Energy*, Energy, Environment, and Sustainability, https://doi.org/10.1007/978-981-10-7206-2_5

Keywords Direct absorption · Solar thermal · Nanofluid · Surface absorption Solar selective

Nomenclature

English Symbols

A_{sa}	Solar-weighted absorption coefficient
C_{ratio}	Solar concentration ratio
D	Diameter of the nanoparticle [nm]
$d\lambda$	Spectral interval [μm]
E_x	Magnitude of electric field vector [NC^{-1}]
G_{surf}	Solar irradiance incident on the surface [Wm^{-2}]
K	Optical coefficient [m^{-1}]
S_λ	Spectral solar irradiance [$Wm^{-2}\mu m^{-1}$]
T	Temperature [K]
y	Thickness of the fluid layer [m]

Greek Symbols

ε	Dielectric function [Fm^{-1}]
ε_{eff}	Effective emissivity
ε_λ	Spectral emissivity
λ	Wavelength [μm]

Abbreviations

CNTs	Carbon nanotubes
HCE	Heat collector element
MWCNTs	Multi-walled carbon nanotubes
SAS	Surface absorption system
VAS	Volumetric absorption system

1 Incumbent Solar Thermal Systems

Solar thermal systems can be broadly classified according to the way in which solar thermal energy is harnessed. Either the incident solar irradiance is absorbed by a solar selective surface and subsequently transferred to the working fluid as heat, referred to as 'surface absorption-based system' (SAS), or some additives are added to the working fluid so that the working fluid directly absorbs and converts the solar radiation without the need of any intervening surface. The later is referred to as 'volumetric absorption-based system' (VAS). Most of the incumbent solar thermal systems are SAS type.

1.1 Need for Concentrating Solar Power

Non-concentrating solar thermal systems such as flat plate/evacuated tube collectors are effective for low-temperature applications (such as for solar water/air heating applications). However, in order to expand the horizon of solar thermal systems, there is a need to couple the primary solar thermal cycle (involving conversion of solar irradiance into the thermal energy of the working fluid) to secondary cycles (power/refrigeration cycles). Output from primary solar thermal cycle forms the input to these secondary cycles. Now, in order to ensure efficient operation of these secondary cycles (and consequently higher overall efficiencies of the whole system), heat transfer fluid (HTF) at sufficiently high temperatures is desired. This motivates the design of solar thermal systems which are capable of capturing high magnitude of incident solar irradiance without increasing the effective thermal loss receiver area (which in turn could lead to high thermal losses since they are proportional to surface area). Both of these objectives can be achieved by interposing some concentrating optics (reflectors/refractors) in between the incident solar irradiance and the receiver (see Fig. 1). For instance, in the case of linear parabolic troughs, solar irradiance is firstly incident normally on the aperture of the parabolic mirrors and then gets reflected on the linear cylindrical receiver placed at the focal line of the parabolic trough. Consequently, the flux (on the receiver) has increased by a factor equal to the solar concentration ratio [defined as the ratio of aperture area

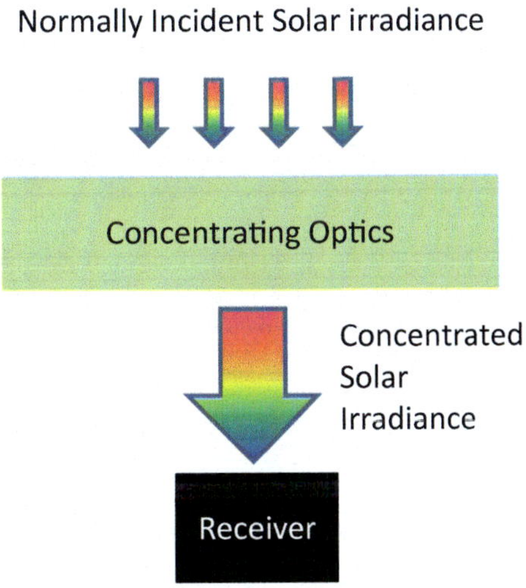

Fig. 1 Schematic showing the process of harnessing concentrating solar power

to the receiver area ($C_{ratio} = A_{aper}/A_{rec}$)] without any corresponding increase in the effective heat loss surface area.

The maximum theoretical achievable (limited by second law of thermodynamics) solar concentration ratio in the case of point and line type concentrators is 45000 and 212, respectively [2]. Some typical values of solar concentration ratios are shown in Table 1.

1.2 Surface Absorption-Based Concentrating Solar Thermal Systems

Incumbent concentrating solar thermal systems predominantly employ surface absorption-based receivers. Here, the solar irradiance is firstly absorbed by a solar selective surface (a surface having high solar irradiance absorptivity and low thermal emissivity) and subsequently this absorbed energy is transferred to the working fluid through conduction and convection modes of heat transfer (see Fig. 2).

Solar selective surfaces. Ideally, a solar selective surface should have absorptivity of 1 for all wavelengths shorter than the cut-off wavelength and an absorptivity of 0 for all other wavelengths. However, real solar selective surfaces do not have such sharp cut-off wavelengths (and extreme absorptivity values). This section details the fundamental limits of ideal solar selective surfaces and intends to explore the present state of technology employing these solar selective surfaces.

The fundamental limit. From a fundamental standpoint, the cut-off wavelength dictates the trade-off between solar irradiance absorption capability and radiative losses from a solar selective surface. Since effective absorptivity and emissivity are not mutually independent of each other, a trade-off needs to be carried out to optimize the performance of a solar selective surface. Figure 3a shows the solar weighted absorptivity (defined by Eq. 1) as a function of cut-off wavelength and similarly, Fig. 3b shows effective emissivity (defined by Eq. 2) as a function of surface temperature for various cut-off wavelengths, respectively.

$$A_{sa} = \frac{\int_0^{\lambda_{cutoff}} \alpha_{s,\lambda} S_\lambda d\lambda}{\int_0^{2.5\mu m} S_\lambda d\lambda}, \quad (1)$$

$$\varepsilon_{eff,T} = \frac{\int \varepsilon_{s,\lambda} E_{\lambda,T} d\lambda}{\int E_{\lambda,T} d\lambda}, \quad where \quad \varepsilon_{s,\lambda} = 1 \quad for \quad \lambda \leq \lambda_{cutoff} \quad (2)$$

$$\varepsilon_{s,\lambda} = 0 \quad for \quad \lambda > \lambda_{cutoff}$$

Table 1 Incumbent surface absorption-based concentrating solar thermal technologies

Concentrating solar thermal technology	Capacity range[a] (MW)	Typical solar concentration ratio[a]	Selected commercial installations				
			Name of the project	Installed capacity (MW)	Typical temperatures (°C)	Heat transfer fluid (HTF)	Application
Parabolic trough collectors	10–200	70–80	SEGS (I-IX)	353.8[b]	307–390[b]	Mineral[b] oil, Therminol VP-1	Electricity generation
Linear fresnel reflector systems	10–200	25–100	Puerto Errado 2	30[e]	270[e]	Water[e]	Electricity generation
Power tower surrounded by heliostat fields	10–150	300–1000	PS10	11[d]	250[d]	Water/Steam[d]	Electricity generation
Dish–stirling	0.01–0.4	1000–3000	Vangaurd 1[d]	25[d]	720[d]	Hydrogen[d]	Electricity generation

[a][8], [b][4], [c][21], [d][5], [e][20]

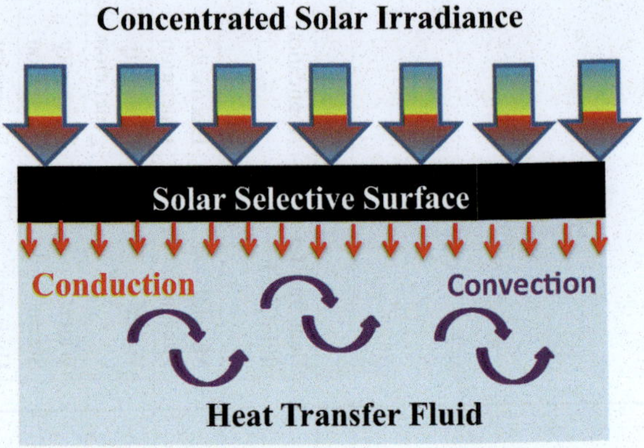

Fig. 2 Heat transfer mechanism in surface absorption-based concentrating solar thermal systems

where S_λ is the spectral solar irradiance, $\alpha_{s,\lambda}$ is the spectral absorptivity of the surface, $E_{\lambda,T}$ is the spectral emissive power for given surface temperature and $\varepsilon_{s,\lambda}$ is the spectral emissivity of the surface.

It should be noted that these plots are for ideal solar selective surfaces and represent the maximum achievable theoretical performance characteristics. These plots essentially point out that there exists a theoretical limit to the achievable solar weighted absorptivity and thermal emissivity.

The present state of technology. Real solar selective surfaces do not have sharp cut-off wavelengths and extreme absorptivity/emissivity values like that for ideal solar selective surfaces. However, they are designed to have properties as close as possible to the ideal solar selective surfaces. Figure 4 shows common solar selective surfaces along with their absorptivity and emissivity values. Recently, Almeco Solar has developed TiNOX-based highly solar selective surfaces employing aluminium and copper substrates [24].

Essentially, TiNOX is a composite coating constituted of multiple layers such as protective layer, antireflective layer, absorber layer, bonding layer and substrate. Each layer has a distinctive role to play; for instance, antireflective layer ensures maximum transmissivity so that most of the incident solar irradiance reaches the absorber layer. Typically, TiNOX coated copper substrate has solar weighted absorptivity of 95% and thermal emissivity of 4% (TiNOX Energy, accessed June 2017).

The absorber layer material is so selected that it can absorb most of the incident solar irradiance. Substrate should effectively be IR reflective in order to ensure low radiative losses. In effect, the entire constituent layers work together to ensure high solar irradiance absorptivity and low IR emissivity.

The use of high concentration ratios as well as solar selective surface serves the overall purpose of harnessing solar energy effectively. Table 1 provides a comparison of the typical operating parameters of various types of surface absorption-based concentrating solar thermal technologies being used at commercial scales.

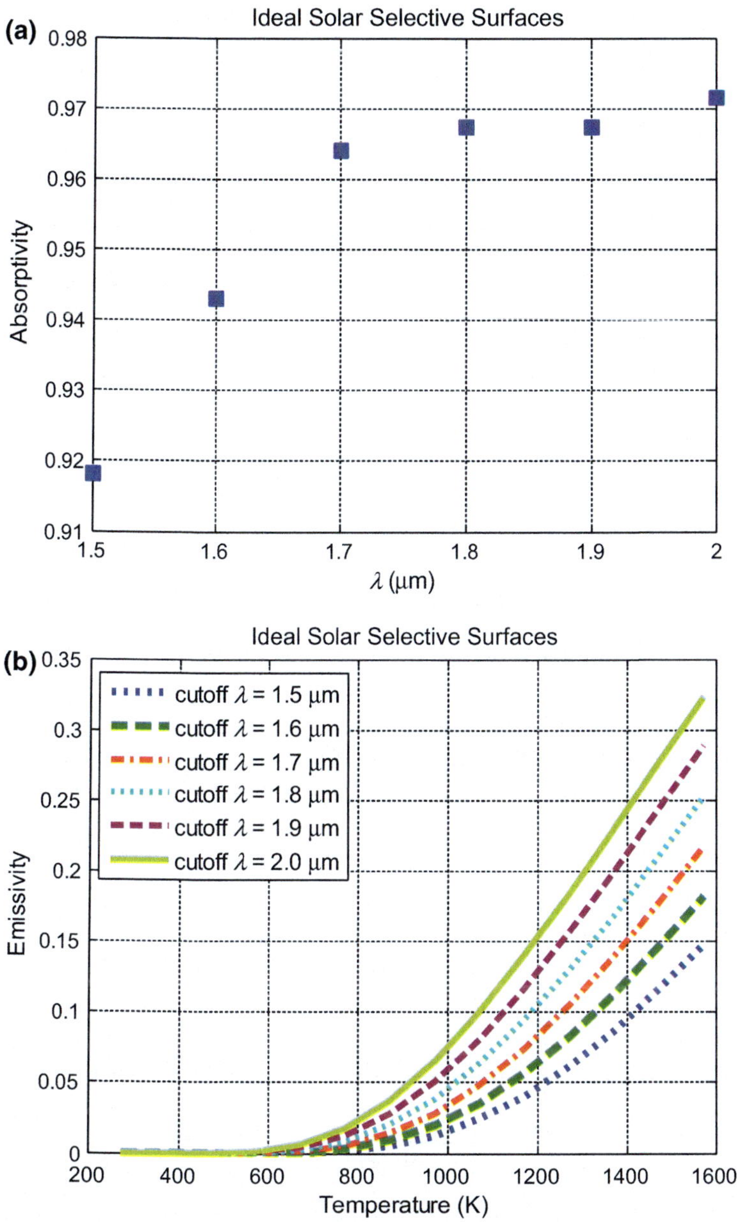

Fig. 3 For ideal solar selective surface, **a** solar weighted absorptivity as a function of cut-off wavelength and **b** emissivity as a function of surface temperature for various cut-off wavelengths

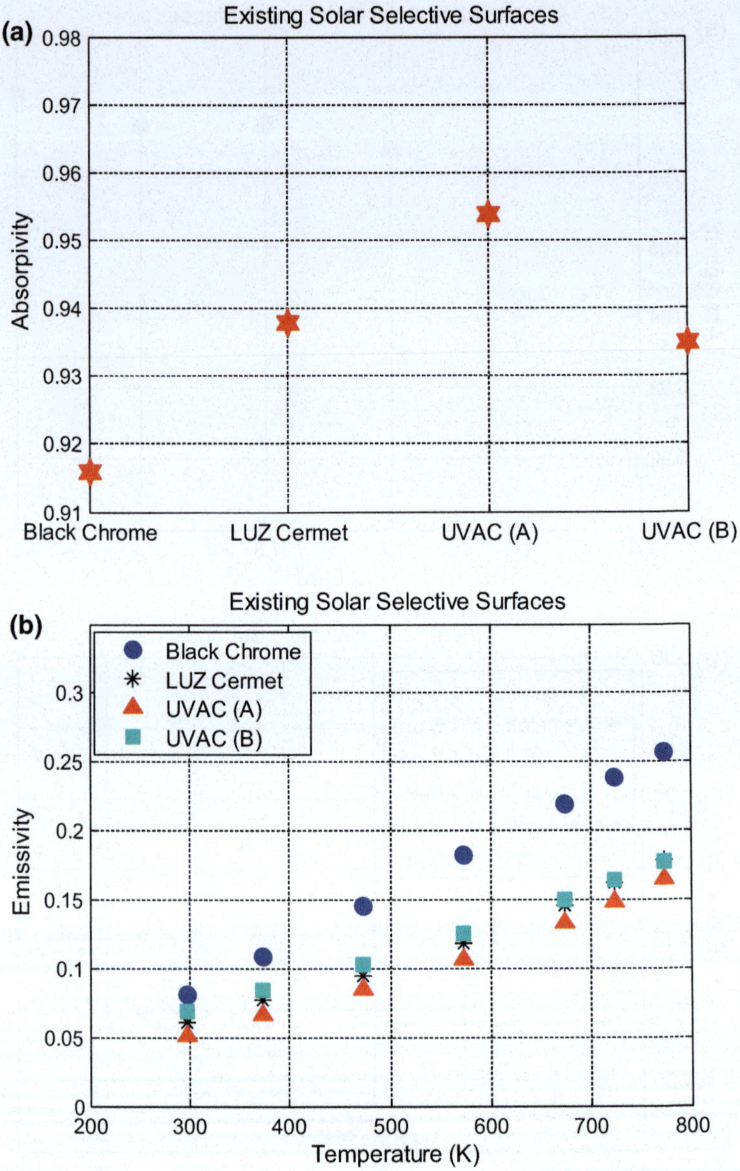

Fig. 4 For commonly used solar selective surfaces, **a** solar weighted absorptivity values and **b** emissivity values as a function of temperature. Data points taken from Kennedy and Price [9]

2 Relevance of Nanofluid-Based Volumetric Absorption Solar Thermal Systems

Limitations such as ineffectiveness of solar selective surfaces (in curbing radiative losses) after a certain temperature and the problem of high overheat temperatures have motivated the exploration of alternative ways to harness solar thermal energy. Among the alternatives, one way is to allow the incident solar irradiance to interact directly with the working fluid (nanofluid) without heating any other structures within the receiver.

2.1 Preparation Techniques and the Stability of the Prepared Nanofluids

Nanofluids (stable dispersions of nanoparticles in the basefluid) could be prepared either in situ through one-step process or could be prepared through two-step process (involving preparation of nanoparticles and their subsequent dispersion into the basefluid).

Nanofluids prepared by one-step process are more stable, but at the same time, careful and precise control of nanoparticle volume fraction is difficult to achieve. Also, this method is difficult for scale-up production of nanofluids. On the other hand, the two-step process has the flexibility of controlling the volume fraction, but stability is an issue in nanofluids prepared by two-step process [16, 23]. Recently, researchers have found surface functionalization as an effective technique to enhance the stability of the nanoparticles in the basefluid. Surface functionalization could be achieved both through physical as well as chemical routes. Surface functionalization essentially involves attaching functional groups onto nanoparticles surfaces so that these nanoparticles could be made hydrophilic and can have better stability characteristics. Nowadays, commercial manufacturers like Sigma-Aldrich, NanoAmor have started manufacturing functionalized nanoparticles; therefore, stable nanofluids of desired volume fractions could be prepared easily through two-step process. Table 2 details the nanofluids used for direct solar thermal conversion applications.

2.2 Measuring Devices or Techniques Employed for Measuring Optical Properties

It is imperative to measure the optical properties of the basefluids/nanofluids for a correct assessment of their suitability in direct volumetric absorption solar thermal systems [18]. Usually, UV–VIS–NIR spectrophotometer and FTIR are employed

Table 2 Selected studies employing nanofluids as direct solar absorbers

Type of study	Nanoparticle			Basefluid	Reference
	Material	Typical size	Shape		
Experimental and numerical	Graphite, magnetite and silver	15–50 nm	Nanosphere, irregular	Water	Gorji TB, and Ranjbar AA [3]
Experimental and numerical	Gold	16 nm	Nanorods	Water	Jeon et al. [6]
Experimental and numerical	MWCNT	——	Nanotube	Water/Therminol 55	Li et al. [15]
Experimental and numerical	Gold	20 nm	Nanosphere	Water	Jin et al. [7]
Experimental, analytical and numerical	MWCNT	D = 20 nm L = 1–25 μm	Nanotube	Water	Lee et al. [12]
Numerical	Au, Ag, Cu, Al	8–10 nm	Nanospheres and nanospheroids	Therminol VP-1	Khullar et al. [11]
Experimental	Amorphous carbon MWCNT	< 50 nm	Irregular and nanotube	Water and ethylene glycol	Khullar et al. [10]
Experimental and numerical	Carbon-coated cobalt	28 nm	Nanosphere	Therminol VP-1	Lenert and Wang [14]
Experimental and numerical	Graphite, Cu, Ag, Al, TiO_2	20–150 nm	Nanosphere	Water and therminol VP-1	Taylor et al. [22]
Experimental and numerical	Graphite, carbon nanotube, Ag	6–30 nm	Nanosphere and nanotube	Water	Otanicar et al. [19]
Numerical	Al	5 nm	Nanosphere	Water	Tyagi et al. [25]

for measurement of spectral characteristics (absorbance/transmittance/reflectance) of the basefluids/nanofluids.

Essentially, two distinct wavelength bands—namely solar irradiance wavelength band (0.3–2.5 μm) and mid-infrared wavelength band (2.5–30 μm) are of interest.

To qualify as good candidate for direct solar absorption, the HTF should have high solar weighted absorptivity as well as low IR emission. Although, inherently nanofluids are bound to have high emission in the IR region [10]; however, it is possible to engineer the desired solar-weighted absorptivity through careful control of the nanoparticle material, size and shape and the basefluid [17]. The nanoparticles should either have broad absorption peaks (such as carbon-based nanoparticles) or should have absorption peaks coincident with the solar irradiance peaks (mostly metal nanoparticles).

For a given material, size and shape of nanoparticles, it is ultimately the volume fraction of the nanoparticles that dictate the solar absorption capability. It should be noted that solar weighted absorptivity increases with increase in volume fraction up to a certain value only, and then any further increase in volume fraction does not have any significant effect on the solar weighted absorptivity. Moreover, a higher value than the optimum value tends to decrease the overall conversion efficiency of the direct volumetric absorption solar thermal systems. This could be understood from the fact that at very high volume fractions, the absorption essentially takes place in the top layer only, thus it behaves like a surface absorption case with emission on order of a typical blackbody.

2.3 Limitations of Selective Surfaces in Curbing Radiative Losses

In order to ascertain the importance of solar selectivity in surface absorption-based concentrated solar thermal systems, radiative losses (%) [defined by Eq. 3] for a perfectly black surface were evaluated as a function of surface temperature (for various solar concentration ratios) and compared to that for an ideal solar selective surface (having a cut-off wavelength value of about 1.7 μm).

$$\begin{aligned} &\textit{Radiative losses}(\%) = \frac{\varepsilon_{s,T}\sigma T^4}{C_{\text{ratio}} G_{\text{surf}}} \times 100, \\ &\textit{where} \quad \varepsilon_{s,T} = 1 \quad \textit{for} \quad \textit{blackbody} \\ &\varepsilon_{s,T} = \varepsilon_{\text{eff},T}\ \textit{for} \quad \textit{solar selective surface} \end{aligned} \tag{3}$$

where $\varepsilon_{s,T}$ is the total hemispherical emissivity of the surface for a given temperature T (K), C_{ratio} is the solar concentration ratio and G_{surf} is the solar irradiance (Wm^{-2}) incident on the surface. This cut-off wavelength has been chosen as higher cut-off wavelengths do not result in significant increase in the absorptivity values. Figure 5 clearly depicts that as the solar concentration ratio increases; the solar selectivity becomes increasingly important in order to reduce radiative losses. However, a close look at Fig. 5 also reveals that through usage of selective coatings it is only possible to delay the onset of typical blackbody behaviour in terms of radiative losses. Hence, solar selective surfaces can be effective up to a certain temperature in relation to ensuring low radiative losses.

2.4 Limitations of Selective Surfaces in Transferring the Absorbed Energy to the Working Fluid

These selectively coated surfaces are designed to have high absorptivity across the incident solar spectrum and low emissivity in the mid-infrared region to ensure high efficiency. Effective transfer of this absorbed energy to the working fluid requires a

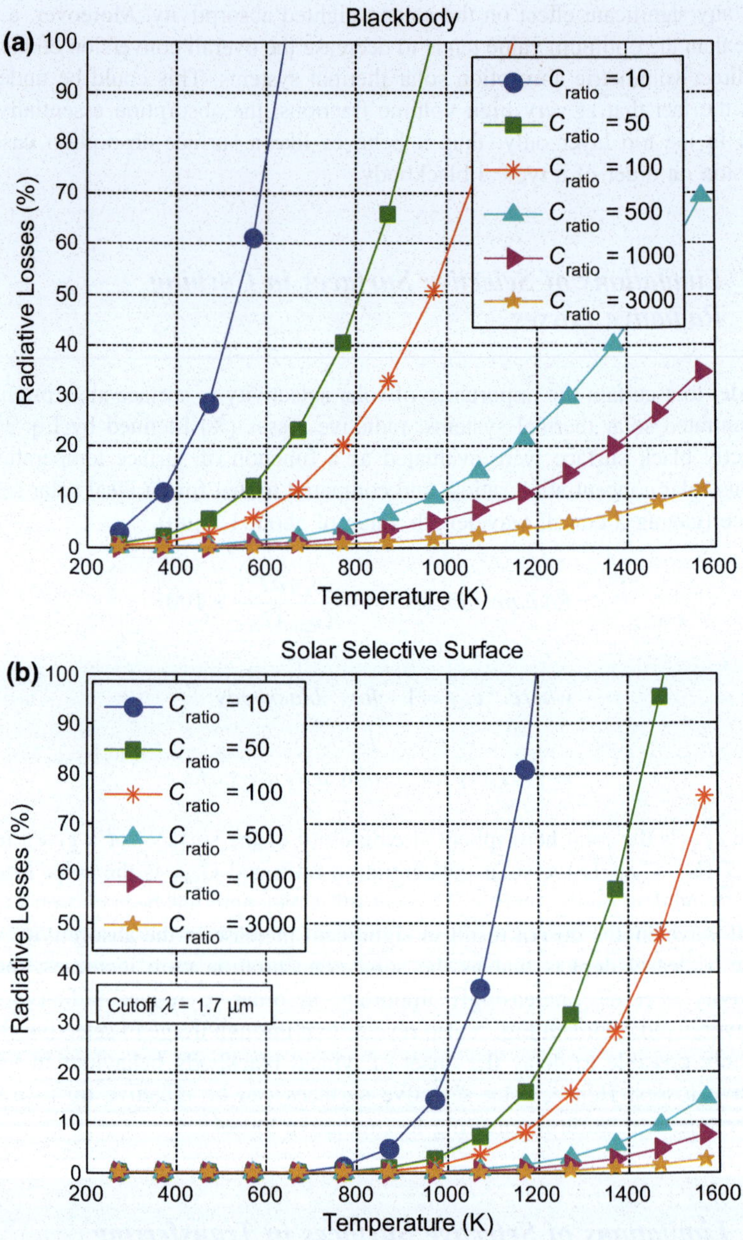

Fig. 5 Radiative losses (%) as a function of surface temperature for various solar concentration ratios **a** surface having a perfect blackbody characteristics and **b** solar selective surface having cut-off wavelength of 1.7 μm

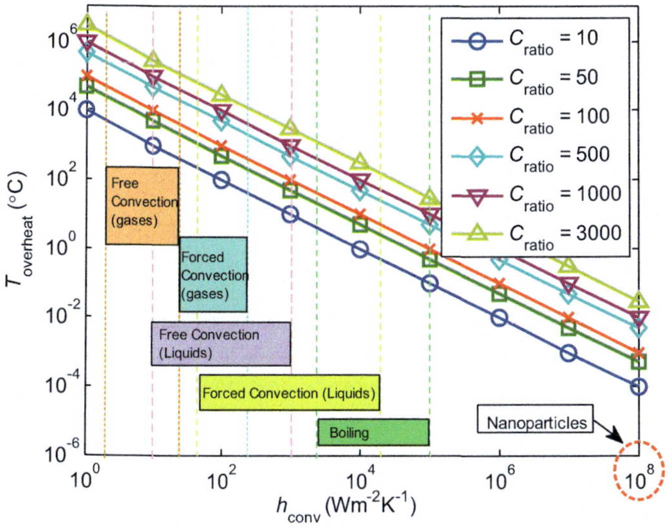

Fig. 6 Overheat temperature (defined as the temperature difference between the solid and the heat transfer fluid) as a function of the convective heat transfer coefficient for various solar concentration ratios. Typical values of heat transfer coefficients have been taken from [1] and [13]

strong convective heat transfer mechanism. In the absence of which the surface temperatures could excessively rise relative to the working fluid (defined as the overheat temperature) and resulting in high radiative losses (owing to the fourth-power dependence of radiative losses on temperature) and also stringent material requirements. Typical values of heat transfer coefficients suggest that most of the convective heat transfer modes are not effective enough to ensure efficient heat transfer at very high values of solar concentration. In other words, overheat temperature increases quite rapidly as the flux that has to be transferred across the surface increases (see Fig. 6).

This thermal resistance is inversely proportional to the heat transfer coefficient. As most of the solar thermal applications come under free/forced convection regimes, heat transfer coefficients typically of these regimes are not sufficient enough to establish an effective heat transfer between the heated solar selective surface and the working fluid at high solar concentration ratios.

2.5 *Overheat Temperatures in Nanoparticle Dispersions*

Absorption mechanism and hence the spatial temperature profile depend on the spatial nanoparticle distribution. Nanoparticles owing to their very high heat transfer coefficients are highly efficient in transferring absorbed energy to the fluid in immediate contact. Heat transfer coefficient as applicable to nanoparticles means heat transfer from the individual nanoparticles to the surrounding medium (basefluid). Metallic nanoparticle dispersed in a basefluid can be safely approximated as a lumped system owing to its extremely small size and high thermal conductivity. Now we can work out the effective convective heat transfer from the nanoparticle to the basefluid by invoking the Biot number relationship as

$$Bi = \frac{h_{\mathrm{conv}} D}{k}, \tag{4}$$

Biot number should be less than 0.1 for a lumped system, and also conservative estimates for thermal conductivity ($k = 100\ \mathrm{Wm^{-1}K^{-1}}$) for an approximately 100 nm size nanoparticle yields convective heat transfer coefficient (h_{conv}) on the order of $10^8\ \mathrm{Wm^{-2}K^{-1}}$ [13, 22]. This is true even at high solar concentration ratios (bottom right corner of Fig. 6). Furthermore, optical properties of nanoparticles are dependent on its shape, size, constituent material and the surrounding dielectric medium. This characteristic lends nanoparticles to be optically tuned for solar thermal applications. In the backdrop of the aforementioned characteristics, it is worthwhile to explore the candidature of nanofluid-based volumetric absorption solar thermal systems.

3 Representative Experimental Study

Detailed theoretical analysis reveals that it is indeed possible (at least in principle) to enhance the overall solar energy conversion process through direct absorption of the solar irradiance by the nanofluid, i.e. nanofluid-based volumetric absorption solar thermal technologies could actually outperform their surface absorption-based counterparts under certain controlled conditions.

Next, a representative experimental investigation has been carried out under low solar concentration ratio (a condition which is more amenable to surface absorption-based technologies).

For the experimental investigation, two identical parabolic trough collectors have been employed. In the first collector, a surface absorption-based receiver has been simulated by rolling a solar selective metallic sheet into a glass tube (with nanofluid inside rolled sheet), whereas in the second collector, a volumetric absorption-based receiver design has been simulated by simply housing the nanofluid in the glass tube [as shown in Fig. 7a].

The two parabolic troughs (receiver designs) have been tested under similar conditions (solar irradiance and wind speed). It was found that under similar operating conditions, the volumetric absorption-based receiver design records higher temperature relative to the surface absorption-based receiver design [see

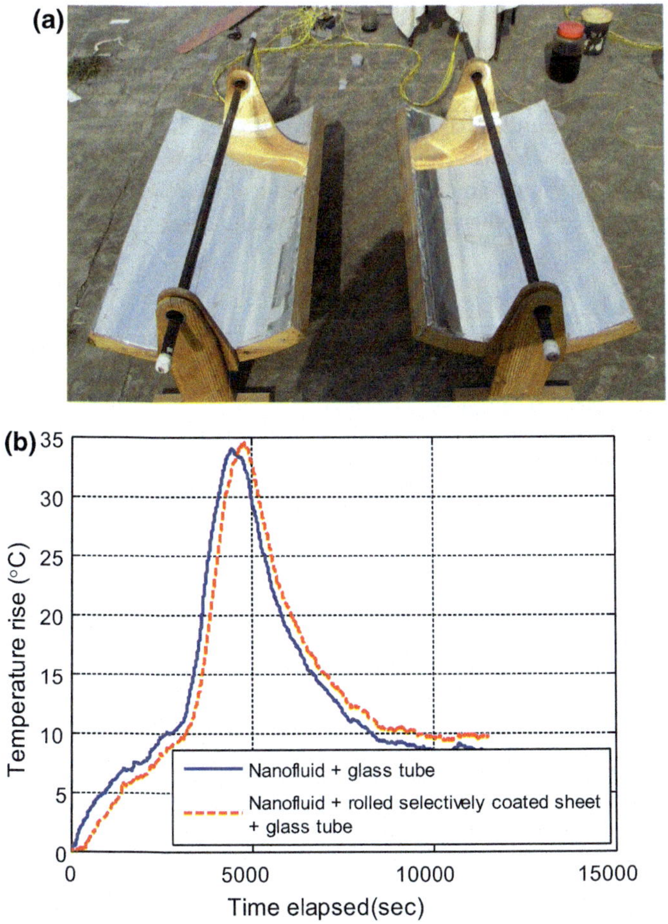

Fig. 7 **a** Photograph of the parabolic trough collectors employing surface absorption and volumetric absorption-based receivers and **b** temperature rise as a function of time in the two receiver designs

Fig. 7b]. The nonlinear temperature rise trend could be understood from the fact that the two collectors were fixed in orientation shown in Fig. 7a (i.e. non-tracking condition) and as the relative position of the sun and the collector changed with time, the magnitude of the concentrated solar irradiance changed from a minimum value to a maximum value and then to a minimum value. In other words, it was only for a particular duration that two receivers received the concentrated sunlight.

Moreover, as predicated theoretically, it is expected that the volumetric receiver designs shall perform still much better when operating at high solar concentration ratios.

4 Conclusions

Theoretical as well as experimental investigation clearly point out that surface absorption-based receivers have inherent limitations at high temperatures and solar concentration ratios. However, under the aforementioned conditions, volumetric absorption-based systems could actually be more efficient in converting the incident solar energy into useful thermal energy gain of the working fluid.

References

1. Cengel YA (2003) Heat transfer: a practical approach. MacGraw-Hill
2. Duffie JA, Beckman WA (2006) Solar engineering of thermal processes. Wiley
3. Gorji TB, Ranjbar AA (2016) A numerical and experimental investigation on the performance of a low-flux direct absorption solar collector (DASC) using graphite, magnetite and silver nanofluids. Sol Energy 135:493–505
4. Goswami DY (ed.). (2005) Advances in solar energy. An annual review of research and development. Fuel and Energy Abstracts (Vol. 36). Boulder: American Solar Energy Society
5. Goswami DY, Kreith F (2007). Handbook of energy efficiency and renewable energy, vol 20070849. CRC Press
6. Jeon J, Park S, Lee BJ (2016) Analysis on the performance of a flat-plate volumetric solar collector using blended plasmonic nanofluid. Sol Energy 135:493–505
7. Jin H, Lin G, Bai L, Amjad M, Filho EPB, Wena D (2016) Photothermal conversion efficiency of nanofluids: An experimental and numerical study. Sol Energy 139:278–289
8. Kalogirou SA (2009) Solar energy engineering: processes and systems. Academic Press
9. Kennedy CE, Price H (2005) Progress in development of high-temperature solar-selective coating. In: Proceedings of ISEC2005 international solar energy conference. Orlando, Florida
10. Khullar V, Tyagi H, Hordy N, Otanicar TP, Hewakuruppu Y, Modi P, Taylor R (2014) Harvesting solar thermal energy through nanofluid-based volumetric absorption systems. Int J Heat Mass Transf 77:377–384
11. Khullar V, Tyagi H, Otanicar T, Hewakuruppu Y, Taylor R (2016) Solar selective volumetric receivers for harnessing solar thermal energy, paper no. IMECE2016-66599, ASME international mechanical engineering congress and exposition, Phoenix, Arizona, USA
12. Lee S-H, Choi TJ, Jang SP (2016) Thermal efficiency comparison: surface-based solar receivers with conventional fluids and volumetric solar receivers with nanofluids. Energy 115:404–417
13. Lenert A (2010) Nanofluid-based receivers for high-temperature, high-flux direct solar collectors. Doctoral dissertation, Massachusetts Institute of Technology. Accessed from 12 April 2012. http://dspace.mit.edu/bitstream/handle/1721.1/61881/706146087.pdf
14. Lenert A, Wang EN (2012) Optimization of nanofluid volumetric receivers for solar thermal energy conversion. Sol Energy 86(1):253–265
15. Li Q, Zheng C, Mesgari S, Hewkuruppu YL, Hjerrild N, Crisostomo N, Rosengarten G, Scott d JA, Taylor RA (2016) Experimental and numerical investigation of volumetric versus surface solar absorbers for a concentrated solar thermal collector. Sol Energy 136:349–364
16. Otanicar T, Hoyt J, Fahar M, Jiang X, Taylor RA (2013) Experimental and numerical study on the optical properties and agglomeration of nanoparticle suspensions. J Nanopart Res 15:2039
17. Otanicar TP, Chowdhury I, Prasher R, Phelan PE (2011) Band-gap tuned direct absorption for a hybrid concentrating solar photovoltaic/thermal system. J Sol Energy Eng 133(4):41014

18. Otanicar TP, Golden JS, Phelan PE (2009) Optical properties of liquids for direct absorption solar thermal energy systems. Sol Energy 83(7):969–977
19. Otanicar TP, Phelan PE, Prasher RS, Rosengarten G, Taylor RA (2010) Nanofluid-based direct absorption solar collector. J Renew Sustain Energ 2(3):033102
20. Puerto Errado 2 thermosolar power plant (2013) Accessed May 26, 2014. from http://www.nrel.gov/csp/solarpaces/project_detail.cfm/projectID=159
21. Rabl A (1985) Active solar collectors and their applications. Oxford University Press
22. Taylor RA, Phelan PE, Otanicar TP, Adrian R, Prasher R (2011) Nanofluid optical property characterization: towards efficient direct absorption solar collectors. Nanoscale Res Lett 6 (1):225
23. Taylor R, Coulombe S, Otanicar T, Phelan P, Gunawan A, Lv W, Tyagi H (2013) Small particles, big impacts: a review of the diverse applications of nanofluids. J Appl Phys 113 (1):011301
24. TiNOX Energy (2013) http://www.almecogroup.com/uploads/1004%20ALMECO_TinoxEnergy_ENG_S402_05_2013_mail.pdf. (n.d.). Almeco Solar. Accessed from 01 Jan 2014. http://www.almecogroup.com/uploads/1004-ALMECO_TinoxEnergy_ENG_S402_05_2013_mail.pdf
25. Tyagi H, Phelan P, Prasher R (2009) Predicted efficiency of a low-temperature nanofluid-based direct absorption solar collector. J Sol Energy Eng 131(4):041004

Solar Thermal Energy: Use of Volumetric Absorption in Domestic Applications

Vishal Bhalla, Vikrant Khullar, Harjit Singh and Himanshu Tyagi

Abstract Solar thermal systems are one of the renewable energy systems used in the residential buildings for the heating purpose, and with these systems, the usage of non-renewable energy resources decreases. To improve the performance of solar collectors, engineers and scientists are regularly working on it. Direct absorption-based solar thermal collectors (DASTC) are kind of solar collectors in which the fluid can be heated directly (without any absorption surface). The present study deals with numerical model of direct absorption-based solar collector which can be used for residential purposes. The absorbed energy fraction, effect of the height, length of collector, and mass flow rate on the collector efficiency have been determined. The analysis shows that collector efficiency increases with the increase of mass flow rate when the height of the fluid in the collector is same and the efficiency of the collector deceases with the increase of channel length. Further, it has been observed that it is beneficial to use an optimum volume fraction of the nanoparticles in DASTC because at an optimum volume fraction, the collector achieved maximum efficiency.

Keywords Direct absorption solar thermal collector · Nanofluid
Volume fraction · Heat transfer

V. Bhalla (✉) · H. Tyagi
Department of Mechanical Engineering, Indian Institute of Technology Ropar,
Nangal Road, Rupnagar 140001, Punjab, India
e-mail: vishal.bhalla@iitrpr.ac.in

V. Khullar
Mechanical Engineering Department, Thapar University,
Patiala 147004, Punjab, India

H. Singh
College of Engineering, Design and Physical Sciences, Brunel University London,
Uxbridge, MDX UB8 3PH, UK

H. Tyagi et al. (eds.), *Applications of Solar Energy*, Energy, Environment, and Sustainability, https://doi.org/10.1007/978-981-10-7206-2_6

Nomenclature

A	Area
C_p	Specific heat
D	Diameter
f_v	Volume fraction
G_T	Incident solar flux on the collector, 1000 W/m^2
H	Height
HTE	Heat transfer equation
h	Convective heat transfer coefficient
I	Intensity
k	Thermal conductivity
K	Radiative constant
L	Length
$\dot{m}$	Mass flow rate
$p(\Omega' \rightarrow \Omega)$	Phase function
q_r	Radiative heat flux
RTE	Radiative transfer equation
T	Temperature
T_{solar}	Black body temperature
y	Depth of penetration
η	Collector efficiency
ρ	Density
Ω	Solid angle

Subscripts

a	Absorption
b	Blackbody
e	Extinction
λ	Spectral
s	Scattering
∞	Ambient
in	Inlet
out	Outlet
min	Minimum
max	Maximum

1 Introduction

The demand of energy is increasing day by day due to which the dependence on fossil fuels (coal, oil, and natural gas) is increasing. With the regular usage of fossil fuels, emission of carbon dioxide takes place, which is deteriorating the environment. According to Dincer [1], there have many environmental issues with the usage of fossil fuels and have proposed that in order to overcome the environmental issues, the usage of renewable energy resources like solar energy, wind energy, ocean energy should be encouraged.

Solar thermal systems are one of the renewable energy systems, used in the residential buildings for the heating purposes, and with these systems, the usage of fossil fuels is decreasing. The solar thermal collectors can be differentiated by the way solar irradiation has been absorbed by them. According to the absorption mechanism, these solar thermal collectors are (a) surface absorption-based solar thermal collector (SASTC) and (b) direct absorption-based solar thermal collector (DASTC).

In SASTC, a black painted metallic surface has been used on the top side of the collector (see Fig. 1a) which absorbs the incident solar radiation and converts into thermal energy. The converted thermal energy is transferred to the working fluid by conduction and convection heat transfer mechanism. The working mechanism of SASTC is as shown in Fig. 1a. Some challenges are associated with SASTC like (a) at high flux high emissive losses from the surface (b) at high flux, a huge temperature difference rise in between surface and the mean fluid temperature results in reduction of thermal efficiency of the solar collector (c) weather influences the internal parts of SASTC.

In order to overcome these challenges, scientists and researchers suggest the solar energy should be directly absorbed by the working fluid. In the recent years, direct absorption-based solar collectors have been introduced [2–4] in which the solar radiation is directly absorbed by the heat transfer fluid (water, ethylene glycol, propylene glycol, Therminol VP-1 etc.). Since the absorptivity of the heat transfer fluids is very low in the visible region, some external particles (nanoparticles) are added in the heat transfer fluid, which improve its absorptivity. The working mechanism of DASTC is as shown in Fig. 1b. In this DASTC, the nanofluid is entering from left side of the channel and flowing with velocity U and the outlet of the nanofluid is on the right side of the channel. The solar irradiation incident on the top side of the channel as shown in Fig. 1b and the nanofluid will absorb the incident radiation. A glass is placed on the top side of the channel in order to reduce the radiative and convective losses from the nanofluid. The optical properties of the glass play an important role in DASTC. The optical properties of the glass should

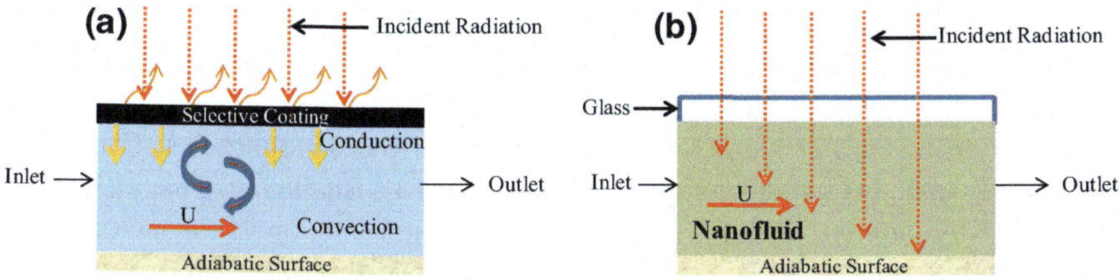

Fig. 1 **a** Schematic of surface absorption-based solar thermal collector (SASTC) and **b** direct absorption-based solar thermal collector (DASTC)

be in such a way that it will allow the incident radiation to pass through it, but do not allow the infrared radiation (which will generate due to heated nanofluid) to pass through it. In this way, glass helps in thermal trapping. In DASTC, the solar irradiation absorption capability depends on two parameters: (a) spectral absorption coefficient of the material ($K_{a\lambda}$) and (b) depth of the solar collector (y). When the solar irradiation incident on the nanofluid, it get absorbed by the nanofluid. The absorbed energy is given by Eq. 1.

$$E_{absorbed} = I_o[1 - \exp(-K_{a\lambda} \cdot y)] \quad (1)$$

The spectral absorption coefficient ($K_{a\lambda}$) depends on the volume fraction of the nanoparticles. So, the performance of the DASTC depends on the volume fraction of the nanoparticles (f_v) and the depth of the channel (y).

Tyagi et al. [2] have proposed a numerical model for a low-temperature direct absorption-based solar thermal collector. They have found that with aluminum nanoparticles ($D = 5$ nm), the absorptivity of base fluid has been increased by nine times as compared to pure fluid (water). Further, they have compared the collector efficiency of direct absorption-based solar thermal collector with convectional flat plate collector under same working conditions, and they have showed that under same operating conditions, the efficiency of direct absorption solar collector is 10% higher (on an absolute basis) than the conventional flat plate solar collector. Otanicar et al. [3] have experimentally concluded that there has 5% improvement in the efficiency of the solar thermal collectors when different types of nanoparticles have been used. Lenert and Wang [5] have prototyped a nanofluid-based cylindrical receiver which was optimized with carbon-coated nanoparticles suspended in Therminol VP-1. They have showed that the optical thickness plays an important role in the efficiency of the solar collector and have concluded that at high optical thickness, the losses to the environment increases. Chen et al. [6] have experimentally found that at very low volume fraction of gold nanoparticles, photo-thermal efficiency of gold nanoparticles is more than water. Luo et al. [7] have showed that the absorption film (place on the top surface) has more emission losses to the environment than the direct absorption-based solar thermal collector. From the numerical analysis, they have concluded that direct absorption-based solar thermal collector is a good choice for the harnessing of solar energy. Khullar et al. [8], Bhalla, and Tyagi [9] have shown experimentally that under similar working conditions, the temperature rise with DASTC is more than SASTC and it is beneficial to use nanoparticles-laden fluid-based solar thermal collector.

The performance of the direct absorption-based solar thermal collector (DASTC) mainly depends on the following factors: (a) material of the nanoparticles (b) volume fraction of the nanoparticles, and (c) shape of the nanoparticles. Gorji et al. [10], Hordy et al. [11], Sani et al. [12], Chen et al. [13], Bhalla and Tyagi [9], Tyagi et al. [2], Khullar et al. [8] have suggested that volume fraction of the nanoparticles play an important role in enhancing the efficiency of the direct absorption-based solar collector. It is very significant to use an optimum volume fraction of the nanoparticles in the direct absorption-based solar thermal collector. The optimum volume fraction also depends on the material of the nanoparticles. The material of the nanoparticles can be of metals, metal oxides, graphites. Each of the aforementioned materials has a different optical signature which depends on the area in which these collectors will be used.

It is very important to have awareness about the factors and properties of the nanoparticles which affect the toxicity of the nanoparticles. Sufian et al. [14] have presented that the toxicity of the nanoparticles depends on the size, material, shape, surface chemistry and charge, medium of synthesis, aggregation, storage time, mobility, and reactivity. But in actual working conditions, the heated nanofluid will allow to pass through the heat exchanger where it will heat the pure water (used for domestic applications) and that pure water will be used for heating purposes.

In the present study, a numerical model has been studied for a flat-shape direct absorption-based solar collector (nanoparticles: aluminum (D = 10 nm) and base fluid: water) which can be used for domestic applications. In the numerical model, aluminum nanoparticles have been used because these are metallic nanoparticles, and due to local surface plasmon resonance (LSPR) effect, these nanoparticles can absorb very good amount of incident solar irradiation. In the present study, firstly, the solar weighted absorptivity of the nanoparticles-laden fluid has been evaluated, and then, the effect of different parameters like (a) mass flow rate of the nanoparticles-laden fluid, (b) height of the solar collector, and (c) the length of the solar collector on the collector efficiency has been studied.

2 Theoretical Model

Modeling of radiative and thermal nature of direct absorption-based solar thermal collector is a challenging task, because for the modeling, radiative transfer equation (RTE) and heat transfer equation (HTE) need to be solved simultaneously and both these equations are of different nature. Where, RTE describes the ballistic transport of the photons inside a medium and HTE describes the diffusion of heat due to molecular and lattice vibration. In this numerical model, the coupled RTE and HTE model is developed to predict the behavior of direct absorption-based solar thermal collector (DASTC).

In order to evaluate the performance of the DASTC, few assumptions have been used: (1) The top surface is exposed to the ambient air at T_{∞} = 20 °C, and heat transfer coefficient is 10 W/m^2 K, so losses from top surface are due to convection only. (2) Bottom surface is adiabatic and transparent, i.e., no heat flux can pass through it. (3) Emission losses from the top surface are negligible because the temperature rise from the collector is not expected to be very high. (4) Scattering by the nanoparticles is an independent scattering because the volume fraction of the nanoparticles is very small. (5) Spectrally distributed solar irradiance is incident normally to the solar collector with an intensity of 1000 W/m^2 (no atmospheric absorption). (6) Thermo-physical properties of the nanoparticles are independent of the temperature rise of the fluid.

Table 1 Different values of the parameters used in numerical analysis

Sr. no.	Length of the collector (m)	Mass flow rate (kg/s)	Volume fraction (%)	Height of the collector (m)
1	0.3	0.003	0.01	0.002
2	0.4	0.0035	0.05	0.004
3	0.5	0.004	0.1	0.006
4	0.6	0.0045	0.5	0.008
5	0.7	0.005	1	0.01
6	0.8	–	5	0.012
7	0.9	–	–	0.014
8	1.0	–	–	0.016
9	–	–	–	0.018
10	–	–	–	0.02

The parameters which majorly affect the performance of the DASTC are (1) length of the collector, (2) mass flow rate of the nanofluid, (3) volume fraction of the nanoparticles, and (4) height of the collector. In order to observe these parameter's affect on the performance of DASTC, different values have been considered in the numerical analysis and are as given in Table 1.

The schematic of the flat-shape direct absorption-based solar thermal collector is as shown in Fig. 2.

In order to simulate the intensity distribution within the channel (along the height of the channel), Eq. (2) which is radiative transfer equation (RTE) has been used.

$$\frac{dI_\lambda(\Omega)}{dy} = -(K_{s_\lambda} + K_{a_\lambda})I_\lambda(\Omega) + K_{a_\lambda}I_{b_\lambda} + \frac{K_{s_\lambda}}{4\pi}\int_{4\pi} I_\lambda(\Omega')p(\Omega' \to \Omega)d\Omega \quad (2)$$

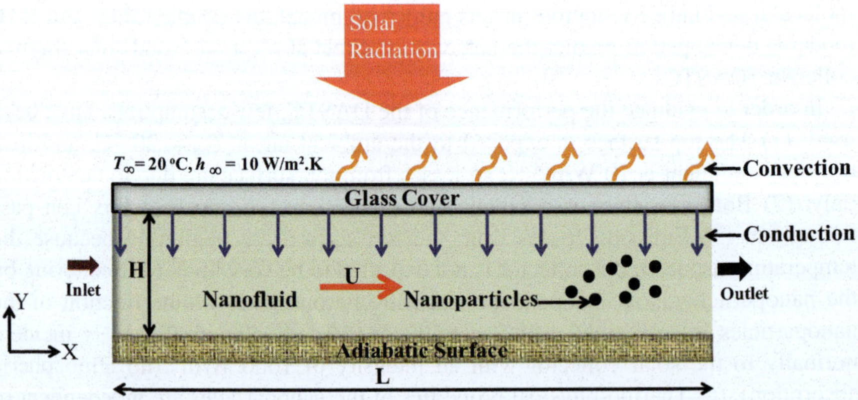

Fig. 2 Schematic of theoretical model of flat-shape direct absorption-based solar thermal collector

Here I is the incident intensity; I_b is the black body intensity; Ω is solid angle; $p(\Omega' \rightarrow \Omega)$ is phase function; $K_{a\lambda}$ and $K_{s\lambda}$ are the spectral absorption and scattering coefficients.

Equations (2a–9) and procedures for solving the intensity distribution, radiative heat flux, heat transfer equation, and the collector efficiency have been adapted from Tyagi et al. [2]. For pure fluid (water), scattering can be neglected and the attenuation of the incident radiation can be only by absorption. So, the absorption coefficient for pure fluid can be evaluated from Eq. 2.

$$K_{a\lambda,\,water} = 4\pi\kappa/\lambda \tag{2a}$$

where κ is index of absorption.

Hence, the RTE for pure fluid can be represented by Eq. 3, which is given as

$$\frac{dI_\lambda}{dy} = -\left(K_{a_\lambda,\,water}\right)I_\lambda \tag{3}$$

But with the addition of nanoparticles in the fluid, the absorption and scattering by nanoparticles take place and some complex relations need to be used for absorption ($K_{a\lambda,\ \mathrm{nanoparticle}}$)and scattering coefficients ($K_{s\lambda,\ \mathrm{nanoparticle}}$) of nanoparticles which are given in Eqs. 4 and 5.

$$K_{a\lambda,\,nanoparticle} = \frac{6\pi f_v}{\lambda}\mathrm{Im}\left\{\frac{m^2-1}{m^2+2}\left[1+\frac{\pi^2D^2}{15\lambda^2}\left(\frac{m^2-1}{m^2+2}\right)\frac{m^4+27m^2+38}{2m^2+3}\right]\right\} \tag{4}$$

$$K_{s\lambda,\,nanoparticle} = \frac{4\pi^4D^3f_v}{\lambda^4}\left|\frac{m^2-1}{m^2+2}\right|^2 \tag{5}$$

and

$$\begin{aligned} K_{e\lambda,\,nanoparticle} = {} & \frac{6\pi f_v}{\lambda}\mathrm{Im}\left\{\frac{m^2-1}{m^2+2}\left[1+\frac{\pi^2D^2}{15\lambda^2}\left(\frac{m^2-1}{m^2+2}\right)\frac{m^4+27m^2+38}{2m^2+3}\right]\right\} \\ & + \frac{4\pi^4D^3f_v}{\lambda^4}\left|\frac{m^2-1}{m^2+2}\right|^2 \end{aligned} \tag{6}$$

where $K_{e\lambda}$ is extinction coefficient, which is sum of absorption and scattering coefficient; D is the diameter of the nanoparticle; m is complex refractive index $\left(m = m_{particles}/n_{fluid}\right)$; f_v is volume fraction of nanoparticles in the base fluid.

Finally, the attenuation of incident light for nanoparticles-laden fluid can be evaluated by Eq. 7

$$\frac{\partial I_\lambda}{\partial y} = \left[-K_{a\lambda,\,water} + \left(-K_{e\lambda,\,nanoparticles}\right)\right]I_\lambda \tag{7}$$

Since, in this model, we assume independent scattering, hence, we can add two intensities.

Finally, to find out the temperature gain ($\Delta T = T_{out} - T_{in}$; $T_{in} = 20$ °C) in the fluid, RTE is solved simultaneously with the two-dimensional energy equation. The energy equation is given in Eq. 8.

$$k\frac{\partial^2 T}{\partial y^2} - \frac{\partial q_r}{\partial y} = \rho C_p \frac{\partial T}{\partial t} \tag{8}$$

where k is the thermal conductivity of the nanoparticles-laden fluid; q_r is radiative heat flux, and C_p is the specific heat of the fluid.

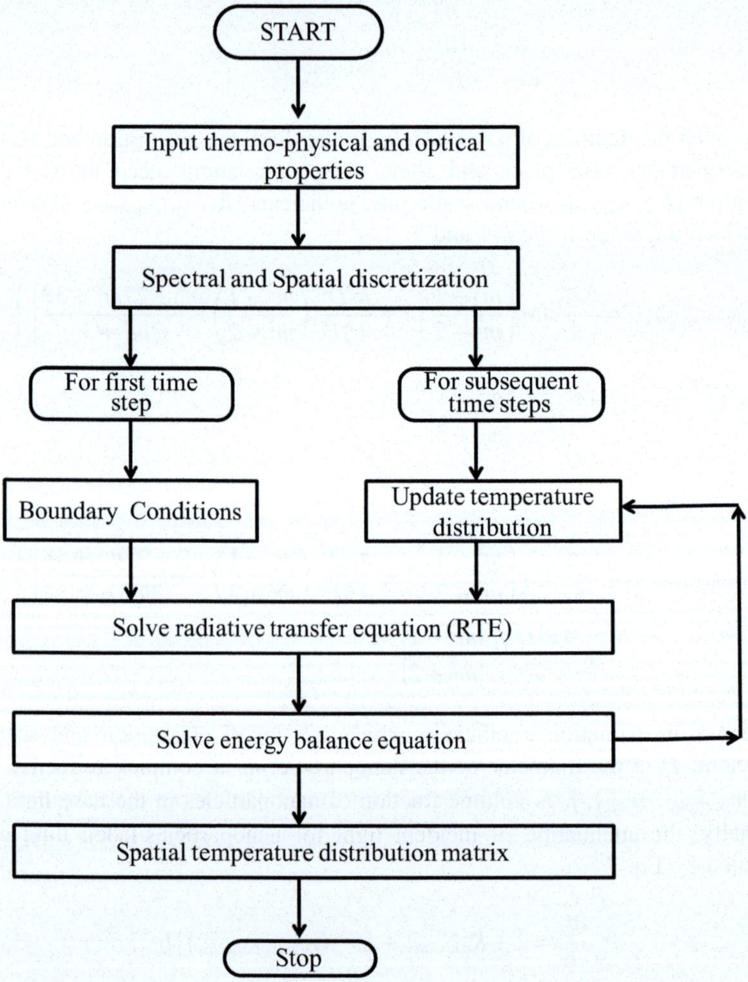

Fig. 3 Algorithm for spatial temperature distribution in direct absorption-based solar thermal collector (DASTC)

Finally, the collector efficiency has been found by using Eq. 9

$$\eta = \frac{\dot{m} Cp \Delta T}{A G_T} \tag{9}$$

where $\dot{m}$ is mass flow rate; A is top cover area of the collector; G_T is the incident solar flux (1000 W/m^2).

The algorithm used to evaluate the spatial temperature distribution for direct absorption-based solar thermal collector (DASTC) is as shown in Fig. 3.

3 Results and Discussions

To understand the potential of nanofluid in the direct absorption-based solar collector (DASTC), it is necessary to understand how much irradiation is getting absorbed by the fluid. To evaluate the absorbing capability of the nanoparticles-laden fluid, we calculated the absorbed energy fraction F, which has been calculated by using Eq. 10. This absorbed energy fraction depends on the volume fraction (f_v) of the nanoparticles and the depth of penetration (y) [15].

$$F = \frac{1 - \int_{\lambda_{\min}}^{\lambda_{\max}} I_{b\lambda}(\lambda, \mathrm{T}_{solar}) e^{-K_{e\lambda} y} d\lambda}{\int_{\lambda_{\min}}^{\lambda_{\max}} I_{b\lambda} d\lambda} \tag{10}$$

where $I_{b\lambda}$ is the black body spectral intensity; $K_{e\lambda}$ is the extinction coefficient; y is depth of penetration; T_{solar} is black body temperature (=5800 K).

Figure 4 shows the absorbed energy fraction F for various volume fractions of nanoparticles (0, 0.01, 0.05, 0.1 and 0.5%) at different depth of penetration. It shows that water has very low absorbed energy fraction and the absorbed energy fraction is directly proportional to the volume fraction of the nanoparticles. Figure 4 shows that when the volume fraction is increasing, the depth of penetration is decreasing,

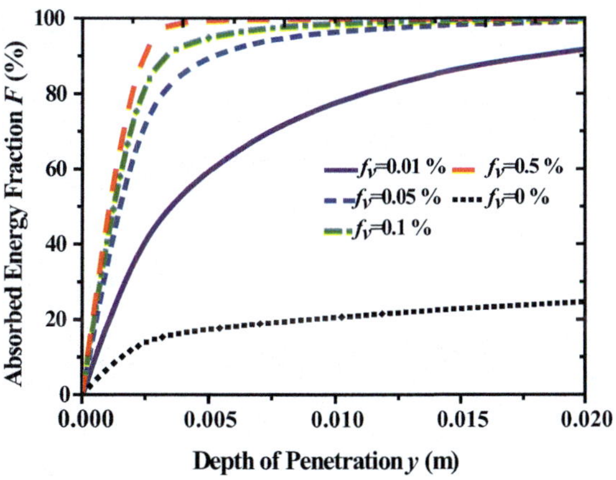

Fig. 4 Absorbed energy fraction versus the depth of penetration for different volume fractions of nanofluid

i.e., irradiation gets absorbed in a very thin layer of nanoparticles-laden fluid and maximum amount of energy get absorbed within very small depth of penetration. For example, at low volume fraction (0.01%), approximately 90% energy get absorbed at the depth of 0.02 m, whereas at high volume fraction (0.5%), 100% energy get absorbed at the depth of 0.005 m. This information is very helpful while designing the direct absorption-based solar collector, and according to this, we can adjust the height of the nanoparticles-laden fluid in the direct absorption-based solar thermal collector (DASTC).

The temperature rise of the fluid is an important parameter in the solar collector. Figure 5a, b shows effect of mass flow rate on the temperature rise of pure fluid and nanoparticles-laden fluid.

Figure 5a, b shows the temperature gain by water at different mass flow rates. Figure 6a, b shows the efficiency of the direct absorption solar collector using water

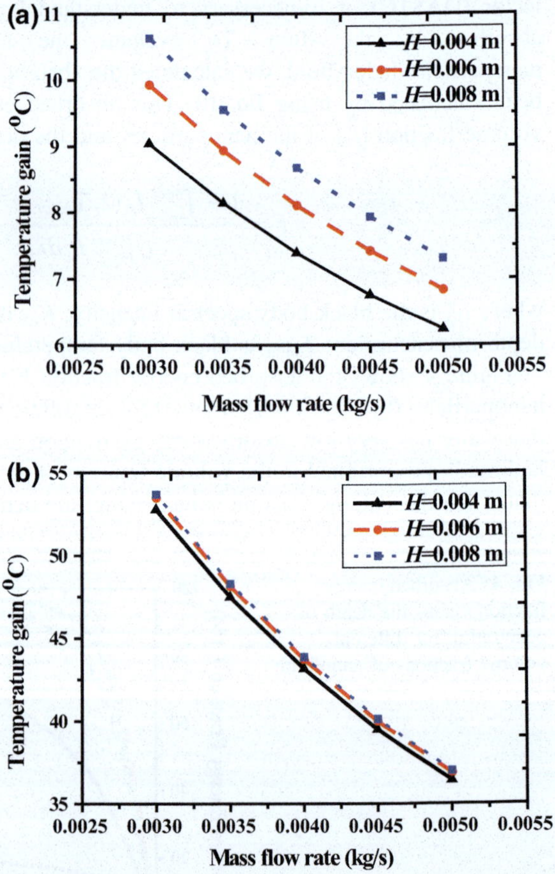

Fig. 5 **a** Temperature gain of water and **b** temperature gain of nanoparticles-laden fluid at different mass flow rates and different height of the DASTC channel for constant volume fraction of nanoparticles ($f_v = 0.2\%$). Length of the collector is 1 m

and nanofluid at different mass flow rates and at different mass flow rates. The trend of temperature rise decreases with the increase of mass flow rate as shown in Fig. 5a, b. The reason for this decreasing trend is that when the mass flow rate is high, average velocity of the fluid is very larger, results in less exposure time of the incident solar radiation. The conversion of the incident solar radiation to heat take some time, results in less temperature rise of the fluid.

Whereas, on the other hand, when the depth of the channel increases and the mass flow rate is constant, the average velocity of the fluid decreases that results in increase of temperature rise of the fluid. Secondly, one more observation can be interpreted in such a way that when the height of the channel (height of the fluid) increases, the attenuation of incident sunlight increases. Due to increasing attenuation, the fluid gets heated and gives high temperature gain. Further, on comparing the temperature rise of pure fluid and nanoparticles-laden fluid shows that temperature gain is very high with nanoparticles.

Figure 6a, b shows the effect of mass flow rate on the collector efficiency at different height of the DASTC channel for constant volume fraction of nanoparticles ($f_v = 0.2\%$). Figure 6a, b shows that the collector efficiency increases with the increase of mass flow rate $\dot{m}$ and height of the collector H. It can be understood in such a way that η is directly proportional to both $\dot{m}$ and ΔT. The rate of increase of mass flow rate is larger than the decrease in temperature gain for all values of H; thus, the collector efficiency can be increased. From Fig. 6b, we can conclude that the collector efficiency with nanofluid is more than the pure water.

Another parameter which affects the performance of the solar collector is the length of the solar collector. Figure 7 shows the effect of channel length on the temperature gain and the collector efficiency. From Fig. 7, it can be concluded that with the increase of length of the collector the temperature gain increases because the amount of sunlight absorbed by the flowing fluid increases.

But similarly, the collector efficiency decreases, and it can be interpreted as when the length of the collector is small, the temperature gain is less and correspondingly the losses to the atmosphere through convection are less due to which the collector efficiency is high. Simultaneously, on increasing the length of the collector, the convective losses increases and due to which the collector efficiency decreases.

The volume fraction of the nanoparticles plays an important role in the direct absorption solar collector. Figure 8 shows the effect of the volume fraction on the temperature gain of the nanofluid. Figure 8 shows an optimum volume fraction of the nanoparticles, i.e., the temperature below and above the optimum volume fraction is less. The trend of the temperature gain shows that the system is highly dependent on the volume fraction of the nanoparticles. Increasing the volume fraction increases the attenuation of the sunlight. This attenuation varies exponentially with the extinction coefficient and hence also with the volume fraction of the nanoparticles. So, there has an optimum volume fraction of the nanoparticles where the temperature gain is maximum. For this optimum volume fraction, height of the collector was 0.006 m and the mass flow rate was 0.003 kg/s.

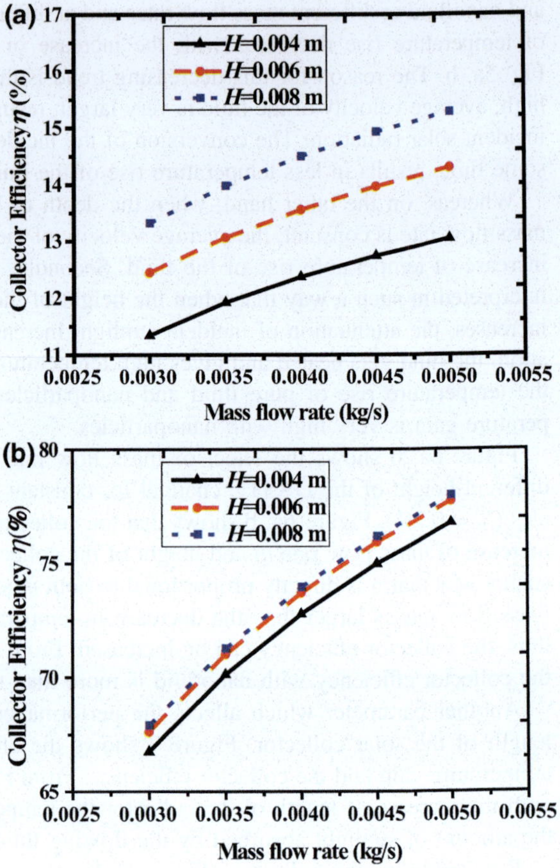

Fig. 6 **a** Collector efficiency for pure fluid and **b** collector efficiency for nanoparticles-laden fluid at different mass flow rates and different height of the DASTC channel for constant volume fraction of nanoparticles ($f_v = 0.2\%$). Length of the collector is 1 m

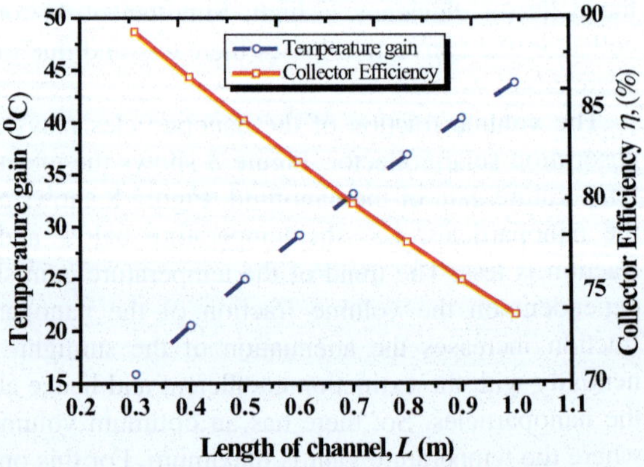

Fig. 7 Effect of channel length on the temperature gain and the collector efficiency for the nanofluid-based solar collector

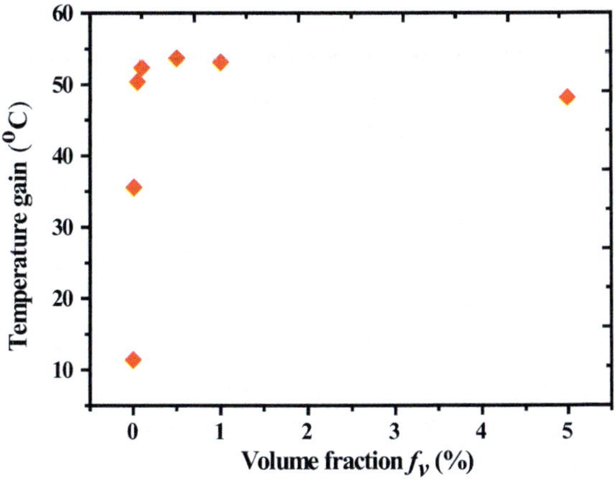

Fig. 8 Effect of volume fraction on the temperature gain

4 Conclusion

The analysis shows that collector efficiency increases with the increase of mass flow rate when the height of the fluid in the collector is same and the efficiency of the collector deceases with the increase of channel length. The results show that with the nanofluid, the temperature gain is 42 °C higher than pure water. The results also show that it is instrumental to use an optimum volume fraction of the nanoparticles in DASTC.

References

1. Dincer I (2000) Renewable energy and sustainable development: a crucial review. Renew Sustain Energy Rev 4(2):157–175
2. Tyagi H, Phelan P, Prasher R (2009) Predicted efficiency of a low-temperature nanofluid-based direct absorption solar collector. J Sol Energy Eng 131(4):41004
3. Otanicar TP, Phelan PE, Prasher RS, Rosengarten G, Taylor RA (2010) Nanofluid-based direct absorption solar collector. J Renew Sustain Energy 2(3)
4. Taylor RA, Phelan PE, Otanicar TP, Walker CA, Nguyen M, Trimble S, Prasher R (2011) Applicability of nanofluids in high flux solar collectors. J Renew Sustain Energy 3(2):23104
5. Lenert A, Wang EN (2012) Optimization of nanofluid volumetric receivers for solar thermal energy conversion. Sol Energy 86(1):253–265
6. Chen M, He Y, Zhu J, Kim DR (2016) Enhancement of photo-thermal conversion using gold nanofluids with different particle sizes. Energy Convers Manag 112:21–30
7. Luo Z, Wang C, Wei W, Xiao G, Ni M (2014) Performance improvement of a nanofluid solar collector based on direct absorption collection (DAC) concepts. Int J Heat Mass Transf 75:262–271
8. Khullar V, Tyagi H, Hordy N, Otanicar TP, Hewakuruppu Y, Modi P, Taylor RA (2014) Harvesting solar thermal energy through nanofluid-based volumetric absorption systems. Int J Heat Mass Transf 77:377–384

9. Bhalla V, Tyagi H (2017) Solar energy harvesting by cobalt oxide nanoparticles, a nanofluid absorption based system. Sustain Energy Technol Assessments 1–10
10. Gorji TB, Ranjbar AA (2016) A numerical and experimental investigation on the performance of a low-flux direct absorption solar collector (DASC) using graphite, magnetite and silver nanofluids. Sol Energy 135:493–505
11. Hordy N, Rabilloud D, Meunier JL, Coulombe S (2014) High temperature and long-term stability of carbon nanotube nanofluids for direct absorption solar thermal collectors. Sol Energy 105:82–90
12. Sani E, Mercatelli L, Barison S, Pagura C, Agresti F, Colla L, Sansoni P (2011) Potential of carbon nanohorn-based suspensions for solar thermal collectors. Sol Energy Mater Sol Cells 95(11):2994–3000
13. Chen N, Ma H, Li Y, Cheng J, Zhang C, Wu D, Zhu H (2017) Complementary optical absorption and enhanced solar thermal conversion of CuO-ATO nanofluids. Sol Energy Mater Sol Cells 162(Nov 2016):83–92
14. Sufian MM, Khattak JZK, Yousaf S, Rana MS (2017) Safety issues associated with the use of nanoparticles in human body. Photodiagnosis Photodyn Ther 19:67–72
15. Drotning WD (1978) Optical properties of solar-absorbing oxide particles suspended in a molten salt heat transfer fluid. Sol Energy 20(4):313–319

Thermal and Materials Perspective on the Design of Open Volumetric Air Receiver for Process Heat Applications

Gurveer Singh, Rajesh Kumar, Ambesh Dixit and Laltu Chandra

Abstract The concentrated solar thermal technologies (CST) are versatile in view of their multi-faceted applications, such as, process heat, cooling, and electricity generation. These are of line and point-focusing types with the later having much higher flux concentration (in Suns). This allows achieving a temperature in excess of 1200 K using, for instance, the open volumetric air receiver (OVAR). Such a high temperature is useful for applications, like the one which is developed at IIT Jodhpur, namely the solar convective furnace for heat treatment of aluminum. This requires a temperature of up to 750 K in the first phase of development. Thus, a suitable solar selective coating withstanding such a high temperature and having a thermal conductivity close to the base material for operating in an open atmosphere is desirable. Because of its atmospheric exposure, air and dust-induced degradation is inevitable, which may lead to its failure. These challenges are to be addressed for adapting such high-temperature CST technologies in arid deserts of India, the Middle-East, and Africa. In view of such challenges, the following details and foreseen developments are discussed in the paper:

(a) design of OVAR including various sub-components;
(b) flow-stability and the effect of heat-flux distribution on an absorber pore;
(c) the developed coating and its characterization for OVAR.

G. Singh · L. Chandra (✉)
Department of Mechanical Engineering, Indian Institute of Technology Jodhpur, Jodhpur 342011, India
e-mail: chandra@iitj.ac.in

R. Kumar · A. Dixit (✉)
Department of Physics, Indian Institute of Technology Jodhpur, Jodhpur 342011, India
e-mail: ambesh@iitj.ac.in

G. Singh · R. Kumar · A. Dixit · L. Chandra
Centre for Solar Energy, Indian Institute of Technology Jodhpur, Jodhpur 342011, India

H. Tyagi et al. (eds.), *Applications of Solar Energy*, Energy, Environment, and Sustainability, https://doi.org/10.1007/978-981-10-7206-2_7

Nomenclature

Latin

A_H	Total hole area in perforated plate (m^2)
A_P	Cross-section area of perforated plate (m^2)
B	Plank's distribution of blackbody radiation ($W/m^2\mu m$)
B_c	Coefficient
c_{pf}	Specific heat (kJ/kgK)
D	Diameter of hole in perforated plate (m)
d_p	Diameter of pore (m)
d_f	Diameter of foot-piece (m)
F	Friction factor
G	Spectral irradiance ($W/m^2\mu m$)
I	Heat-flux along the z-direction (W/m^2)
I_0	Heat-flux on the inlet surface of absorber (W/m^2)
K	Nozzle loss coefficient
k_f	Thermal conductivity of air (W/mK)
k_s	Thermal conductivity of solid (W/mK)
k_p	Derived pressure-drop coefficient for absorber
K_{la}	Theoretical loss coefficient for a thick plate
K_{lp}	Loss coefficient based on v_p
L	Length of the absorber (m)
P	Static pressure (Pa)
q_s''	Incident power on a pore with absorber material (W)
R	Reflectance
Re_p	Reynolds number in a pore
T	Perforated plate thickness (m)
T	Temperature (K)
T_o	Inlet temperature od air (K)
T_{out}	Outlet temperature od air (K)
u_f	Superficial speed (m/s)
V	Velocity (m/s)
v_f	Average pipe fluid velocity (m/s)
v_p	Average speed in an absorber pore (m/s)

Greek

α	Absorptivity
β	Geometry-based parameter
γ	Vena contracta contraction ratio
Δp	Pressure-drop (Pa)
Δp^2	Difference in quadratic pressures (Pa^2)
∇	Gradient operator
$\in$	Porosity

ε	Emissivity
μ	Absolute viscosity (kg/m.s)
λ	Wavelength (μm)
ν	Kinematic viscosity (m^2/s)
ξ	Modified extinction coefficient (m^{-1})
ρ	Density (kg/m^3)
ρ_a	Reflectivity of the absorber material
θ	Incidence angle
Φ	Contraction coefficient
η_{ov}	Overall efficiency
η_{th}	Thermal efficiency

1 Introduction

Rajasthan and Gujarat in India receive abundant direct normal irradiance (DNI), which is desirable for implementing the concentrated solar thermal (CST) technologies [1–3]. These are classified under line and point focusing. For example, the heliostat based central receiver system (CRS) is a point-focusing technology. The CRS is capable of achieving a high temperature in excess of 1200 K [4–7]. For evaluating the sub-systems of CRS, the solar air tower simulator (SATS) facility is installed at IIT Jodhpur; see Fig. 1a. One of the designed open volumetric air receivers (OVARs) with circular absorbers is shown in Fig. 1b. To improve the absorbers' durability in open dusty ambient conditions, a suitable solar selective

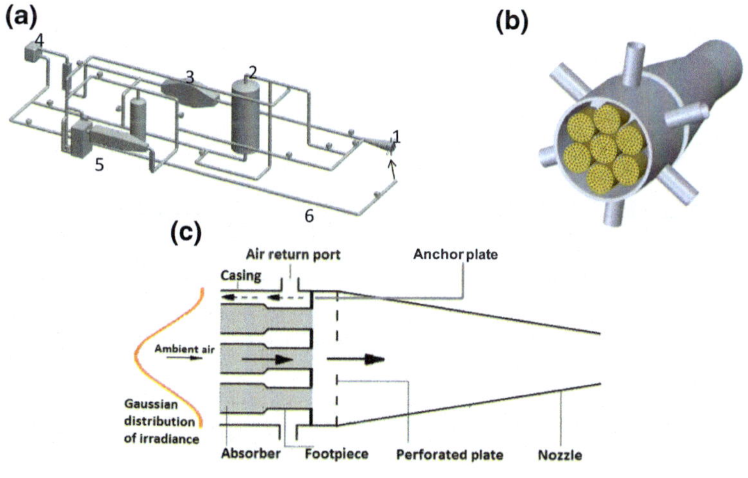

Fig. 1 **a** Solar air tower simulator (SATS) facility showing—1: OVAR, 2: TES, 3: heat exchanger, 4: blower, 5: SCF, 6: return-air line; **b** designed open volumetric air receiver (OVAR) **c** a section of the designed OVAR

coating (SSC) should be deposited on these porous structures. These coatings should sustain a temperature >750 K beside the mentioned challenging properties. However, most of the reported SSCs are developed for vacuum or inert atmosphere [8]. The SSCs should exhibit an absorptivity of ≥ 0.95 in the solar spectral range viz. 0.3–2.5 μm and emissivity of ≤ 0.05 in the near/far infrared spectral range viz. 2.5–25 μm. The ceramic metal (cermet)-based spectrally selective absorber structures are commonly used due to their relatively high spectral selectivity and the ease of depositing metal in ceramic matrix [9]. These absorbers are exposed to the concentrated solar irradiance (CSI) with an aim to achieve volumetric heating [10]. The same was demonstrated in laboratory with peripheral Joule heating of porous absorbers [11, 12]. Simultaneously, the generated heat is removed by the suction of air through absorber pores by virtue of forced convection. The foot-piece ensures radial thermal uniformity at the absorber outlet. To allow removal of dust and cleaning of receiver, a space is provided between the foot-piece and the anchor plate. Moreover, for homogenization of air temperature at the OVAR outlet, the nozzle with perforated plate is provided. Hence, the objectives of the designed OVAR are (a) achieving thermal uniformity at the OVAR outlet and (b) safe and stable operation in desert regions. Further details and evaluation of OVAR were reported [11, 12]. This is developed in view of a futuristic application viz. metals processing in solar convective furnace (SCF) [13]. This requires an air temperature of ca. 750 K, which is the upper limit at the first stage of development. The advantages of OVAR are reported by, e.g., [6, 10]. Some of the developed OVAR designs are listed in Table 1 [11, 14]. This describes their geometrical, absorber material, and operating parameters along with thermal efficiencies.

The associated challenges or uncertainties with the operation of an OVAR include (a) failure of absorbers due to thermally induced flow instability,

Table 1 Some details of the reported open volumetric air receivers

Name	Material	Absorber/ Pore	Efficiency (%)	Average outlet air temperature (°C)	Diameter (m)	Thickness (m)
TSA	Inconel 601	Coiled knit wires	79	700	0.280	0.050
Bechtel 2	Nichrome	Wire mesh	66	563	0.875	–
Catrec 2	Stainless steel	Honeycomb foil	70	460	0.756	0.090
SANDIA FOAM	Alumina	Circular	54	550	0.875	0.030
Selective receiver	SiSiC	Straight pores	62	620	0.835	0.092 + 0.080
HiTRec II	re-SiC		72	700	0.14 (H)	–
SOLAIR 3000	re-SiC		75	750	0.14 (S)	–
IIT Jodhpur	Brass		70–80	325	0.0254	0.0254

H: diagonal of hexagon, S: side of square

(b) thermal stresses on absorber, and (c) a lower outlet air temperature and thermal efficiency than expected; see, e.g., [14–18]. Some of these are attributed to (a) Gaussian distribution of CSI on the OVAR aperture and (b) non-uniform heat-flux distribution along the pore length of absorber. This is converted to power on aperture (PoA) of OVAR for its assessment; see, e.g., [19]. The radial non-uniformity of heat-flux distribution leads to the highest temperature of centrally located absorbers. Moreover, the increasing kinematic viscosity of air with temperature results in a relatively higher pressure-drop. As a consequence of its radial invariance in an absorber, a lower mass flow rate (MFR) is expected in the centrally located absorbers than its peripheral counterparts. A probable outcome is damaged or failure of OVAR at the end and even the flow instability [20]. There is another challenge while operating an OVAR in arid desert region viz. dust deposition. Indian desert with known frequent dust/sand-storm is prone to dust-induced failure of absorber [21]. In addition, blockage of pores with the deposited dust having a low thermal conductivity (1–2 W/mK) may lead to an accident or failure on-the-field [22, 23]. Some of these following practical aspects are addressed for operating an OVAR in arid deserts:

(a) stable flow regime through absorber pores at different conditions;
(b) the deposited coating on absorber surface, which should be resistant to high temperature, air-based corrosion besides dust-related abrasion.

Considering the above-mentioned aspects, the paper is divided into two sections viz. thermal-hydraulic analysis and coating development and characterization.

2 Thermal-Hydraulic Analysis

In this section, the analysis of pressure-drop, flow-stability, and heat-flux distribution for the designed straight pore absorber-based OVAR is presented. Pressure-drop depicts the offered resistance to fluid-flow inside the pore. High values of pressure-drop imply large capacity of blower requirement and reduction in OVAR performance. It should be noted that all the parts of OVAR contribute to this pressure-drop, which is depicted in Fig. 1b and c. The available expressions for the estimation of pressure-drop in these parts of OVAR are listed in Table 2.

The total pressure-drop (Δp_O) in OVAR provides the required blower power during its operation. This must be addressed while analyzing the efficiency of OVAR. Generally, the expression for thermal efficiency of OVAR does not include this parasitic loss. Thus, a new parameter viz. overall efficiency is introduced [27], which is given as follows:

$$\eta ov = \left(\frac{\eta th \times PoA/MFR}{PoA/MFR + \Delta p_o/\rho} \right) \quad (1)$$

Table 2 Estimation of pressure-drop of OVAR components

Component	Source	Expression for pressure-drop
Absorber (foam)	Ergun Eqn. [24]	$\frac{\Delta p}{l} = \frac{150\mu(1-\in)^2 u_f}{d_p^2 \in^3} + \frac{1.75\rho(1-\in)u_f^2}{d_p \in^3}$
Absorber (straight pores)	Hagen–Poiseuille Eqn. [25]	$\Delta p = \frac{32\mu v_p l}{d_p^2}$
Foot-piece (straight pipe)	Darcy–Weisbach Eqn. [25]	$\Delta p = \frac{\rho f l}{d_f} \cdot \frac{v^2}{2}$
Perforated plate	Holt et al. [26]	$\Delta p = K_{lp} \frac{\rho v_f^2}{2}$ $K_{la} = 1 - 2(A_p/A_H) + 2(A_p/A_H)^2(1 - 1/\gamma + 1/2\gamma^2)$ $K_{lp} = (2.9 - 3.79\varphi + 1.79\varphi^2)K_{la}$ if $\varphi \leq 0.9$ $K_{lp} = (0.876 + 0.069\varphi)K_{la}$ if $\varphi > 0.9$
Nozzle	White [25]	$\Delta p = k.\frac{\rho v^2}{2}$ where k = 0.04

To analyze the thermally induced flow-stability in the designed OVAR, the pressure-drop in an absorber pore is required. A correlation for the estimation of pressure-drop across a straight circular pore-based absorber is reported in [20], which is given as follows:

$$\Delta p = k_p \times \frac{\rho v_p^2}{2} \quad \text{with } k_p = 179.25 \text{Re}_p^{-0.588} \tag{2}$$

The correlation as in Eq. (2) is used to derive the difference in quadratic pressures (Δp^2) across an absorber pore [20]. An unexpected oscillatory behavior of Δp^2 with respect to the outlet air temperature depicts flow instability; see, e.g., [18]. The final expression of Δp^2 is as follows:

$$\Delta p^2 = p_{inlet}^2 - p_{outlet}^2 = B' \cdot \left(\frac{q_s'' / \varepsilon T_{out} - \Sigma \beta \sigma T_{out}^5}{(T_{out} - T_0)} \right)^{1.412} \text{where } B' = \frac{B_c}{c_{pf}^{1.412}} \tag{3}$$

The Eq. (3) is analyzed using two cases with different flux concentrations (FC). A validated-two-dimensional axis-symmetry-based approach solving continuity, momentum, and energy equations is employed for numerical simulation. Pressure outlet and mass flow inlet are defined in the geometry. The details are summarized in Table 3.

2.1 Results and Discussions

A comparison between the numerical and Eq. (3) based values of Δp^2 is shown in Fig. 2b. The different values of PoA viz. 1.03 and 2.06 W are applied on a single pore of OVAR corresponding to FC of 167 and 335 on the front surface of

Table 3 Numerical setup for two-dimensional analysis

Governing equations	Mesh	Numerical scheme	Convergence
Fluid (air) $\nabla \cdot \vec{V} = 0$ $(\vec{V} \cdot \nabla)\vec{V} = -\frac{\nabla p}{\rho} + \nu \nabla^2 \vec{V}$ $\left(\nabla \cdot \rho \vec{V} H\right) = \nabla \cdot \left(\frac{k_f}{c_{pf}} \nabla H_f\right)$ Solid (brass absorber) $\nabla \cdot (k_s \nabla T) = 0$	Type: rectangle Size: 0.015–0.05 mm Aspect ratio: 5.5–7.5 Orthogonality: 0.8-0.99	First order upwind with SIMPLE algorithm	10^{-6}

absorber, respectively. For a comparison, 200 $\leq$ *PoA/MFR* (kJ/kg) $\leq$ 1000 is considered. Here, the heat-flux is uniformly distributed over the entire pore length. This depicts volumetric heating, and the following are inferred:

(a) the expression of Δp^2 as in Eq. (3) is consistent with two-dimensional simulation for *PoA/MFR* > 600 (kJ/kg) within an uncertainty of ca. ±10%;
(b) the decreasing Δp^2 with outlet air temperature T_{out} depicts the usual behavior during forced convection-based heat transfer in the pore.

Thus, it is concluded that the derived expression in Eq. (3) with some assumptions is reasonable and that flow instability is not foreseen in the designed OVAR. This implies a stable operation of the designed SCF. Flow instability may be triggered with a non-uniform heat-flux distribution, due to the elevated flow resistance in lieu of differential kinematic viscosity. This aspect will be implicitly addressed while comparing the effect of heat-flux distribution on the air-side in OVAR. Interestingly, for the same value of *PoA/MFR* (kJ/kg), a relatively higher T_{out} is observed with FC of 335. Moreover, the differences increase with *PoA/MFR*, which is attributed to a higher Re_p or heat transfer coefficient. The same is depicted in Fig. 2b.

In laboratory, a uniform heat-flux on the absorber may be achieved under a controlled condition. However, on-the-field conditions, the heat-flux distribution

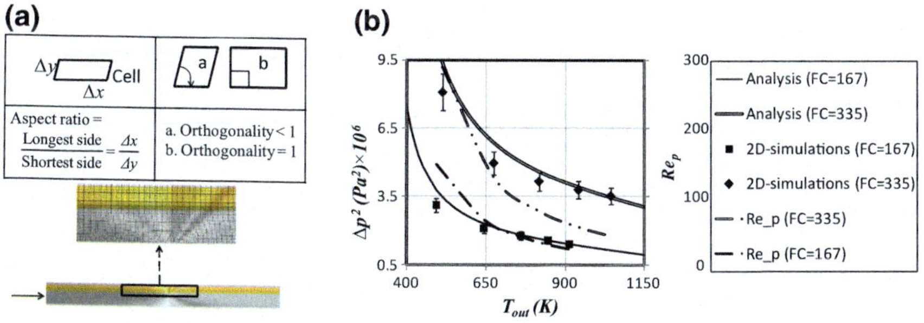

Fig. 2 **a** Mesh employed for two-dimensional simulations and **b** difference in quadratic pressure across the cylindrical absorber of OVAR (Fig. 1)

owing to influx condition will be non-uniform along the length [19, 28], which is given by

$$\left.\begin{aligned} I(z) &= I_0 e^{-\xi z} \\ \xi_{2d} &= \frac{-\ln \rho_a}{d_p \tan\theta} \\ \xi_{3d} &= \frac{3(1-\epsilon)}{d_p} \end{aligned}\right\} \quad (4)$$

Eq. (4) shows its dependence on the extinction coefficient (ξ), which is different for two (2d)- and three (3d)-dimensional pores. A straight pore is two-dimensional, whereas other types are generally classified as three-dimensional; see Table 1. For analyzing the effect of heat-flux distribution, a PoA of 1.03 W is considered. This results in flux concentration of ca. 167/1700 *Suns* on the absorber surface for a uniform/non-uniform heat-flux distribution along the pore length (z). This is due to the exponential decay of the irradiance, for instance in a three-dimensional pore, along the z-direction given by

$$I(z) = \frac{q_s'' d_p \xi}{4(1-e^{-\xi L})} e^{-\xi z} \quad (5)$$

In Fig. 3a, the variation of the centerline air temperature is shown along the flow direction for a uniform and a non-uniform heat-flux distribution with ξ = 720. This figure depicts comparable values of centerline air temperature, and thus, as in a straight pore, stable flow is expected. Similar trend is not observed with the absorber material temperature along the pore length ($z/L \leq 0.25$) under these different conditions; see Fig. 3b. Because of the volumetric heating, the temperature at the inlet surface of absorber is lower than that of the outlet with uniform heat-flux distribution [8]. With non-uniform heat-flux distribution, the highest temperature is found at the front surface [7, 28, 29]. A high value of the absorber front surface temperature will lead to substantial radiation-based heat loss. Thus, volumetric heating is indeed desirable to achieve higher overall efficiency. However, on-the-field condition, achieving volumetric heating is not possible, and thus installation of field is of prime importance to reduce ξ.

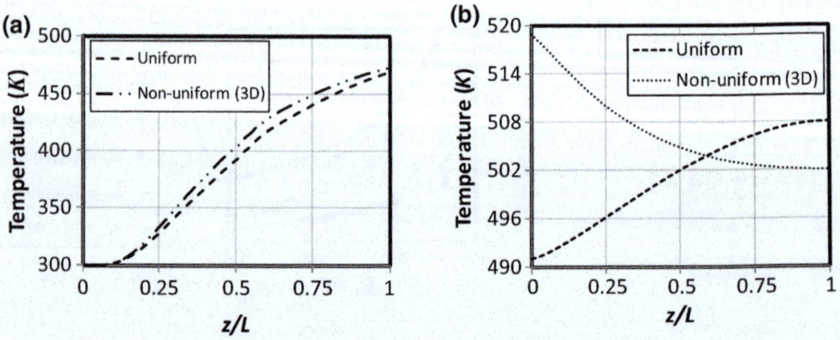

Fig. 3 Temperature distribution along the length for **a** air at the centerline **b** solid

3 Coating Development and Characterization

A ceramic metal (cermet)-based SSC is being developed for an OVAR to maximize the absorption of CSI on its aperture. There are various physical and chemical deposition processes explored for synthesizing theses cermet structures [8, 9, 30–32]. However, the sol-gel coating process is practically suitable for irregular surfaces in comparison to other deposition techniques [33, 34]. This process allows coating of the desired materials on wide variety of substrates including glasses, metals, and ceramics. The manipulation of relative material fraction can be easily controlled by using the suitable dopants in the base solution. Considering high-temperature operating conditions, silver metal-rich $BaTiO_3$ ternary oxide SSC has been developed on plane stainless steel and copper substrates [35–37].

3.1 *Experimental Details*

As a starting point, SSC is deposited on the stainless steel 304 (SS*304*) substrates, using the sol-gel dip-coating process. The dimensions of SS*304* flat substrate are 30 mm × 30 mm × 1 mm. These are metallographically polished with 2000 grade silicon carbide (SiC) abrasive paper to remove any surface—roughness and impurities. Subsequently, these polished substrates are chemically cleaned in trichloroethylene and acetone at 353 K for 5–10 min for removing organic impurities. Finally, the substrates are rinsed with deionized water and dried using nitrogen gas jets. Tetraethoxysilane, methyltriethoxysilane, acetic acid, barium acetate (from Alfa Aesar); titanium tetraisopropoxide (from Sigma Aldrich); isopropanol (from Vtec), with purity >98% are used for preparing $BaTiO_3$ sol. Silver nitrate with high purity ~99.9% (from Alfa Aesar) is used for introducing the metal fraction in $BaTiO_3$ ceramic SSC.

Acetate-derived $BaTiO_3$ sol is prepared using the glacial acetic acid, barium acetate, titanium tetraisopropyl alkoxide (TTIP), 2-propanol, and deionized water. The molar ratio of barium acetate to TTIP is maintained at the same level (Ba:Ti molar ratio = 1:1). First, barium acetate is dissolved in acetic acid at 353 K, and further, cooled to the room temperature. TTIP is added drop-wise in the desired molar ratio. The molar ratio of acetic acid is maintained seven times the barium acetate with DI water. The continuously stirred sol is refluxed to avoid the acid loses during sol preparation. Finally, a transparent and colorless sol is synthesized for $BaTiO_3$ ternary ceramic oxide. Further, silver nitrate solution is added $BaTiO_3$ solution to prepare Ag-$BaTiO_3$ cermet SSC. Silica sol is synthesized using hydrolysis and condensation of methyltriethoxysilane and tetraethoxysilane in 4:1 molar ratio in acidic medium by employing 0.1 M HCl solution.

Ag-$BaTiO_3$ followed by alternating $BaTiO_3$ and SiO_2 layers are dip coated on SS*304* substrates at a dipping speed of 2 mm/s and withdrawal speed of 1 mm/s (see Fig. 4a, b, and c). The process is repeated twice to achieve the desired stacking,

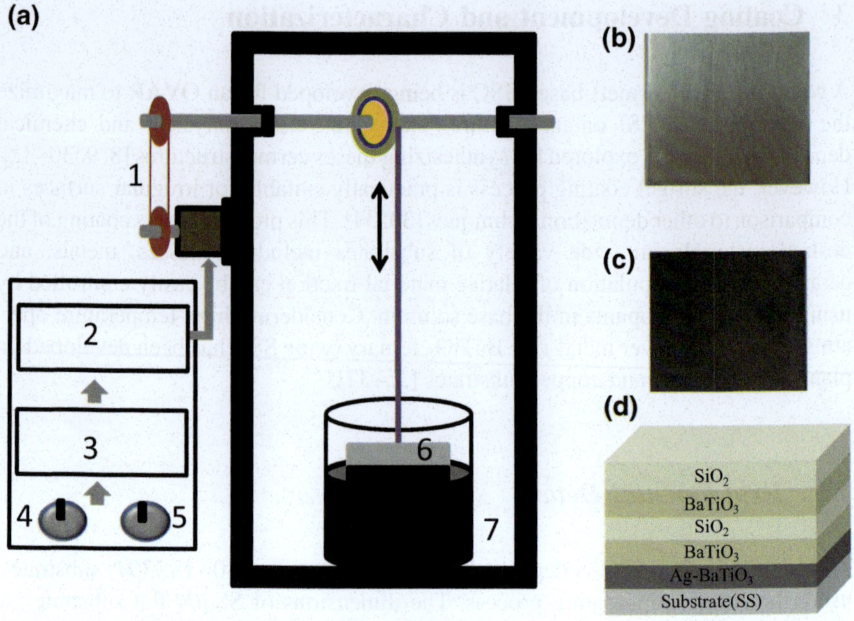

Fig. 4 **a** Schematic of a dip-coating system, with respective sub-components showing: 1: chain pulley arrangement, 2: stepper motor drive, 3: pic controller unit, 4: speed knob,5: direction knob, 6: substrate, 7: coating solution **b** optical image of uncoated SS*304* **c** and coated SSC **d** schematic of different SSC stacks on SS substrate

as schematically shown in Fig. 4d. The individual layers are dried in air at 353 K for 10 min. The Ag-$BaTiO_3$ layer is cured using UV lamp (from matrix instrument) to neutralize the charged Ag particles. All these layers are deposited at a temperature of 303 ± 2 K. The synthesized five-layered SSC is thermally cured at 1073 K for 1 h under argon ambient.

3.2 *Characterization*

X-ray diffraction (XRD) measurements are performed to understand the development of crystal structures of these absorber structures using Bruker D8-advance XRD system in logged couple mode using copper $K\alpha$ (λ = 1.5406 Å) monochromatic radiation source operating at 40 kV and 40 mA. The diffraction measurements are carried out within 20–80° at an angular speed of $0.02°s^{-1}$. The microstructure and surface properties are investigated using Carl Zeiss EVO 18-special edition scanning electron microscope (SEM). The elemental compositions are measured using an energy dispersive X-ray (EDX) (OXFORD instrument, an accessory integrated with SEM system). The thicknesses of these films are measured using Dektak XT stylus surface profiler (Bruker make). The optical reflectance of the SSC is measured using 110 mm-integrated sphere-based diffuse reflectance accessory with a Cary 4000 UV–Vis spectrophotometer in 300–900 nm. A polytetrafluoroethylene (PTFE) sample is used as the reference for this purpose. FTIR spectrophotometer (model Bruker Vertex 70 V) is used for reflectance measurements in 2.5–25 μm with the standard gold mirror as reference. These UV–Vis and FTIR reflectance measurements are used to calculate the room temperature solar absorptance $\alpha(\lambda)$ and thermal emittance $\varepsilon(\lambda)$ in the respective wavelength range. The solar absorptance is defined as the fraction of the total irradiance, which is absorbed by the SSC. Considering an opaque material with Kirchhoff's law, the spectral absorptance can be expressed in terms of reflectance $R(\lambda)$,

$$\alpha(\lambda) = 1 - R(\lambda) \tag{6}$$

where λ is the wavelength and $R(\lambda)$ includes the diffuse and collimated reflectance, which also depends on the temperature T. The solar absorptivity is calculated as follows:

$$\alpha = \frac{\int_{\lambda_1}^{\lambda_2} [1 - R(\lambda)G(\lambda)d\lambda}{\int_{\lambda_1}^{\lambda_2} G(\lambda)d\lambda} \tag{7}$$

$$\varepsilon(\lambda, T) = 1 - R(\lambda, T) \tag{8}$$

where $G(\lambda)$ is the spectral irradiance at AM1.5 [38]; λ_1, λ_2 are the minimum and the maximum wavelengths, respectively. The thermal emittance characterizes the performance of SSC in infrared region. This is defined as the fraction of emitted intensity from a hot body with respect to the blackbody at the same temperature. As the thermal radiation is emitted along all the directions, the spectral hemispherical thermal emittance is measured for characterization of SSC [39]. Using Kirchhoff's law, for an opaque object, the spectral emittance and total emissivity are given as follows:

$$\varepsilon(T) = \frac{\int_{\lambda_1}^{\lambda_2} [1 - R(\lambda, T)B(\lambda, T)d\lambda}{\int_{\lambda_1}^{\lambda_2} B(\lambda, T)d\lambda} \tag{9}$$

$$B(\lambda, T) = c_1\lambda^{-5}(e^{\frac{c_2}{\lambda T}} - 1)^{-1} \tag{10}$$

where $B(\lambda, T)$ is the Plank's distribution of blackbody radiation at temperature T [40]. In Eq. 5, the parameters $c_l = 3.743 \times 10^{-16}$ Wμm^4/m^2 and $c_2 = 1.4387 \times 10^{-16}$ μmK are first and second constant.

3.3 Results and Discussions

The deposited structure is investigated using XRD, and the diffractogram is shown in Fig. 5a. The different planer diffractions for SS*304* substrate are (111), (200), and (220), which corresponds to $Fe_{0.64}Ni_{0.36}$ and that of (110) depicts Fe–Cr phase. In contrast, no diffraction peaks are observed for $BaTiO_3$ and SiO_2 layers. This may be attributed to a thickness ~50 nm and the absence of crystalline phases for these materials. Scanning electron microscopy (SEM) image of the deposited SSC is shown in Fig. 5b. The synthesized structure shows homogeneous granular morphology with a diameter of ~0.5 μm. The deposited $SiO_2/BaTiO_3/SiO_2/BaTiO_3$/Ag-$BaTiO_3$/SS*304* SSC is homogeneous over the entire surface, as shown in Fig. 5b. The thickness is around 300 nm for $SiO_2/BaTiO_3/SiO_2/BaTiO_3$/Ag-$BaTiO_3$/SS*304* SSC. The atomic fractions of SSC are estimated using EDX and are summarized as the inset in Fig. 5b. These measurements show the presence of Ti, Ba, Si, and O atoms for the deposited SSC and are consistent with XRD observations.

The optical performance of the developed SSC is evaluated using UV-Vis and FTIR measurements. The measured infrared reflectance is shown in Fig. 5c, with

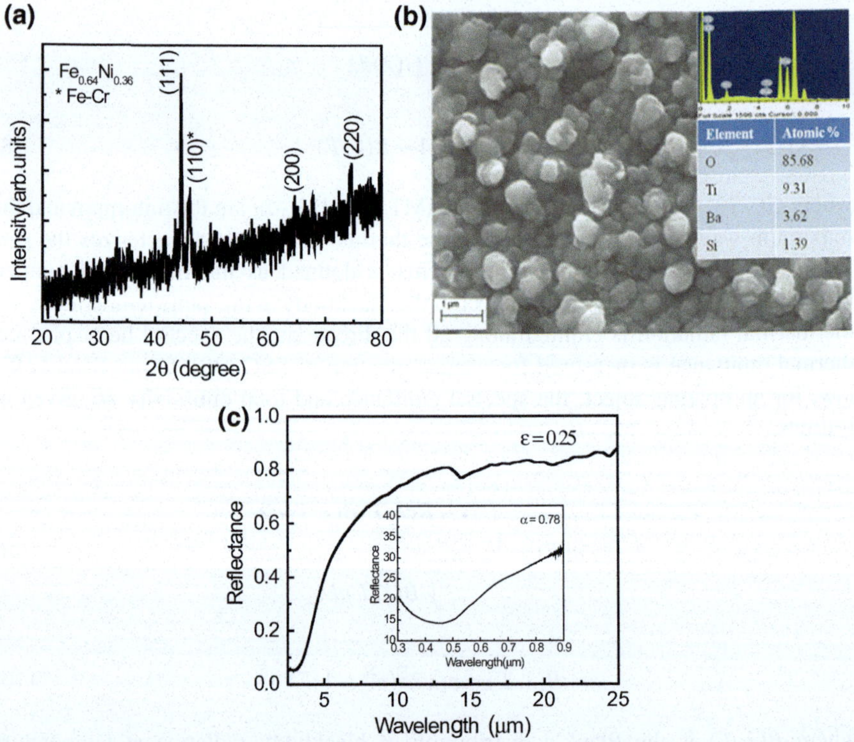

Fig. 5 **a** X-ray diffraction spectrum **b** SEM with EDX and **c** reflectance as a function of wavelength of deposited $SiO_2/BaTiO_3/SiO_2/BaTiO_3$/Ag-$BaTiO_3$/SS structure

inset showing the for UV-Vis spectral region. These are used to estimate the absorptivity and emissivity using Eqs. (7) and (9), respectively. The standard AM1.5 solar irradiance in ~0.3–2.5 μm and spectral black-body radiation in the desired spectral range are used for these measurements [38]. The obtained absorptivity and emissivity for the developed $SiO_2/BaTiO_3/SiO_2/BaTiO_3/$ Ag-$BaTiO_3$/SS*304* SSC are ~0.8 ± 0.05 in 0.3–0.9 μm and 0.25 ± 0.02 in 2.5–25 μm, respectively. These values are relatively poor, and therefore further optimization is in progress. The absorptivity and emissivity of SSC can be improved by varying the metal content in $BaTiO_3$ ceramic layer, multilayer stacking numbers, and the thickness of individual layers. However, lower thickness is desired to ensure the good adhesive property and low emissivity on the absorber (substrates) of OVAR. As desired, for such receivers, the developed oxide-based SSC will be stable up to 750 K in open atmosphere. However, the dust-related degradation issues will be revisited during the optimization process. Therefore, it may be safely stated that the selected approach is reasonable for OVAR-based process heat applications.

4 Conclusion

In this paper, some of the thermal and materials aspects pertaining to open volumetric air receivers (OVARs) are discussed. One of the designed OVAR at IIT Jodhpur is reported. The analysis of difference between quadratic pressure (Δp^2) at the inlet and outlet of OVAR depicts stable flow regime up to a *PoA/MFR* = 1000 kJ/kg. This includes the desired air temperature of 750 K for solar convective furnace. Furthermore, a comparative assessment of the derived expression for Δp^2 at two different flux concentrations revealed its independence on operating conditions. The two-dimensional numerical analysis showed that the volumetric heating will lead to a lower radiation-based heat loss in comparison to a non-uniform heat-flux distribution as on-the-field. For maximizing the energy absorption and sustainable operation of OVAR, an oxide-based solar selective coating (SSC) viz. $SiO_2/BaTiO_3/SiO_2/BaTiO_3$/Ag-$BaTiO_3$ is developed and characterized using stainless steel 304 (SS*304*) as the substrate. The measurement revealed the absorptivity and emissivity of about 0.8 ± 0.05 in 0.3–0.9 μm and 0.25 ± 0.02 in 2.5–25 μm, respectively. Strategies to improve the optical properties of this SSC are being discussed for future. Finally, it is concluded that the developed OVAR with SSC will allow a stable and safe operation of solar convective furnace system in arid deserts.

Acknowledgements The authors gratefully acknowledge the infrastructural support by IIT Jodhpur and the financial support from the Ministry of New and Renewable Energy, Govt. of India through grant no. 15/40/2010-11/ST.

References

1. Purohit I, Purohit P (2010) Techno-economic evaluation of concentrating solar power generation in India. Energy Policy 38(6):3015–3029
2. Pandey S, Singh VS, Gangwar NP, Vijayvergia MM, Prakash C, Pandey DN (2012) Determinants of success for promoting solar energy in Rajasthan. India. Renew Sustain Energy Rev 16(6):3593–3598
3. Bishoyi D, Sudhakar K (2017) Modeling and performance simulation of 100 MW LFR based solar thermal power plant in Udaipur India. Resour Effic Technol (in press) https://doi.org/10.1016/j.reffit.2017.02.002
4. Romero M, Buck R, Pacheco JE (2002) An update on solar central receiver systems, projects, and technologies. J Sol Energy Eng 124(2):98–108
5. Bai F (2010) One dimensional thermal analysis of silicon carbide ceramic foam used for solar air receiver. Int J Therm Sci 49(12):2400–2404
6. Behar O, Khellaf A, Mohammedi K (2013) A review of studies on central receiver solar thermal power plants. Renew Sustain Energy Rev 23:12–39
7. Ho CK (2017) Advances in central receivers for concentrating solar applications. Sol Energy (in press). https://doi.org/10.1016/j.solener.2017.03.048
8. Selvakumar N, Barshilia HC (2012) Review of physical vapor deposited (PVD) spectrally selective coatings for mid-and high-temperature solar thermal applications. Sol Energy Mater Sol Cells 98:1–23
9. Kennedy CE (2002) Review of mid-to high temperature solar selective absorber materials. NREL/TP-520–31267
10. Hoffschmidt B, Tellez FM, Valverde A, Fernández J, Fernández V (2003) Performance evaluation of the 200-kWth HiTRec-II open volumetric air receiver. J Sol Energy Eng 125 (1):87–94
11. Sharma P, Sarma R, Chandra L, Shekhar R, Ghoshdastidar PS (2015) Solar tower based aluminum heat treatment system: part I. Design and evaluation of an open volumetric air receiver. Sol Energy 111:135–150
12. Sharma P, Sarma R, Chandra L, Shekhar R, Ghoshdastidar PS (2015) On the design and evaluation of open volumetric air receiver for process heat applications. Sol Energy 121:41–55
13. Patidar D, Tiwari S, Sharma P, Pardeshi R, Chandra L, Shekhar R (2015) Solar Convective Furnace for Metals Processing. J Miner Met Mater Soc 67(11):2696–2704
14. Avila-Marin AL (2011) Volumetric receivers in solar thermal power plants with central receiver system technology: a review. Sol Energy 85(5):891–910
15. Kribus A, Ries H, Spirkl W (1996) Inherent limitations of volumetric solar receivers. J Sol Energy Eng 118:151
16. Pitz-Paal R, Hoffschmidt B, Böhmer M, Becker M (1997) Experimental and numerical evaluation of the performance and flow stability of different types of open volumetric absorbers under non-homogeneous irradiation. Sol Energy 60(3):135–150
17. Fend T, Pitz-Paal R, Hoffschmidt B, Reutter O (2005) Solar radiation conversion. In: Scheffler M, Colombo P (eds) Cellular ceramics: structure, manufacturing, properties and applications. Wiley-VCH Verkag GmbH & Co. KGaA, Weinheim, p 523–546
18. Becker M, Fend T, Hoffschmidt B, Pitz-Paal R, Reutter O, Stamatov V, Trimis D (2006) Theoretical and numerical investigation of flow stability in porous materials applied as volumetric solar receivers. Sol Energy 80(10):1241–1248
19. Roldán MI, Smirnova O, Fend T, Casas JL, Zarza E (2014) Thermal analysis and design of a volumetric solar absorber depending on the porosity. Renew Energy 62:116–128
20. Singh G, Chandra L. On the flow stability, Heat-flux distribution and dust deposition in an open volumetric air receiver for solar convective furnace system, Solar energy (under review)
21. Shao Y, Dong CH (2006) A review on East Asian dust storm climate, modelling and monitoring. Glob Planet Chang 52(1):1–22

22. Agrafiotis CC, Mavroidis I, Konstandopoulos AG, Hoffschmidt B, Stobbe P, Romero M, Fernandez-Quero V (2007) Evaluation of porous silicon carbide monolithic honeycombs as volumetric receivers/collectors of concentrated solar radiation. Sol Energy Mater Sol Cells 91 (6):474–488
23. Schmücker M (2005) Einfluss von Mineralstäuben auf keramische Solarabsorb-er. http://www.dlr.de/Portaldata/73/Resources/dokumente/soko/soko2012/Schmuecker_2012Soko_Mineralstaeuben.pdf. Accessed 30 May 2017
24. Ergun S (1952) Fluid flow through packed columns. Chem Eng Prog 48:89–94
25. White FM (2015) Fluid mechanics, 7th edn. McGraw Hill education, New Delhi
26. Holt GJ, Maynes D, Blotter J (2011) Cavitation at sharp edge multi-hole baffle plates. In: ASME 2011 international mechanical engineering congress and exposition american society of mechanical engineers. pp 401–410
27. Boddupalli N, Singh G, Chandra L, Bandyopadhyay B (2017) Dealing with dust–Some challenges and solutions for enabling solar energy in desert regions. Sol Energy 150:166–176
28. Wu Z, Caliot C, Flamant G, Wang Z (2011) Coupled radiation and flow modeling in ceramic foam volumetric solar air receivers. Sol Energy 85(9):2374–2385
29. Wang F, Shuai Y, Tan H, Yu C (2013) Thermal performance analysis of porous media receiver with concentrated solar irradiation. Int J Heat Mass Transf 62:247–254
30. Barshilia HC, Kumar P, Rajam KS, Biswas A (2011) Structure and optical properties of Ag–Al_2O_3 nanocermet solar selective coatings prepared using unbalanced magnetron sputtering. Sol Energy Mater Sol Cells 95(7):1707–1715
31. Usmani B, Dixit A (2016) Spectrally selective response of ZrOx/ZrC–ZrN/Zr absorber–reflector tandem structures on stainless steel and copper substrates for high temperature solar thermal applications. Sol Energy 134:353–365
32. Kumar R, Saha AK, Usmani B, Dixit A. Optimization and structure-property correlation of black chrome solar selective coating on Copper and Nickel plated copper substrates, materials today proceedings (accepted)
33. Brinker J, Scherer GW (1990) Sol-gel science. Academic Press, p 790
34. Kickelbick, G (2007) Hybrid materials: synthesis, characterization, and applications. Wiley
35. A Mathews P, RC Kalidindi S, Bhardwaj S, Subasri R (2013) Sol-gel functional coatings for solar thermal applications: a review of recent patent literature. Recent Pat Mater Sci 6 (3):195–213
36. Traversa E, Di Vona ML, Nunziante P, Licoccia S, Sasaki T, Koshizaki N (2000) Sol-gel preparation and characterization of Ag-TiO_2 nanocomposite thin films. J Sol-Gel Sci Technol 19(1):733–736
37. Subasri R, Raju KS, Reddy DS, Hebalkar NY, Padmanabham G (2016) Sol–gel derived solar selective coatings on SS 321 substrates for solar thermal applications. Thin Solid Films 598:46–53
38. http://rredc.nrel.gov/solar/spectra/am1.5/. Accessed 05 June 2017
39. Tesfamicharel T (2000) Characterization of selective solar absorbers: experimental and theoretical modeling. ACTA Universitatis Upsaliensis Uppsala, p 4–79
40. Duffie JA, Beckman WA (1991) Solar Engineering of Thermal Process, 2nd edn. Wiley Interscience, New York

Part IV
Thermal Storage of Solar Energy

Solar Thermal Energy Storage

Aniket D. Monde, Amit Shrivastava and Prodyut R. Chakraborty

Abstract Over the past few decades, considerable research efforts have been devoted to improve the usage of renewable energy resources. Till date, major energy demands have been addressed by fossil fuels, and the limited resource of these precious fuels is continuously depleting at an alarming rate. Increase in the energy demands, deficiency of fossil fuels, and influence of pollution on the environment have forced us to opt for renewable energy resources. Solar energy is a natural source of energy that is not depleted by its use. It is a promising option for replacing conventional energy resources partially or totally, but it is transient, intermittent, and unpredictable in nature. Because of this sporadic nature of solar energy across a given interval of hours, days, and season, various practical problems arise. Variable DNI causes power plants to shut down for few hours of the day or to run at part load most of the time. This creates a demand for an effective subsystem which is capable of storing energy when available solar energy overshoots the demand during the interval of radiant sunshine, and to make it accessible during night or season. A similar problem arises for waste heat recovery systems where accessibility of waste heat and usage period are not the same, and thus creates a need for thermal energy storage (TES) for energy conservation. TES has tempted a lot of researchers to improve its high energy storage capacity and efficiency. If solar energy system is not run with TES, a considerable section of energy demand has to depend on conventional resources which in result reduce the annual solar fraction. TES helps to reduce dependency over conventional resource by minimizing energy waste. TES is mainly described by the parameters like capacity, power, efficiency, storage period, charge and discharge time, and cost. There are different storage mechanisms by which energy can be stored: sensible, latent, and chemical reactions. In sensible-type storage, energy is stored by increasing the temperature of solid or liquid storage media (e.g., sand-rock minerals, concrete, oils, and liquid sodium). These materials have excellent thermal conductivity and are cheaper, but due to low heat capacity, it increases system size. In latent-type

A. D. Monde · A. Shrivastava · P. R. Chakraborty (✉)
Department of Mechanical Engineering, Indian Institute of Technology Jodhpur, Jodhpur, India
e-mail: pchakraborty@iitj.ac.in

H. Tyagi et al. (eds.), *Applications of Solar Energy*, Energy, Environment, and Sustainability, https://doi.org/10.1007/978-981-10-7206-2_8

storage, energy is stored/released during phase change; thus, it has higher storage capacity than sensible, but suffers from the issue of low thermal conductivity. As the solid–liquid phase change process of pure or eutectic substances is isothermal in nature, it is beneficial for the application having limitations with working temperature. In chemical-type TES, heat is absorbed/released due to breakdown or formation of chemical bonds. The technology is not much developed and has limited application due to possibility of degradation over time and chemical instability. TES can also be classified as active and passive depending upon the solid or liquid energy storage medium. Active TES is further classified as direct active and indirect active depending on whether the storage fluid and the heat transfer fluid (HTF) are same or some other HTF is required to extract heat from solar field. The discussion in this chapter includes basic heat transfer models, along with experimental studies by different research groups on various TES. Finally, methods and design criteria that can improve the system performance are discussed.

1 Classification of Thermal Energy Storage

Whenever there exists a mismatch between the energy availability from a source and load requirements in terms of rate and time, energy storage becomes an obvious option. The energy storage device stores energy when energy from the source is in surplus amount than its requirement and delivers it to the load when energy from the source may not be sufficient to run the load or may be absent all together. Thermal energy is stored within TES in the form of sensible, latent, and thermochemical. Classification of TES based on different criteria is shown in Figs. 1 and 2.

Brief description of each of these TES is given as follows:

Buffer: Buffer storage systems are storage systems of small capacity, typically designed for few hours only. The load demand for such cases extends over few hours. Their utility can be seen where time interval of load requirement and energy available from source coincides, but at any particular instant of time, there may be a mismatch between the quantities of two energies, as depicted in Fig. 3a. One example of such system is buffer TES for a solar Brayton engine [1].

Diurnal: The load demand extends over a day, but energy from the source may not be available for the load demand as in Fig. 3b. Hence, a storage system of capacity more than that of buffer storage is required, although it will still be a short-term storage system like buffer storage. Examples of such system are oil/rock thermal energy storage, combined molten salt and oil/rock thermal energy storage,

Fig. 1 Classification based on timescale

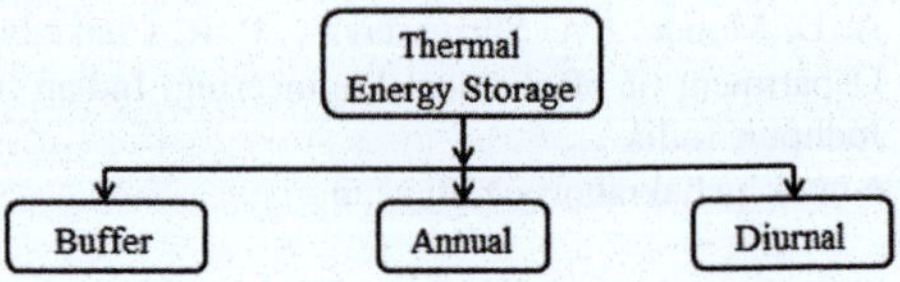

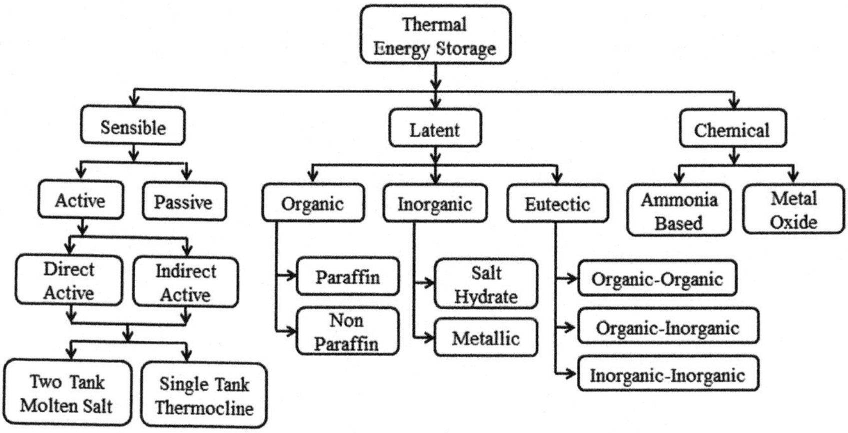

Fig. 2 Classification based on mode of storage

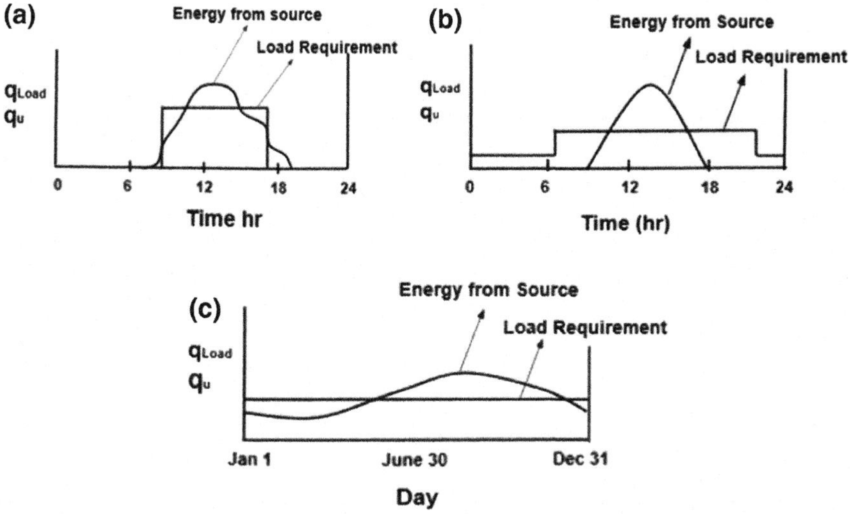

Fig. 3 **a** Buffer; **b** diurnal; **c** annual

and Hitec salts thermal energy storage [2]. Oil/rock thermal energy storage is a good option for intermediate temperature applications.

Annual: Annual storage systems are storage systems of large capacity where the load demand extends over the year. During summer when the energy from the source is more than the load requirement, it is stored in the storage, and during winter, this stored excess energy is delivered. Such large storage systems are called "annual storage systems" shown in Fig. 3c. Example of one such system is large insulated water tanks [3]. Such large size of water tanks for thermal storage can be built using materials like steel, reinforced concrete. Water tanks can also be placed within the earth itself [3]. The upward thermal loss is prevented by placing insulation layer, and lateral heat loss is curtailed by the thermal resistance of earth surrounding it. Dry ground can also be used for annual TES, as low thermal conductivity of solid helps to prevent heat losses.

Sensible Heat Storage: When the storage material (usually liquid or solid) is heated, it does not change its phase. The amount of energy stored in the storage is proportional to the change in temperature of the material and is shown as:

$$E = m \int_{T_1}^{T_2} c_p dT \quad (1)$$

where m is mass of the storage material, c_p is the specific heat, T_1 and T_2 are the temperature limits between which the storage system works, and the difference between temperature limits is known as temperature swing.

Latent Heat Storage: The storage medium undergoes phase change process during heat storage process. The energy stored can be expressed as follows:

$$E = m^*(L) \quad (2)$$

where m and L are mass and latent heat of storage material, respectively. In the above case, the storage is expected to operate almost isothermally at close proximity to melting point temperature of the heat storage medium.

If the system works within a range of temperature T_1 to T_2 such that melting point T_m appears somewhere in between the two limiting temperatures, then energy stored will be given by the following:

$$E = m \int_{T_1}^{T_m} c_{ps} dT + m^*(L) + m \int_{T_m}^{T_2} c_{pl} dT \quad (3)$$

where c_{ps} and c_{pl} represent specific heat of solid and liquid, respectively, and T_m is the melting temperature.

Thermochemical Reaction: Thermochemical heat storage involves reversible reactions.

$$A + \Delta H_r \Leftrightarrow B + C \quad (4)$$

A, B, and C are the reactants, and ΔH_r is the reaction enthalpy (kJ/mole)

During heat storage process, the endothermic reaction takes place, and chemical reactant A dissociates into B and C at the expense of thermal energy. During heat release process, an exothermic reaction takes place, products of the endothermic reaction are mixed in such a way so as to form the initial reactant A again, and heat energy is liberated. However, the temperature at which reverse and forward reactions occur is usually different. The storage operates isothermally during the reaction, and the product obtained after the reaction can be stored.

The thermal energy stored in thermochemical storage medium can be expressed as follows:

$$Q = n_A \Delta H_r \tag{5}$$

where n_A is the number of moles of the reactant A (mol).

A simplified scheme of TES system based on chemical reactions is shown in Fig. 4.

1.1 Selection Criteria for Thermal Storage

Following are the factors upon which selection of storage type depends [5]:

(a) Application for which TES system is to be used.
(b) Range of temperature within which TES is to be operated.
(c) Storage capacity.
(d) Thermal losses taking place from the storage.
(e) The time period for which storage cycle is to be operated.
(f) Cost of the storage unit (including operating cost).

The next three sections are attributed to the three important methods of thermal energy storage, viz. sensible, latent, and thermochemical heat storage.

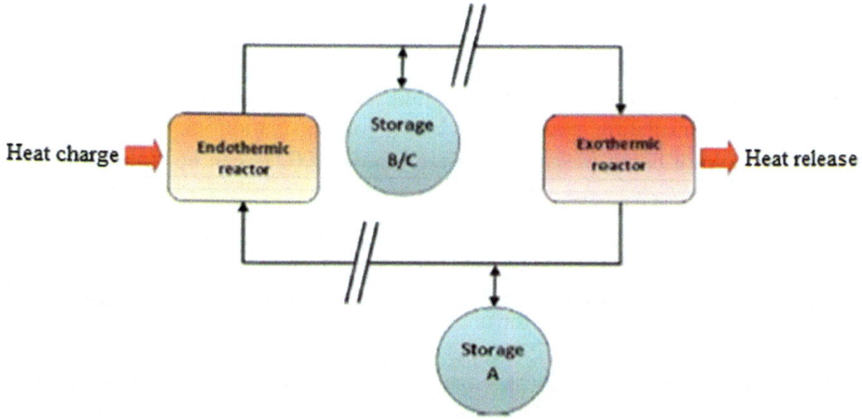

Fig. 4 Simplified scheme of a TES system based on chemical reactions [4]

2 Sensible Heat Storage

In sensible heat storage, the storage medium (liquid or solid) is heated such that its temperature rises. Some common materials used for storing sensible heat are water, oil, organic molten salts, rocks, pebbles, and refractory materials [5]. When solids are used for storing sensible heat, they are packed in such a way so as to form a porous bed through which hot fluid such as air may pass, and exchanging heat with the porous solid material. Choice of storage material is dominated mainly by the temperature level at which heat is to be stored. Further, sensible heat storage system is explained based on the material used and construction or storage configuration.

Disadvantages of sensible heat storage system are as follows:

a. Large volume requirement.
b. Incompetent to store or deliver energy at constant temperature.

2.1 Based on Material Used for Sensible Heat Storage System

2.1.1 Liquid Water

Water is one of the most commonly chosen media for storing sensible heat, especially for low-temperature applications, as it is cheap, easily available, and stable at operating temperature. Thermal management of commercial and residential buildings can be effectively achieved by using hot-water tanks where water is naturally stratified which improves the performance of the system. Steel, concrete, and fiber glass are the materials used for making water storage tanks. In order to reduce the heat loss from hot water, insulation is provided. The thickness of the insulation is usually large (10–20 cm) which becomes a significant part of the total cost. Shelton [6] has studied an underground tank surrounded by the earth that has sufficient value insulation, and this provides a good insulation thickness. If the used tank is capable of being pressurized, water above 100 °C can be stored [5].

2.1.2 Aquifer Thermal Energy Storage [ATES]

The important techniques used for underground thermal energy storage are aquifer thermal energy storage [ATES] (open loop), borehole thermal energy storage [BTES] (closed loop), and cool thermal energy storage [CTES]. Among the three, ATES has proved to be most promising and is commercially applied in many countries like Belgium, the Netherlands, Sweden, Denmark, and UK.

An aquifer is made up of saturated rock that helps water to move through it easily. Aquifers should be porous and permeable in nature and include rock types as sandstone, limestone, sand, and gravel [7]. Using aquifer eliminates the construction cost of storage tank since it uses the available ground water in the aquifer.

An ATES consists of two or more wells used for injection and pumping out water from the aquifer. Aquifer stores thermal energy by injecting hot water into the aquifer and extracting cold water simultaneously from the other well. ATES has two different operating modes: cyclic mode in which pumping wells and injection wells are interchanged with a change in season and continuous mode as shown in Fig. 5. The cyclic mode stores thermal energy in aquifer, thus increases system performance but has a complication in system design. On the other hand, the continuous mode has a simple design, but due to limited temperature range, the system has low performance.

Open ATES are of two types, namely doublet and mono-well. The most common among the two is doublet (Fig. 6) where the cold and warm wells are located in the same aquifer about 100 m apart. In case of mono-well, a single well is drilled, with two separated screens in the same aquifer. Warm water is stored in the upper screens of the well, and cold water is stored in the lower screen. If the cold water is stored in the upper screen, then the cold and hot water will mix due to buoyancy. In mono-well, this interference is avoided by keeping the vertical distance between warm and cold screens considerably large.

The basic elements of an ATES system are as follows: (i) source of thermal energy, (ii) delivery system, (iii) aquifer store, and (iv) thermal loads (Fig. 7).

Thermal Loads: Application of ATES can be commonly found in heating and cooling of buildings which have a high level of utilization like hospitals, academic institutions, shopping complex, offices, and industry. System performance can be optimized if the load demand includes both heating and cooling. At least a thermal

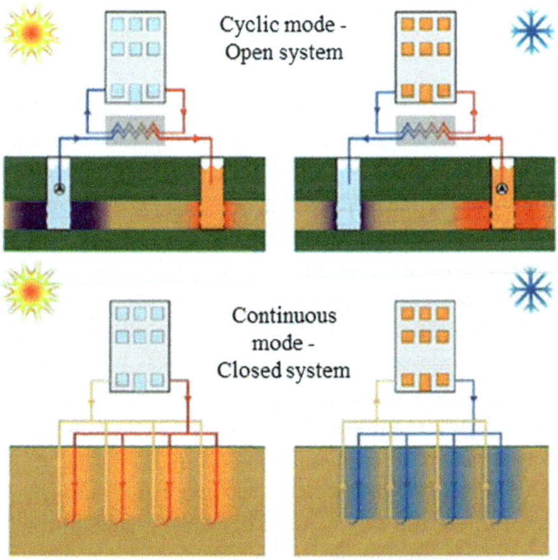

Fig. 5 Aquifer thermal energy storage (open and closed system) [8]

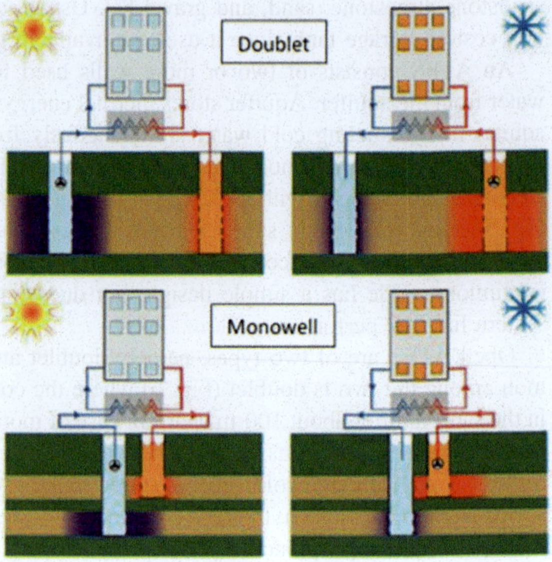

Fig. 6 Aquifer thermal energy storage (doublet and mono-well) [8]

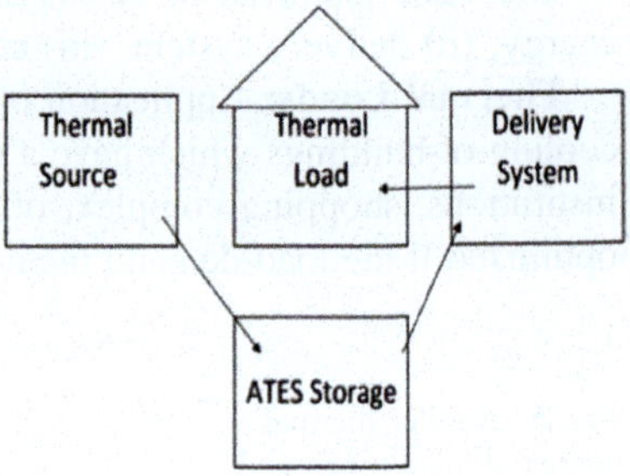

Fig. 7 Basic elements of ATES [9]

load of ~50 tons is required for seasonal thermal storage to work efficiently; otherwise, storage volume will be too low resulting in higher loss factor for stored thermal energy. For residential applications (where the load is typically below ~50 tons), cluster of houses together can be provided with the ATES [9].

Thermal Cold Source: Most convenient way to create a cold reservoir in the aquifer is to use cold air outside the building during winter. A heat exchanger can be conveniently used to cool the water of cold reservoir of the aquifer by exchanging heat between water and cold winter air. Waste cold water from a heat pump (while generating heat) can also be collected. Cold water can also be collected from water bodies like ocean, estuary, river, or lake. If chilled water is stored in the aquifer, a minimum of ~5 °C is recommended below which freezing of water might occur [9].

Thermal Heat Source: To store thermal energy in the hot reservoir part of aquifer, various sources can be used such as hot water from heat pumps in summer months, waste heat from industrial processes, and solar thermal energy collected during summer months [9].

Delivery System: The delivery system delivers hot and cold water for heating and cooling the building (load). It transports fluid from one place to other for the purpose of energy transfer and ultimately storage in the aquifer. It mainly consists of piping, pumps, and heat exchangers. If the HVAC system fulfills the purpose of both heating and cooling, then the system may include heat pump as well [9].

ATES Store: As the name implies, it stores hot and cold water in the aquifer. The energy supplied to the aquifer is partly stored in the ground water and remaining in the grains (or rocks mass) that form the aquifer. As the hot water passes through thc grains, thermal fronts are formed which move in a radial direction from well while charging aquifer and move back while discharging. Thermal interference of warm and cold energy kept in aquifer determines the performance of ATES system. The increase in thermal interference will decrease the efficiency of the system. The thermal interference is essentially influenced by the distance between borehole, the conductivity, and the rate at which water is pumped or injected. ATES system can be designed for both base and peak thermal loads, but it should not affect the properties of surrounding [9].

2.1.3 Other Liquids for TES

Some other liquids for TES are heat transfer oils, inorganic molten salts, and liquid metals. Heat transfer oils are used to store heat in the range of 100–300 °C. Some of them are servotherm, Caloria HT43, and Therminol T66. These are high-temperature applications, and so the oil provided should have good thermal stability to withstand oxidation, carbon deposits, and low volatility but should have good heat transfer characteristic.

Apart from heat transfer oils, molten salts and liquid metals are used to store sensible heat for higher temperature ranges (above 300 °C). Such a salt is eutectic mixture of 53% KNO_3 + 40% $NaNO_2$ + 7% $NaNO_3$ by weight and commercially known as “Hitec” having melting temperature of only 143 °C and maximum usage temperature of 425 °C. If temperature of the salt rises above the recommended temperature, decomposition and oxidation start to occur (Table 1).

Focusing on increasing key parameters of energetic efficiency and reducing cost, a new kind of storage system is being developed based on liquid metals to store sensible heat. These liquid metals can be classified as follows: alkali metals, heavy metals, and fusible metals. Liquid metals have a broad working range and good thermal transference, but their heat capacity is low compared to other material used for the same purpose. Alkali metals have good thermophysical properties, average density, melting point, heat capacity, and thermal conductivity. Some alkali metals are K, Na, and Li. Reactivity and cost of sodium are low, and it can reach up to a temperature of 883 °C, but they react with water to release hydrogen and heat. Heavy metals have good thermal stability and acceptable density values, but low

Table 1 Properties of some sensible heat storage liquids [10]

Liquid	Temperature range (°C)	Density (kg/m^3)	Heat capacity (J/kg K)	Thermal conductivity (W/m K)
Water	0–100	1000	4190	0.63 at 38 °C
Water-ethylene glycol 50/50	–	1050	3470	–
Caloria HT43	−10 to 315	–	2300	–
Dowtherms	12–260	867	2200	0.112 at 260 °C
Therminol 55	−18 to 315	–	2400	–
Therminol 66	−9 to 343	750	2100	0.106 at 343 °C
Ethylene glycol	–	1116	2382	0.249 at 20 °C
Hitec (molten salt)	141–540	1680	1560	0.61
Engine oil	Up to 160	888	1880	0.145
Draw salt (molten salt)	220–540	1733	1550	0.57
Lithium	180–1300	510	4190	38.1
Sodium	100–760	960	1300	67.5
Ethanol (organic liquid)	Up to 78	790	2400	–
Propanol (organic liquid)	Up to 97	800	2500	–
Butanol (organic liquid)	Up to 118	809	2400	–
Isobutanol (organic liquid)	Up to 100	808	3000	–
Isopentanol (organic liquid)	Up to 148	831	2200	–
Octane	Up to 126	704	2400	–

heat capacity and thermal conductivity (lower than alkaline metals). Temperature range of Pb–Bi (LBE) alloy is comparatively large from a low melting point of 125 °C and maximum working temperature up to 1670 °C. In addition to this, it is also inert in air and water but contains lead which is toxic.

Fusible metals have low melting points and high thermal stability, but they are expensive and corrosive materials. The main fusible metal is liquid Sn which has been used in operations at high temperatures in the glass industry (Table 2).

Solids: Sensible heat storage is simple in design and inexpensive and is stored in rocks, pebbles, or refractory materials. The combination of liquid and solid sensible heat storage media is also used where solid is porous and granular with liquid passing through the hollow spaces. The filler refractory material can be magnesium oxide, aluminum oxide, or silicon oxide.

Table 2 Thermophysical properties of liquid metals [11]

HTF	T_{min}	T_{max}	Cp (kJ/kg K)	K (W/m K)	ρ (kg/m^3)	μ (mPa s)
Alkali metals						
Na–K eutectic	−12	785	0.87	26.2	750	0.18
K	64	766	0.76	34.9	705	0.15
Na	98	883	1.25	46	808	0.21
Li	180	1342	4.16	49.7	475	0.34
Heavy metals						
Pb–Bi eutectic	125	1533	0.15	12.8	9660	1.08
Bi	271	1670	0.15	16.3	9940	1.17
Pb	327	1743	0.15	18.8	10.32	1.55
Fusible metals						
Ga	30	2237	0.36	50.0	6090	0.77
In	157	2072	0.24	47.2	6670	0.75
Sn	232	2687	0.24	33.8	6330	1.01

2.2 *Based on Construction or Storage Configuration*

2.2.1 Liquid Storage Tank

Case 1: Direct Active Arrangement: Consider an insulated liquid tank where the liquid is well mixed. As a result, uniform temperature T_L is maintained in the tank which varies only with time. The working fluid which receives energy in the collector is used to store energy in the storage tank. Such arrangement is called direct arrangement and is shown in Fig. 8.

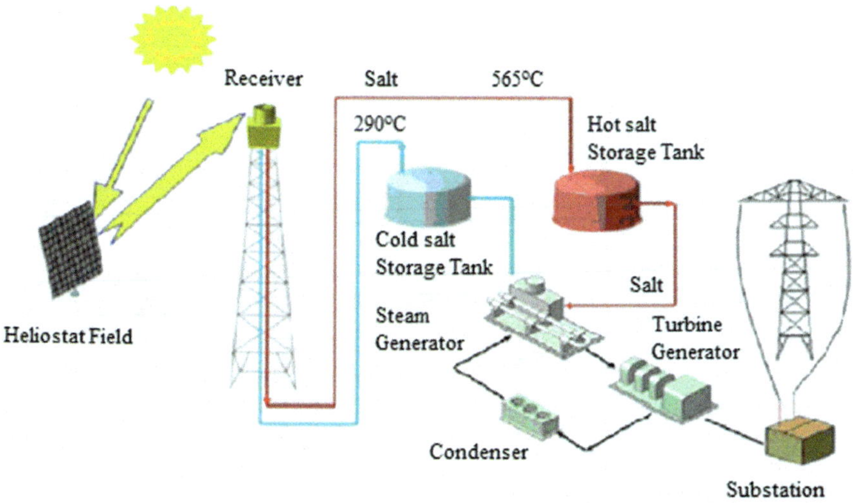

Fig. 8 Schematic of two-tank direct [12]

CASE 2: Indirect Active Arrangement: Consider an insulated liquid tank, where liquid in the tank is always well mixed as a consequence uniform temperature is maintained in the tank T_L which varies only with time. A HTF is used to receive energy from the collector and delivers it to thermal storage media which is different from HTF in this case. Such arrangement is called indirect arrangement. Such kind of arrangement may be required in situations where corrosion is a significant problem at collector end or temperature can go below freezing. Heat transfer between HTF and storage fluid may take place by using coil-immersed heat exchanger in the storage tank.

CASE 3: Two-Tank Molten Salt TES System: To store heat energy, this is the most mature technology commonly used in the concentrated solar plant (CSP). Storage in two-tank system is done by sending fluid from cold fluid tank to solar field directly or to the heat exchanger depending upon the direct active or indirect active arrangement. Thus, fluid is heated and is stored in the hot-fluid tank.

In case of direct active, HTF and energy storage fluid (ESF) are same. During energy charging process, HTF absorbs heat from the collector field and stores it in the hot storage tank. Irrespective of the fact whether solar irradiance is in excess, adequate or less than the requirement of the load, energy received from the collector field will always be stored in the hot tank and load will always receive an adequate amount of energy from the storage tank.

In most of the CSP systems, oil is used as HTF. However, the cost of oil is much higher for increasing volume of TES system. A less expensive alternative is to use oil as HTF in solar field and a secondary ESF fluid in storage like molten nitrate salt. Such systems are indirect active systems, but such indirect system requires an intermediate heat exchanger between the two fluids to transfer heat as shown in Fig. 9.

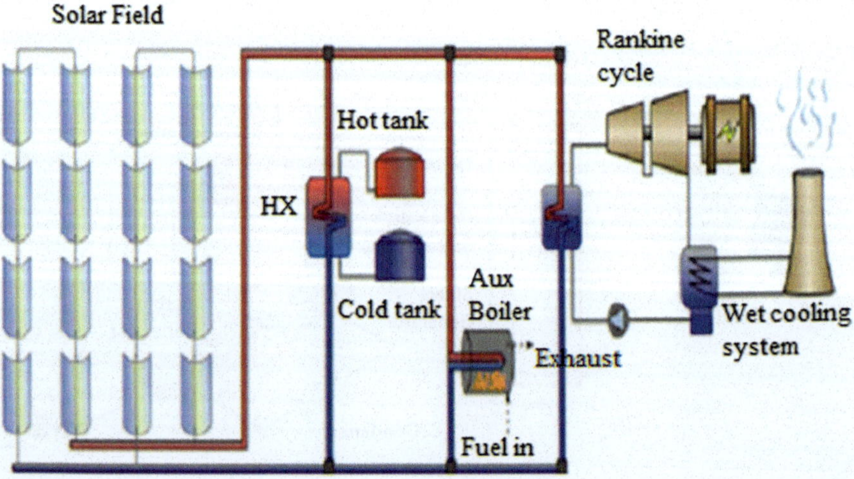

Fig. 9 Schematic of indirect active two-tank system [13]

During charging process, HTF absorbs heat from solar field, part of HTF is used to run the power plant system, and the part is delivered to the heat exchanger. In heat exchanger, cold fluid pumped from the cold tank (blue in the figure) extracts heat from HTF and is stored in the hot storage tank. While during discharging process, fluid from the hot storage tank is pumped into the heat exchanger and rejects heat to HTF, and thus gets cooled and is stored in cold storage tank.

Auxiliary Heater: Auxiliary heater is added into the system so as to supply thermal energy to the fluid when there is no solar resource or storage cannot meet the requirement. Energy is being given to HTF by burning conventional fuel until the fluid reaches the maximum temperature or heater reaches its maximum heat production rate.

Thermocline TES: In a two-tank TES system, the amount and cost of thermal storage fluid used are huge (e.g., 28500 tonnes of molten salt (Table 3), 60% NaNO3, and 40% KNO3, to operate 50 MW turbine for nearly 7.5 h in Andasol plant [14, 15]). Since the heat storage fluid is generally expensive, to reduce the cost of the TES, two-tank configuration can be replaced by a thermocline tank which consists of only one tank. In thermocline, due to density difference, hot fluid remains at the top and cold fluid is present at the bottom. The process of stratification of fluid by maintaining a temperature gradient is mentioned as the thermocline. It exploits buoyancy forces to promote thermal stratification (Fig. 10).

Thermocline can be classified as single-media thermocline or dual-media thermocline (Fig. 11).

In thermocline, charging is done by flowing hot fluid into the tank from the top, and at the same time cold fluid is pumped out from the bottom of the tank and is allowed to absorb heat from the solar field. During discharging process, hot fluid is pumped out from the top of the tank and releases its heat to power plant before it returns to tank from bottom.

Dual-Media Thermocline: Cost can be reduced further by using solid filler material within thermocline where fluid acts only as energy-carrying medium. The filler material (Table 4) can be rocks, salts, concrete, sands, etc. Charging and discharging process is carried out by heat transfer between fluid and filler material. The filler material dampens velocities in the tank which can de-stratify hot and cold HTF regions.

Thermal ratcheting is a significant problem associated with dual-media thermocline (DMT). During charging and discharging process, thermal expansion and contraction take place in both filler material and tank wall. If thermal expansion in the wall is greater than filler material, it may lead to the formation of annular gaps between both. In this gap, the unconsolidated filler material may redistribute, and during cycles, this unconsolidated filler material may not allow the tank wall to contract to its original dimensions which may produce residual stress. This stress may go accumulating with cycles until failure takes place. This complete phenomenon is called ratcheting. Since single-media thermocline does not have any filler material, they do not suffer from any such problem.

Parameters used to capture the performance of thermocline are as follows [18]:

Table 3 Physical properties of molten salts [17]

Name	Solar salt (60% NaNO3, 40%KNO3)	Hitec (7% NaNO3, 53% KNO3, 40% NaNO2)	Hitec (42% Ca(NO3)2, 43% NaNO3, 15% KNO3)
Density (kg/m^3)	$\rho_l = 2090 - 0.64 \times T_l$ (°C)	$\rho_l = 2084 - 0.73 \times T_l$ (°C)	$\rho_l = 2240 - 0.83 \times T_l$ (°C)
Specific heat capacity (J/kg K)	$Cp_l = 1443 - 0.17 \times T_l$ (°C)	$Cp_l = 1561.7$	$Cp_l = 1545.55$
Thermal conductivity (W/m K)	$k_l = 0.44 + 1.9 \times 10^{-4} \times T_l$ (°C)	$k_l = 0.42 - 6.53 \times 10^{-4}(T_l - 260)$	$k_l = 0.52$
Viscosity (kg/ms)	$\mu_l = [22.72 - 0.12 \times T_l + 2.28 \times 10^{-4} \times T_l^2 - 1.47 \times 10^{-7} \times T_l^3 \times 10^{-3}$	$\mu_l = exp[-4.34 - 2.1 \times (lnT_l - 5.1]$	$\mu_l = 10^{6.14} \times T_l^{-3.36}$
Melting point (°C)	221	142	130–140
Stability limit (°C)	600	450–538	550

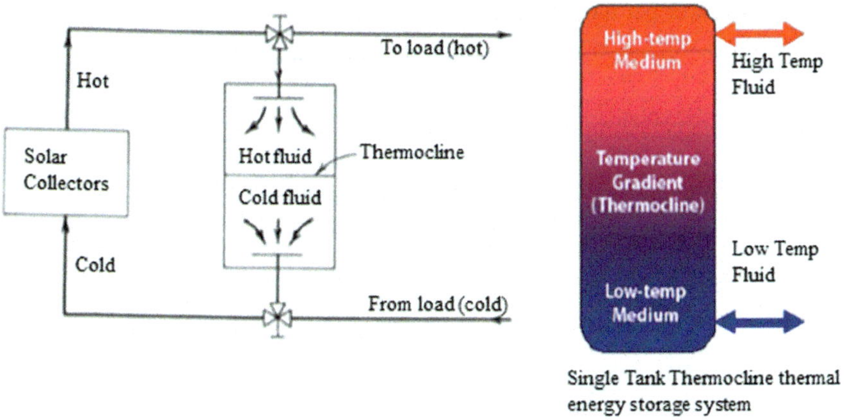

Fig. 10 Schematic of thermocline direct storage [16]

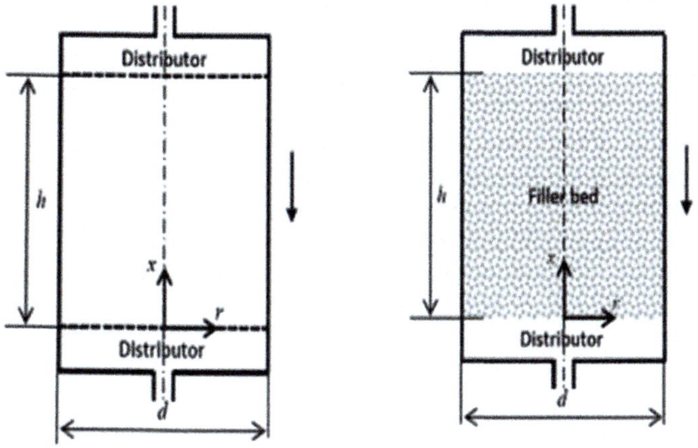

Fig. 11 Single- and dual-media thermocline tanks

Table 4 Physical properties of solid materials [17]

Name	Quartzite rock	Rock (granite)	Taconite	Cofalit
Density (kg/m^3)	2500	2893	3200	3120
Average specific heat capacity (J/kg K)	830	845	800	860
Thermal conductivity (W/m K)	5.69	3	30	2.7

1. Output power per unit cross-sectional area

$$\frac{P}{A} = u_m \rho_{l,c} C_{p_{l,c}} (T_h - T_c) = k_{l,c} \frac{T_h - T_c}{d_s} RePr \quad (6)$$

2. Useful thermal energy per unit cross-sectional area

$$\frac{Q}{A} = \left[\varepsilon \rho_{l,h} C_{p_{l,h}} + (1-\varepsilon)\rho_s C_{p_s}\right](T_h - T_c) \cdot d_s \cdot H \cdot \eta \quad (7)$$

3. Total stored thermal energy per unit cross-sectional area

$$\frac{Q_t}{A} = \left[\varepsilon \rho_{l,h} C_{p_{l,h}} + (1-\varepsilon)\rho_s C_{p_s}\right](T_h - T_c) \cdot d_s \cdot H \quad (8)$$

where

u_m is average speed at inlet to filler bed (m/s).
$\rho_{l,c}$ is density of molten salt phase at low temperature.
$\rho_{l,h}$ is density of molten salt phase at high temperature.
ρ_s is density of solid filler material.
$C_{p_{l,c}}$ is specific heat of molten salt at low temperature.
$C_{p_{l,h}}$ is specific heat of molten salt at high temperature.
T_h is hot temperature at outlet.
T_c is cold temperature at inlet.
$k_{l,c}$ is thermal conductivity of molten salt at low temperature.
d_s is diameter of filler particles (m).
r is porosity.
η is output efficiency.
H is non-dimensional height of tank.

Design Procedure for Thermocline Tank [18]

1. Select tank diameter d and filler particle size d_s according to requirements.
2. Determine the cross-sectional area of the storage tank, and then the output power per unit cross-sectional area (*P/A*) and useful thermal energy per unit cross-sectional area (*Q/A*).
3. Determine non-dimensional output power $Re \cdot Pr$ and useful thermal energy $H. \eta$ using Eqs. 6 and 7.
4. Determine *Re* from the value of $Re \cdot Pr$ and assume $H = H \cdot \eta$.
5. Use *Re* and *H* values to calculate η from Eq. $\eta = 1 - 0.1807(Re)^{0.1801}\left(\frac{H}{100}\right)^m$.

where $m = 0.00234Re^{-0.6151} + 0.00055Re - 0.485$, and Re is Reynolds number. Above formula has a maximum error of 1% for $1 < Re < 50$ and $10 < H < 800$.

6. Determine H by dividing $H.\ \eta$ with η obtained in step 5.
7. Repeat steps 5 and 6 till convergence of H is obtained within 0.1%.
8. The dimensional tank height is calculated as $h = d_s \cdot H$ where d_s is diameter of the filler particle.

To avoid excessive load on foundation, height of thermocline tank should not exceed beyond 39 ft (11.9 m) [19]. Thus, output efficiency increases with tank height H and decreases with an increase in the Re (Fig. 12).

Governing Equation: DMT

Energy Balance for Fluid:

$$\rho_f \varepsilon \pi R^2 U(h_z - h_{z+dz}) + hS_r(T_r - T_f)dz = \rho_f C_f \varepsilon \pi R^2 dz \frac{\partial T_f}{\partial t} \quad (9)$$

$$\frac{hS_r}{\rho_f C_f \varepsilon \pi R^2}(T_r - T_f) = \frac{\partial T_f}{\partial t} + U\frac{\partial T_f}{\partial z} \quad (10)$$

Energy Balance for Filler:

$$hS_r(T_r - T_f)dz = -\rho_r C_r (1-\varepsilon)\pi R^2 dz \frac{\partial T_r}{\partial t} \quad (11)$$

where heat transfer surface area of rocks per unit length of the tank (S_r) is as follows:

$$S_r = \frac{f_s \pi R^2 (1-\varepsilon)}{r};$$

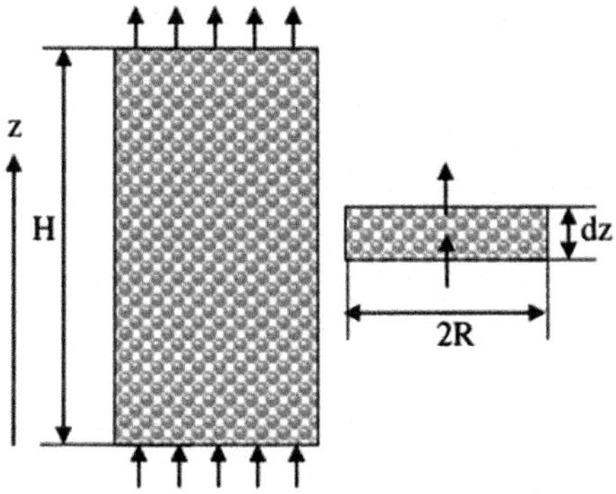

Fig. 12 Dual-media thermocline tank [20]

Heat transfer coefficient between solid and liquid $(h) = 0.191 \frac{mC_f}{\varepsilon \pi R^2} Re^{-0.278} Pr^{-2/3}$;

Reynolds number for porous media $(Re) = \frac{4Gr_{char}}{\mu_f}$;

Mass flux of fluid $(G) = \frac{m}{\varepsilon \pi R^2}$;

Characteristic radius $(r_{char}) = \frac{\varepsilon d_r}{4(1-\varepsilon)}$

Assumptions in Model

1. Temperature in each rock is homogenous.
2. Point contact is assumed between two rocks, and thus heat transfer through conduction can be neglected.
3. It is assumed that lumped heat capacitance method is applicable.
4. Temperature variation in radial and azimuthal direction is considered negligible. Model becomes 1-D.
5. Tank is well insulated.
6. Flow through filler bed is laminar.

Stratification Characterization

There are various methods reported in the literature to characterize thermal stratification in energy storage based on the degree of stratification, first law, second law, and some other basis. Since stratification is one of the key performance parameters of a storage system, it will be extremely useful to characterize it. There are various factors due to which stratification may tend to disappear (thermocline tends to be more uniform in temperature) such as the following:

1. Mixing of water inside the tank due to natural convection. Natural convection can occur if heat is supplied to the bottom of the tank, or heat loss takes place from the top surface of the tank.
2. Mixing due to kinetic energy generated by entering the water at the inlet of the storage tank.
3. Conduction taking place within water or wall material of TES may reduce the temperature difference and thus diminish the stratification.

Stratification may serve as a comparison tool for various TES. There are two approaches to characterize stratification, i.e., "density approach" and "temperature approach" out of which temperature approach is described (Fig. 13).

MIX NUMBER: Mix number was first presented by Davidson et al. [22] and then was improved by Anderson et al. [23]. Mix number is based on the moment of energy which depends on stored energy and distribution of temperature in the storage tank. Moment of energy is the addition of energy stored in packed bed; thus, higher the energy stored in tank higher will be the moment of energy [24].

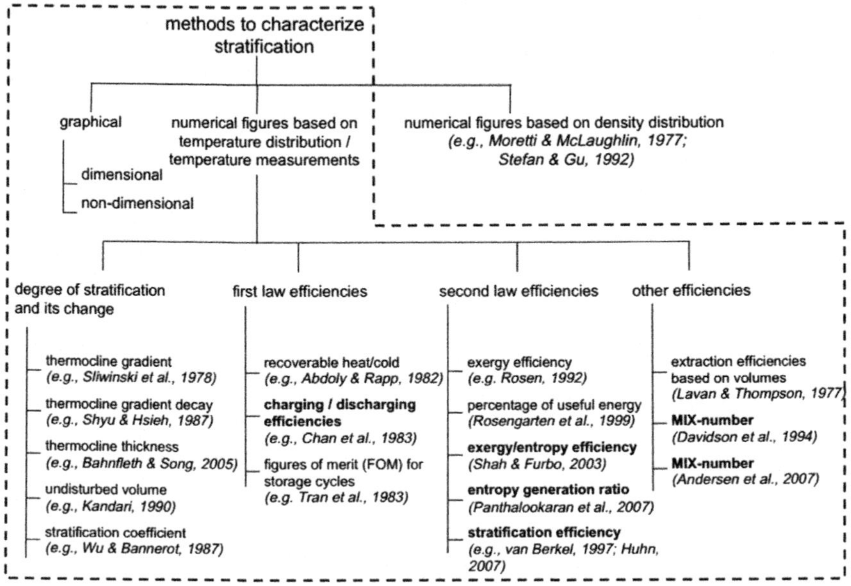

Fig. 13 Stratification characterization [21]

Two TES configurations were derived on the basis of the amount of energy stored in real TES, ideally stratified TES and fully mixed TES. The ideally stratified TES is marked by the certainty of having two adiabatically separated regions, high- and low-temperature regions, and the whole packed bed of the fully mixed TES is at average temperature. *MIX* number is derived by combining the moment of energy of the real TES (M_{exp}) with those of ideally stratified TES (M_{str}) and fully mixed TES (M_{mix}), respectively, as follows [24]:

$$MIX = \frac{M_{str} - M_{exp}}{M_{str} - M_{mix}} \tag{12}$$

$$\eta = 1 - MIX \tag{13}$$

The stratification efficiency ranges from 0 to 1, where moment of energy of an ideally stratified tank is given by 1. The thermocline tank efficiency can be expressed as the ratio of energy stored (or delivered) during a complete charging (or discharging) cycle, over the total energy that can be stored in the particular tank over the operating temperature range:

$$\eta = \frac{\int_0^{\tau_{CHARGE}} \dot{m} C_P (T_{INLET/OUT} - T_{c/h}) dt}{(\rho C_P)_{eff} V_{TANK} (T_h - T_C)} \tag{14}$$

where τ is non-dimensional time, and t is dimensional time.

Qualitatively, a thinner thermocline implies a steeper gradient between the cold- and hot-fluid regions and, therefore, higher thermocline tank efficiency.

3 Latent Heat Storage

Phase change materials (PCM) are itself called as latent heat storage materials. They are capable to release or store a significant quantity of energy while converting from one phase to another phase during melting/solidification or gasification/liquefaction process. Compared to sensible type, latent heat storage has higher phase-transition enthalpy which increases its storage density. There is a wide range of thermal properties of PCMs having eutectic temperature ranging from 100 to 897 °C, and latent heat ranging from 124 to 560 kJ/kg. As latent heat storage works in an isothermal way, it has an advantage over sensible heat storage which has a significant change in temperature difference while storing or releasing energy. Thus, latent heat storage is advantageous for applications with stringent working temperatures. Latent heat storage has low thermal conductivities ranging from 0.2 to 0.7 W/(m K). Thus, a new technology should be adopted which can enhance heat transfer process and overcome this drawback.

Properties

Thermophysical Properties

1. Eutectic temperature of PCM should be within a temperature range of application.
2. PCM should have high latent heat of fusion per unit volume so as to increase storage density.
3. It should have greater specific heat such that it can provide sensible heat storage with latent heat storage.
4. It should have high thermal conductivity in individual phase (solid and liquid) which helps in charging and discharging of energy in storage systems.
5. PCM should melt or freeze at an identical temperature so that constant storage capacity can be attained with each freezing/melting cycle.

Kinetic Properties

1. It should have high nucleation rate, such that super cooling of the liquid phase can be avoided.
2. The system should attain requirement of heat recovery from the storage system, thus should have a significant rate of crystal growth.

Chemical Properties

1. PCM should have high chemical stability.
2. Cycle should be reversible in nature.
3. It should not degrade after number of freeze or melt cycles.
4. It should not corrode construction materials.
5. It should not be toxic, flammable, or explosive in nature.

Classification of PCMs

Melting temperature and latent heat are the important parameters which help to recognize a PCM. Although most of the PCMs do not satisfy the requirement of the storage system, it has to be selected from available materials, and thus poor physical properties should be compensated with good system design. Example, metallic fins are used to enhance the thermal conductivity of PCMs, introducing nucleating agent may suppress super cooling in storage material. Different thermal and chemical properties, which affect the design of latent heat storage systems, are shown in Table 5.

Table 5 Commercial PCMs materials, inorganic salts, and eutectics [25]

Materials	Phase change temperature (°C)	Density (kg/m^3)	Thermal conductivity (W/mK)	Latent heat (kJ/kg)
RT100 (paraffin)	100	880	0.2	124
RT110 (paraffin)	112	–	–	213
E117 (inorganic)	117	1450	0.7	169
A164 (organic)	164	1500	–	306
$NaNO_3$	307	2260	0.5	172
KNO_3	333	2110	0.5	226
KOH	380	2044	0.5	149.7
$AlSi_{12}$	576	2700	1.6	560
$MgCl_2$	714	2140	–	452
NaCl	800	2160	5	492
LiF	850	–	–	–
Na_2CO_3	854	2533	2	275.7
K_2CO_3	897	2290	2	235.8
48%$CaCO_3$–45% KNO_3–7% $NaNO_3$	130	–	–	–
KNO_3–$NaNO_2$–$NaNO_3$	141	–	–	275
$LiNO_3$–$NaNO_3$	195	–	–	252
$MgCl_2$–KCl–NaCl	380	2044	0.5	149.7
$CaCl_2 \cdot 6H_2O$	29–30	1562	0.561	170–192

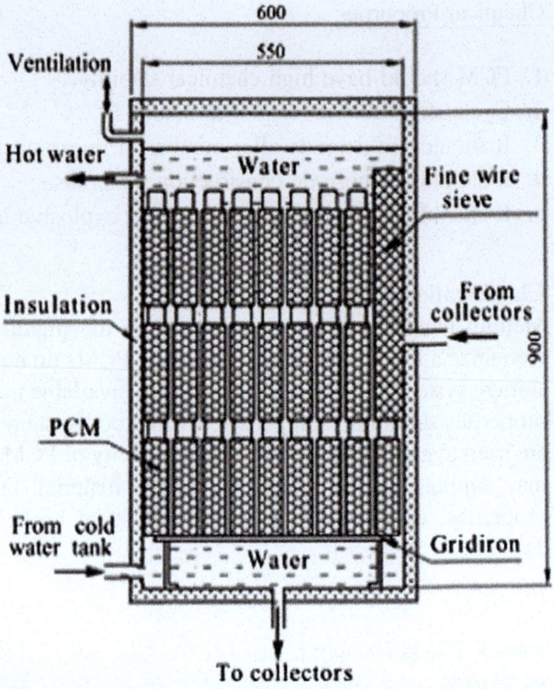

Fig. 14 Detailed cross-sectional view of the cylindrical heat storage tank combined with PCM [26]

Latent Heat Storage Systems

Solar Water-Heating Systems:

Suat et al. [26] studied a conventional open-loop passive solar water-heating system combined with sodium thiosulphate pentahydrate (PCM). Thermal energy storage performance was experimentally investigated and was compared with a conventional system with no PCM. Figure 14 shows a detailed cross-sectional view of heat storage tank used for the experiment by [26]. The solar tank has a volume of 190 l combined with PCM and has an insulation material made up of glass wool. The volume of bottles was intentionally selected a small value, thus minimizes issues mark during solidification and melting of PCM. The volumes of PCM and water in the heat storage tank are equal to 107.8 and 82.2 l, respectively. The total mass and density of PCM used in the heat storage tank are up to 180 kg and 1666 kg/m^3.

The flow of water is in the vertical direction and is taken care by bottles because of their close location. So, cavities are provided at the upper and lower parts of the storage tank, thus gain a homogenous heat distribution in a horizontal direction in the storage tank as water is flowing through vertical cavities although it is known that thermal stratification is higher at the upper part. During day time, part of the water from the collector is used for the household purpose and the part is used to charge PCM. Once charging is done, heat stored in PCM can be utilized during night time to heat water coming from the cold tank.

Fig. 15 Photograph of the air-heating system. A, Collector assembly with energy storage and air-heating subsystems; B, heated space [29]

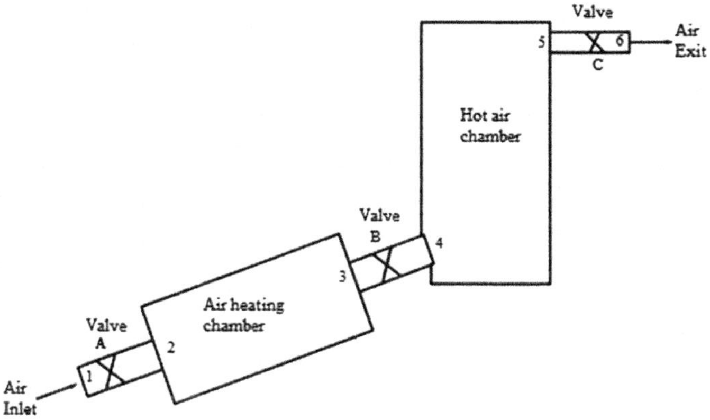

Fig. 16 Schematic view of a natural circulation air heater. The hot air (working chamber) will hold the crops in a cabinet-type crop dryer and will hold the eggs in an egg incubator [29]

Solar Air-Heating Systems:

Morrison et al. [27, 28] in their work studied the performance of solar air-heating systems using energy storage unit. The objective of their study was to (i) evaluate the effect of latent heat and melt temperature on the thermal performance of solar air-heating systems and (ii) to formulate empirical model based on experimental data of phase change energy storage (PCES) units. The conclusion obtained from above study was that the selection of PCM should depend upon its melting temperature and not over its latent heat.

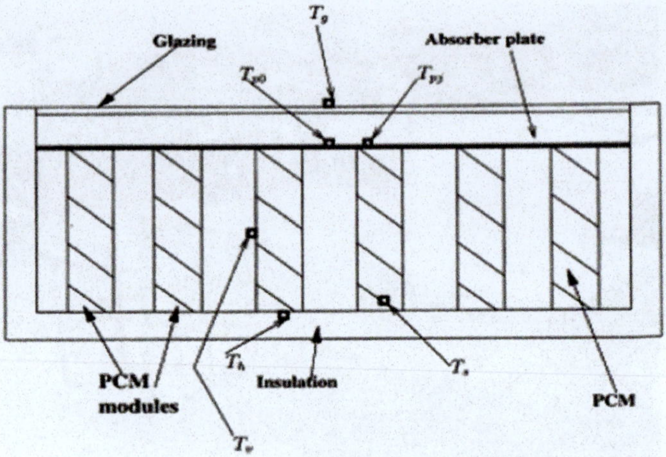

Fig. 17 Cross-sectional view of the collector assembly. The heated air flows perpendicular to the page, and the little boxes indicate approximate locations of thermocouples [29]

Figures 15, 16, and 17 show experimental setup and schematic diagram designed and developed by Enibe [29] to examine the performance of solar air heater for natural convection with PCM energy storage system. It is an assembly which comprises a flat-plate collector combined with the thermal energy storage system. Paraffin wax with known thermal properties is used as PCM for energy storage. PCM is prepared in thin rectangular modules and is enclosed within walls of a rectangular box which acts as fins. The material used for the box has better thermal conductivity, and box is divided into small partitions of same dimensions. These modules are placed within collector as shown in Fig. 17. The empty slot between every module acts as an air heater, and the heaters are organized to the common air inlet and header manifolds.

Surrounding air enters into air-heating chamber by natural convection through control valve *A*, and once air is heated it leaves chamber through control valve *B*. Air coming from the hot chamber is then discharged to ambient through valve *C*. Heat loss is minimized by placing absorber plate and PCM module within insulated box. Also, the solar collector is covered with an opaque screen during nights to reduce the thermal loss coefficient Fig. 16.

The modules including the PCM are made of rectangular boxes. These boxes are welded at the tops to the absorber plates, and the bottom part is allowed to rest over bottom insulation of collector. To reduce air leakage within the system, the absorber plate and module assembly are arranged in such a manner that air enters through inlet chamber and exits from outlet chamber. The design intends to take the advantage put up by the large heat transfer area of the PCM walls. The single-glazed collector area is 1.503 m^2, and it is coated by non-selective steel absorber plate material. The overall mass of PCM is about 65 kg, and overall volume of the air-heating chamber is 0.31 m^3. The volume of the hot air chamber is 0.26 m^3. The thickness of insulation at bottom and sides of the collector is 0.052

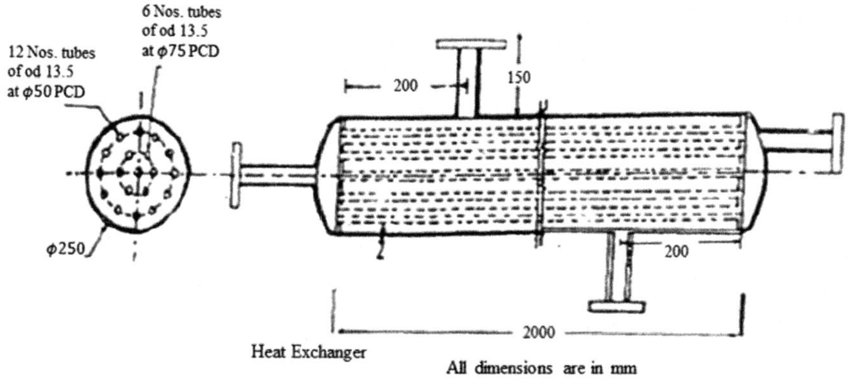

Fig. 18 Shell and tube heat exchanger for low-temperature waste heat recovery [32]

and 0.044 m, respectively, and working chamber is surrounded by insulation of about 0.022 m.

Waste Heat Recovery Systems:

Air-conditioning system rejects thermal energy in both forms, i.e., sensible and latent forms, in which sensible heat reports for about 15–20% of the total heat exhausted [30]. It was observed that if freon is used as a refrigerant, exhaust temperature of the compressor is always higher (<65 °C). This high-temperature exhaust heat can be recovered using a collector and thus attains higher temperature. A heat recovery system was developed by Gu et al. [31] to recover rejected heat from air-conditioning system and thus yields warm water for washing and bathing. It was concluded that recovery system reduces utilization of energy to heat water and heating up of surroundings due to the rejection of heat from systems. The efficiency of the system can become better if all rejected heat from air-conditioning can be recovered.

Buddhi [32] in 1997 manufactured a shell- and tube-type heat exchanger (Fig. 18) which was capable to recover waste heat at low temperature using stearic acid as PCM (59 °C, 198 kJ/kg). To increase the thermal conductivity of tubes, a radial distance of 3–4 cm was maintained. The shell part was filled 50 kg stearic acid. Water was used as HTF for charging and discharging of PCM. It was concluded from experimental data that PCM is feasible for heat recovery system. PCM has low thermal conductivity, and thus fins should be provided to increase the effectiveness of heat exchanger.

Further PCM can be used in solar cookers, latent heat storage exchanger, etc., thus has wide applications.

Heat Transfer Enhancement

In phase change process, the solid–liquid interface moves away from the heat transfer surface. This reduces heat flux from the surface as thermal resistance

Table 6 Materials used as chemical heat storage [25]

Materials	Temperature range (°C)	Enthalpy change during chemical reaction (GJ/m^3)	Chemical reaction
Iron carbonate	180	2.6 GJ/m^3	$FeCO_3 \leftrightarrow FeO + CO_2$
Methanolation–demethanolation	200–250	–	$CH_3OH \leftrightarrow CO + 2H_2$
Metal hydrides	200–300	4 GJ/m^3	(Metal xH_2) $\leftrightarrow$ metal $yH_2 + (x - y)H_2$
Ammonia	400–500	67 kJ/mol	$NH_3 + \Delta H \leftrightarrow 1/2N_2 + 3/2H_2$
Hydroxides	500	3 GJ/m^3	$Ca(OH)_2 \leftrightarrow CaO + H_2O$
Methane/water	500–1000	–	$CH_4 + H_2O \leftrightarrow CO + 3H_2$
Calcium carbonate	800–900	4.4 GJ/m^3	$CaCO_3 \leftrightarrow CaO + CO_2$
Metal oxides (Zn and Fe)	2000–2500	–	Two-step water splitting using Fe_3O_4/FeO redox system
Aluminum ore alumina	2100–2300	–	–

increases due to growing layer of solidified thickness. During solidification, conduction is the only transport mechanism present, but, for melting natural convection with conduction may occur within the liquid region and thus enhances the rate of heat transfer. However, the drawback of low heat transfer rate during solidification can be overcome if appropriate heat transfer enhancement methods are used. Various methods like finned tubes of different arrangements are present that assist to enhance heat transfer in latent heat storage system.

4 Chemical Heat Storage

A large amount of energy is absorbed by chemicals when they break chemical bonds during endothermic reactions and is released when they form chemical bonds during exothermic reactions. This important characteristic of chemical heat is used as one of the storage methods. Materials used for storage can be organic or inorganic provided that their reversible chemical reaction involves absorbing or releasing a large amount of heat. Better chemical reversibility, a large change in chemical enthalpy, and lucid reaction conditions are basic demands required to design a chemical storage system. A list of materials used for chemical heat storage is shown in the Table 6, out of which ammonia-based thermochemical energy storage and metal oxides (Fe_3O_4/FeO) is discussed in detail. Chemical storage has its own problems like the need for sophisticated reactors for particular chemical reactions, poor durability, and chemical instability which limits its application.

Chemical Heat Storage Systems

Ammonia-Based Thermochemical Energy Storage [33]:

In closed cycle chemical storage systems, a certain amount of reactants is circulated by turn between energy-storing and energy-releasing reactors. Heat transfer at each reactor takes place between reactants entering and exiting using counterflow heat exchangers which help to reduce thermal loss by maintaining ambient temperature of storage. NH_3 reaction has its own advantages over other alternative reactions.

$$NH_3 + H \leftrightarrow 1/2N_2 + 3/2H_2$$

It has no side reactions which make it easy to control. Reactions taking place at dissociator are endothermic reaction which runs at a temperature favorable to solar concentrators. As operating pressure is beyond saturation pressure of ammonia, the fraction of ammonia present in storage will be in liquid form. Thus, phase separation is obtained of ammonia and hydrogen/nitrogen which gives the advantage to use a common storage.

Figure 19 shows the experimental setup used. It has 20 l storage container that is able to store un-dissociated NH_3 and 3:1 H_2/N_2 gas mixture. Liquid ammonia from the bottom of the container is supplied using circulation pump to either electric dissociator or solar dissociator. Products obtained in the form of gas are then cooled by flowing it through counterflow heat exchanger and are stored in 20 l energy storage container. Energy recovery is done with the help of ammonia synthesis reactor. Gas is extracted out from the top of storage container and is supplied to 10 l buffer pressure container using circulation pump.

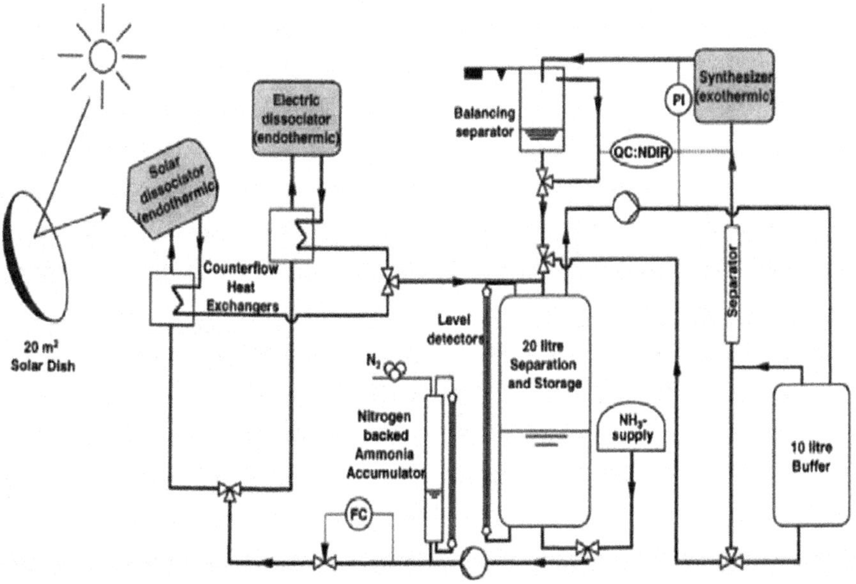

Fig. 19 Solar-driven ammonia-based thermochemical energy storage system [33]

Before it is supplied to synthesis reactor, it is allowed to flow through the cold separator to remove ammonia vapor if present in the gas mixture. In synthesis reactor, exothermic process takes place and energy is recovered. The overall system works at constant volume, and thus pressure increases as NH_3 in the container is dissociated. Level indicator is present to determine liquid indicator.

Two-Step Water Splitting Using Fe_3O_4/FeO Redox System [34]:

This section deals with the following chemical transformation:

$$Fe_3O_4(l) \rightarrow 3FeO(l) + 1/2O_2, \quad T \geq 1875\,K$$

Here sunlight is stored in the form of chemical energy. FeO is allowed to react with water or carbon dioxide exothermally to yield H_2 or C as given in reactions below.

$$3FeO(s) + H_2O(l) \rightarrow Fe_3O_4(s) + H_2$$

$$3FeO(s) + 1/2CO_2 \rightarrow Fe_3O_4(s) + 1/2C$$

The Fe_3O_4 that is obtained in above reactions is recycled to a solar furnace where FeO is again obtained from the reaction. Thus, solar energy is utilized to yield either H_2 from H_2O or C from CO_2. The two-step water-splitting cycle represented by the reactions is schematically shown in Fig. 20.

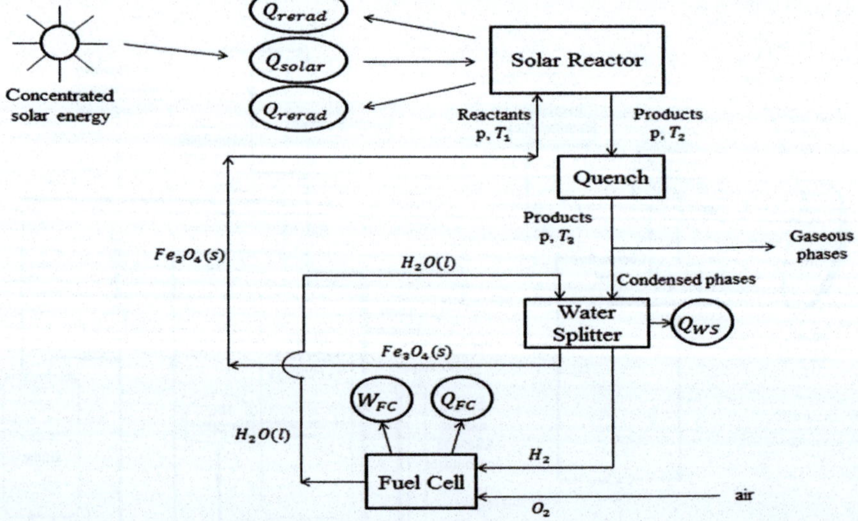

Fig. 20 Process flow diagram of two-step water-splitting thermochemical cycle using Fe_3O_4/FeO and solar energy [34]

Solar reactor:

It is a cavity receiver with a small aperture, which absorbs solar radiation falling over it. Absorption efficiency is defined as the ratio of net energy absorbed to solar energy received from the concentrator (neglect conduction + convection losses).

$$\eta_{absorption} = \frac{Q_{absorbed,\,net}}{Q_{solar}} = 1 - \frac{\sigma \varepsilon_{receiver} T^4_{reactor}}{I^* C} \quad (15)$$

The reactants enter into the reactor at T_1 and undergo chemical reaction as they are heated to $T_2 = T_{reactor}$.

$$Fe_3O_4(l) \rightarrow 3FeO(l) + \frac{1}{2}O_2$$

Thus, net absorbed energy should be equal to change in enthalpy of reaction.

$$Q_{absorbed,\,net} = \Delta H_{reactants T_1, p \rightarrow products T_2, p} \quad (16)$$

Quench:

Products coming out of the reactor is rapidly cooled to surrounding temperature, T3 = 298 K. During quenching process, products may undergo phase change without change in its composition. The product composition remains unchanged, but phase change may occur. Amount of heat energy lost during quenching is as follows:

$$Q_{quench} = -\Delta H_{products T_2, p \rightarrow products T_3, p} \quad (17)$$

Quenching is completely irreversible process which causes significant drop in system efficiency. Irreversibility corresponding to quenching is as follows:

$$Irr_{quench} = (Q_{quench}/298) + \Delta S_{products T_2, p \rightarrow products T_3, p} \quad (18)$$

Water-splitter:

Water is allowed to react with FeO to form H_2 molecule.

$$3FeO(s) + H_2O(1) \rightarrow Fe_3O_4(s) + H_2$$

Heat liberated due to exothermic process can be utilized in reactor to carry out water-splitting reaction beyond surrounding temperature. Heat liberated to surrounding is given as follows:

$$Q_{WS} = -\Delta H_{3FeO(s) + H_2O(l) \rightarrow Fe_3O_4(s) + H_2} \quad (19)$$

Fuel cell (*FC*):

Hydrogen obtained from the water-splitter reactor can be used as fuel in heat combustion engine to produce work or can be used in a fuel cell to produce electricity. Maximum work can be obtained from hydrogen by assuming a reversible fuel cell as shown in Fig. 20. In fuel cell, hydrogen reacts with oxygen which is extracted from atmosphere to form water and thus produces electrical power.

$$W_{FC} = -\Delta G_{H_2 + 0.5O_2 \rightarrow H_2O(l)} \tag{20}$$

Assuming that fuel cell works at constant temperature, the amount of heat loss to surrounding is as follows:

$$Q_{FC} = -298K \times \Delta S_{H_2 + 0.5O_2 \rightarrow H_2O(l)} \tag{21}$$

Overall efficiency of system is evaluated as follows:

$$\eta_{overall} = \frac{W_{FC}}{Q_{solar}} \tag{22}$$

Addition to work output if we consider energy that can be extracted, the overall efficiency is then evaluated as follows:

$$\eta_{overall,\,max} = \frac{W_{FC} + 298 \times Irr_{quench}}{Q_{solar}} \tag{23}$$

The above calculation makes it possible to analyze cyclic system and provides a base to compare the efficiencies of different solar processes.

Closure

A wide variety of solar TES have been described in this chapter, along with detailed description of associated advantages, disadvantages, application areas, and procedures of evaluating performance parameters. A wide range of applications of solar TES vary from small-scale domestic utilities to large-scale power generation. While sensible TES is easy to maintain an inexpensive, they require large storage volume. On the other hand, latent heat and chemical reaction-based TES systems, although poses many design challenges, are very high energy density storage devices requiring small storage volume. For successful implementation of solar thermal power plants in future, TES is going to play the key role, and continuous research effort toward the development of suitable TES for large-scale (GW level) solar power plants is the need of the present to address looming energy crisis of the near future.

References

1. Barr KP (1982) Buffer thermal energy storage for a solar Brayton engine. Mirror
2. Somasundaram S, Brown DR, Kevin Drost M (1997) Diurnal thermal energy storage for cogeneration applications. Cogener Compet Power J 12(2):51–78

3. Givoni B (1977) Underground longterm storage of solar energy—an overview. Sol Energy 19 (6):617–623
4. Pardo P et al (2014) A review on high temperature thermochemical heat energy storage. Renew Sustain Energy Rev 32(2014):591–610
5. Sukhatme K, Sukhatme SP (1996) Solar energy: principles of thermal collection and storage. Tata McGraw-Hill Education
6. Shelton J (1975) Underground storage of heat in solar heating systems. Sol Energy 17 (2):137–143
7. imnh.isu.edu/digitalatlas/hydr/concepts/gwater/aquifer.htm. Accessed 3 July 2017
8. Caljé RJ (2009) Future use of Aquifer thermal energy storage below the historic centre of Amsterdam. Master Study Hydrology Department of Water-Management, Delft University of Technology, December 16th 2009 (5/7/2017)
9. Paksoy H, Snijders A, Stiles L (2009) State-of-the-art review of aquifer thermal energy storage systems for heating and cooling buildings. In: Proceedings of the EFFSTOCK conference, Stockholm, Sweden
10. Ercan Ataer O (2006) Storage of thermal energy, in energy storage systems. In: Gogus YA (ed) Encyclopedia of life support systems (EOLSS). Developed under the Auspices of the UNESCO. Eolss Publishers, Oxford, UK. http://www.eolss.net
11. Ushak S, Fernández AG, Grageda M (2014) Using molten salts and other liquid sensible storage media in thermal energy storage (TES) systems. In: Advances in thermal energy storage systems: methods and applications, 1st ed. Cabeza, LF, Ed, pp 49–63
12. Ortega JI, Ignacio Burgaleta J, TÃŠllez FM (2008) Central receiver system solar power plant using molten salt as heat transfer fluid. J Sol Energy Eng 130(2):024501
13. Jorgenson J, Gilman P, Dobos A (2011) Technical manual for the SAM biomass power generation model. No. NREL/TP-6A20-52688. National Renewable Energy Laboratory (NREL), Golden, CO
14. Solar Millennium (2008) The parabolic trough power plants Andasol 1 to 3. Tech. Solar Millennium AG
15. https://www.nrel.gov/csp/solarpaces/project_detail.cfm/projectID=3. Accessed 5 July 2017
16. Tesfay M, Venkatesan M (2013) Simulation of thermocline thermal energy storage system using C. Int J Innov Appl Stud 3(2):354–364
17. Chang ZS et al (2015) The design and numerical study of a 2 mwh molten salt thermocline tank. Energy Procedia 69:779–789
18. Flueckiger SM, Yang Z, Garimella SV (2013) Review of molten-salt thermocline tank modeling for solar thermal energy storage. Heat Transf Eng 34(10):787–800
19. (2010) Solar thermal storage systems: preliminary design study. EPRI report 1019581
20. Li P et al (2011) Generalized charts of energy storage effectiveness for thermocline heat storage tank design and calibration. Sol Energy 85(9):2130–2143
21. Haller MY et al (2009) Methods to determine stratification efficiency of thermal energy storage processes—review and theoretical comparison. Sol Energy 83(10):1847–1860
22. Davidson JH, Adams DA, Miller JA (1994) A coefficient to characterize mixing in solar water storage tanks. Guide to the records of the Solar Energy Applications Laboratory and Solar Village. http://lib.colostate.edu/archives/findingaids/university/ucsv.html
23. Andersen E, Furbo S, Fan J (2007) Multilayer fabric stratification pipes for solar tanks. Sol Energy 81(10):1219–1226
24. Zavattoni SA et al (2015) Evaluation of thermal stratification of an air-based thermocline TES with low-cost filler material. Energy Procedia 73:289–296
25. Tian Y, Zhao C-Y (2013) A review of solar collectors and thermal energy storage in solar thermal applications. Appl Energy 104:538–553
26. Canbazoğlu S et al (2005) Enhancement of solar thermal energy storage performance using sodium thiosulfate pentahydrate of a conventional solar water-heating system. Energy Build 37(3):235–242
27. Morrison DJ, Abdel-Khalik SI (1978) Effects of phase-change energy storage on the performance of air-based and liquid-based solar heating systems. Sol Energy 20(1):57–67

28. Jurinak JJ, Abdel-Khalik SI (1979) On the performance of air-based solar heating systems utilizing phase-change energy storage. Energy 4(4):503–522
29. Enibe SO (2002) Performance of a natural circulation solar air heating system with phase change material energy storage. Renew Energy 27(1):69–86
30. Zhang J (1999) Energy saving technology of refrigeration devices
31. Gu Z, Liu H, Li Y (2004) Thermal energy recovery of air conditioning system—heat recovery system calculation and phase change materials development. Appl Therm Eng 24 (17):2511–2526
32. Buddhi D (1997) Thermal performance of a shell and tube PCM storage heat exchanger for industrial waste heat recovery. In: Proceedings of the ISES 1997 solar world congress, Taejon, Korea
33. Lovegrove K, Luzzi A, Kreetz H (1999) A solar-driven ammonia-based thermochemical energy storage system. Sol Energy 67(4):309–316
34. Steinfeld A, Sanders S, Palumbo R (1999) Design aspects of solar thermochemical engineering—a case study: two-step water-splitting cycle using the Fe_3O_4/FeO redox system. Sol Energy 65(1):43–53

Review on Integration of Solar Air Heaters with Thermal Energy Storage

Prashant Saini, Dhiraj V. Patil and Satvasheel Powar

Solar radiation on the earth's surface is abundant and truly a zero-carbon energy source. The solar energy needs to be harnessed using various efficient equipments, which has a very low carbon footprint. Various solar thermal energy harvesting techniques have been used which employ solar radiation incident on the optimal area with the help of concentrators. The solar air heater is one of the solar thermal harvesting techniques, which has gained tremendous popularity in recent times. In this chapter, design and operations of various solar air heaters are detailed. Further, due to the intermittent nature of the solar energy, energy storage becomes an integral part of the solar energy harvesting and the solar energy may be stored using latent or sensible heat thermal energy storage. A discussion on various thermal energy storage technologies is then followed. Here, integration of a solar air heater along with low cost or efficient, sensible heat thermal energy storage system is given.

1 Introduction

Energy is important to raise the standard of living of citizens and also contribute substantially in the overall development of the economy. The energy source should be cheap, abundant, renewable, and environment-friendly. Since the industrial revolution, conventional way to obtain electricity is by using the fossil fuels (coal-/oil-fired power plants). Further, transportation contributes heavily to the consumption of liquid

P. Saini · D. V. Patil (✉) · S. Powar
School of Engineering, Indian Institute of Technology Mandi, Mandi 175005, Himachal Pradesh, India
e-mail: dhirajat27@gmail.com; dhiraj@iitmandi.ac.in

H. Tyagi et al. (eds.), *Applications of Solar Energy*, Energy, Environment, and Sustainability, https://doi.org/10.1007/978-981-10-7206-2_9

fossil fuels. These fossil fuel resources are depleting at a very fast rate and by the present depletion rate, they are projected to be depleted soon. Hence, humankind should rely on resources which are available on earth in abundant. Such resources are renewable sources of energy. Currently, fossil fuels provide about 80% of total power consumption all over the world, wherein 31.1% energy is obtained from oil, 28.9% obtained from coal, 21.4% obtained from natural gas [1]. Hydropower is one of the best sources of energy, which contributes about 16% in the world's electricity production [1]. Earth's total natural gas reserves and oil reserves are measured to last long for next 65 years and 45 years, respectively. The coal reserves are measured to last long as little more than 200 years [2].

As the renewable energy is freely available in nature, energy from the various renewable energy resources can hence be applied in a decentralized way, to reduce the overall price of the transportation. After the second half of the eighteenth century, coal was utilization slowly increased the industrialization throughout the world [3, 4]. The awakening of interest for harness of the solar energy started in the beginning of 1970s, because of its three prevalent reasons. The first prevalent reason is the oil crisis, second being the continuous utilization of non-conventional energy resources making irreversible and permanent environmental harm to nature and to irreplaceable cultural monuments. The balance relation should be maintained among, the environment, industries, and agriculture [5–9]. The use of carbon-containing fossil fuel includes various burning processes and subsequently, oxidation products are discharged into the atmosphere such as CO, CO_2, SO_2, NO_x, unburned C_nH_m, toxicants, and substantial metals, dust, ash, and particulate. Further, with the use of desulfurizing or any detoxing element, gypsum or ammonia is formed. Only the carbon dioxide generated from biomass burning is ecologically neutral because the carbon dioxide generated by burning will be obtained from the atmosphere when the biomass made [10–12]. The last prevalent reason behind awakening the interest in the use of solar energy is due to the fact that the nuclear energy is just reluctantly endured in various countries or halted in most of other countries, primarily due to inability to resolve the issue of radioactive waste material disposal. Solar energy utilization has two essential aspects [13] as stated here. The first essential aspect comprises of various types of local utilization which includes compilation, transformation, and application of solar energy, e.g., passive solar energy utilization such as usage in building heating/illumination, heat generation using solar thermal collectors, electricity generation using photovoltaic solar panels, usage of ambient heat in various types of heat pumps, and the hydropower, wind power/biomass into electricity. The commonality in the usage of all these solar energy technologies is that they are not prescribed to only those places in the world, which have good solar radiation.

Large hydropower plants, ocean thermal energy conversion systems, photovoltaic solar power plants, etc., require large area for installation, and these systems need to be installed in places having high direct solar radiation. These systems belonging to a category in which solar energy converted into electricity and this generated electricity are delivered to other secondary places which are many miles far away from the sites.

2 Renewable Energy

Renewable energy is that form of energy which is cultivated from those categories of resources which are available in abundant amount, i.e., sunlight, rain, waves, wind, tides, and geothermal heat. Due to association of various issues with the application of fossil energy, renewable energy is gaining much research attention in recent years. Conventional energy is being replaced by the renewable energy in areas such as electricity generation, space heating, water heating, etc. Usage of these resources facilitates various social, environmental, and economic advantages like energy security, replacing coal, water management, reduced health impact via reduced air pollution, energy access, and bio-energy [14]. In India, a Commission for Additional Sources of Energy (CASE) was formed which is later changed into a Department of Non-Conventional Energy Sources (DNES) in 1982. After a decade, i.e., 1992, this department promoted to the Ministry of Non-Conventional Energy Sources (MNES), and since 2006, it is known as Ministry of New and Renewable Energy (MNRE) [15].

India cultivates its energy requirement from various resources. Manifestly, major furnisher is the conventional source, which provides about 85% of total power to the country. Table 1 depicts the total power generating capacity of India, which shows that 13.17% of total power comes from renewable energy sources. It has been seen that the growth progression of harnessing energy from renewable sources is largest among all main power generating sectors in India.

At present, renewable energy proportion of India is about 35,774 MW, which is approximately 13% of total production and includes mainly biomass energy, wind energy, and solar energy. Currently, India stands at fifth position of total renewable energy capacity and utilization (excluding hydropower) and in wind energy utilization, India also stands at fifth position after China, USA, Germany, and Spain. In solar energy application, India stands at tenth position. Only the USA and India joined CSP facilities in 2014 [16].

Table 1 Region-wise installed power generating capacity of India (Ministry of statistics and programme implementation (MOSPI), 2016)

Sr. no.	Indian geographic region	Hydro (GW)	Thermal (GW)	Nuclear (GW)	New and renewable energy (GW)	Growth rate from 2013–14 to 2014–15
1	North region	17.07	41.61	1.62	7.15	6.68
2	West region	7.45	78.69	1.84	12.79	14.08
3	South region	11.40	34.40	2.32	15.12	10.67
4	East region	4.11	32.28	0.00	0.44	9.70
5	NER	1.24	1.86	0.00	0.26	15.52
Total		41.27	188.90	5.78	35.77	10.79
(%)		15.19%	69.52%	2.12%	13.17%	

Table 2 Renewable power capacity in India (Ministry of statistics and programme implementation (MOSPI), 2016)

Sr. no.	Source	Capacity		Growth rate from 2013–14 to 2014–15
		On 31.3.2014	On 31.3.15	
1	Biomass energy	4013.55	4450.55	10.88
2	Waste to power	106.58	127.08	19.23
3	Wind power	21136.40	25088.19	18.69
4	Small hydropower	3803.70	4176.83	9.82
5	Solar power	2631.96	4878.88	85.37
	Total	31692.18	388821.53	22.49

The comprehensive description of the contribution of all these resources in power generation in India has been shown in Table 2 [17]. It shows that most of the power is produced by wind energy constituting about 65% of total renewable energy capacity of the country. The growth rate of solar power in India is also recorded at highest with 85.37%. Hence, solar power and wind power attracted the attention of the Indian government and these sources are coming in front for stimulating renewable energy to convert into a prominent source of energy over fossil energy in the near future.

Solar energy is available in abundance in India, and it is also environment-friendly in nature. By the strong influence of these two factors, solar energy is becoming attractive in remaining renewable energy sources. Solar power has the capability of indulging energy requirement exhort in such way that the power production methods should be sustainable, less waste producing, and environment-friendly. For the development of new technologies to collect the greatest amount of solar energy, various methods and studies have been carried out around the world.

3 The Sun

The sun is the best source of energy because solar energy is available in abundance. The sun is in sphere shape and made of hot gaseous matter with its diameter around 1.39×10^9 m and situated at the 1.5×10^{11} m distance from earth. It takes about four weeks (27 days by equator and 30 days by polar regions) to rotate about its axis. It has an effective surface temperature of 5777 K.

Figure 1 dictates the structure of the sun. Out of 100%, 90% of the energy generated in the core of the sun, which is 0–0.23 R (R represents the effective radius of the sun). And the core of sun constitutes about 40% of the total mass of the sun, whereas the volume acquired is about 15% of the total volume. As one move toward

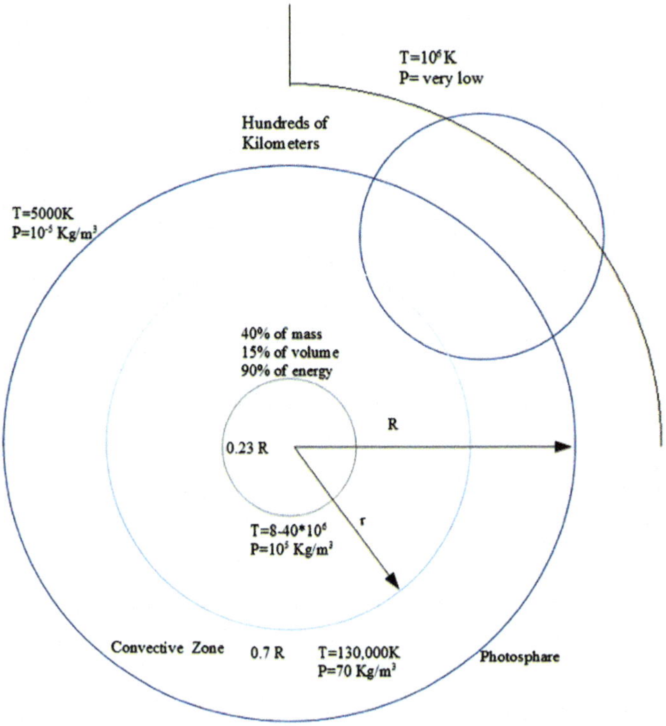

Fig. 1 Structure of the sun

the outer side of the sun, that is distance greater than 0.7 R from the center, temperature is reduced to around 1.3×10^5 K. Distance in between 0.7–1.0 R is called the convective zone of the sun where the temperature is reduced around 5×10^3 K [18]. Outermost region of the sun is known as the photosphere, and all the radiations are generated from this region. Above the photosphere, a region is distributed up to some hundreds of kilometers filled with coolant gases and is called reversing layer. Above this layer, chromosphere situated which is lower in temperature, i.e., 10^6 K as well as in density.

4 Solar Spectrum

The solar radiation obtained on earth is in white light. Indeed, this white light contains a broad wavelength spectrum, which varies from infrared (prolonged than red light) to ultraviolet (lessened than violet light). And these patterns of wavelength in white light depend upon the variation of temperature on the sun's surface [19] (Fig. 2).

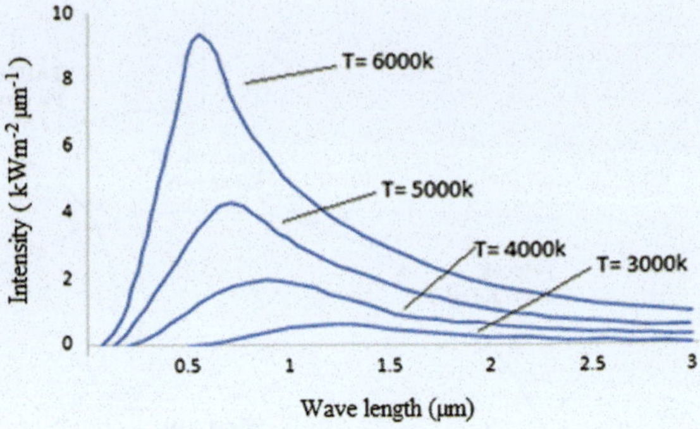

Fig. 2 Spectral distribution of solar radiation

5 Availability of Solar Radiation on Earth

Solar energy received by the earth in the 60 min of time after scattering and absorption of solar radiation by the earth environment would be sufficient to fulfill the energy demand of the entire world for one year. Consequently, the issue is not regarding the accessibility of energy from the sun but regarding the usefulness of converting this energy into various forms of energy which required for the human use. Out of total radiation, 30% instantly reflected to the space by the earth's environment. And the remaining 70% is used for increasing the temperature of oceans (47%), earth's surface, and atmosphere. Further, a small amount of which is captivating for maintaining the evaporation cycle of water. Moderately little portions used for driving the wind, furthermore, in the plant for their photosynthesis. And at the end, total energy after utilization is transmitted back to space from the earth in the form of infrared radiation.

6 Diffuse and Direct Radiations

Earth enclosed in its environment includes distinct types of particulates, for example, dust particles, small solid or liquid particles, and clouds of different types. Hence, there is a reduction in solar radiation intensity as radiation passes through the earth's environment before they reach to the surface of the earth. In a clear weather condition, solar radiation depleted in the three different ways as given below.

(i) Selective absorption of the wavelength occurs when the solar radiation passes through the ozone, water vapors, CO_2, and molecular oxygen.
(ii) Rayleigh scattering of wavelength occurs in the atmosphere as the solar radiation passes through different gasses or dust particles.
(iii) Mie scattering of wavelength occurs when the size of the particulates of different gases and dust particles are large as compared to solar radiation wavelength.

Consequently, there is degradation of solar radiation with true scattering and by particle absorption, in which a part of radiant energy converted into heat. The direction of rceeived radiation on earth is changed by either scattering or by reflection and this changed radiation is called as "Diffused Radiation." With a sunny weather, the magnitude of the diffused radiation varies with an average of 15% of total radiation, which reached on the surface of the earth. On cloudy days, diffused radiation can reach the earth's surface. The radiation received on earth without any change is called as "Direct Radiation." On sunny days, direct radiation (beam radiation) reaching for the surface of the earth has a maximum power density, i.e., 1000 W/m^2 [13]. Both direct and diffused radiations are very useful in the maximum of the solar energy applications, whereas the direct radiations used for generating very high temperatures. However, it is the diffuse radiation that received in the daylight and combinations of both, i.e., diffuse and direct directions are called as "Global Radiation."

The fundamental form of energy that is responsible for the existence of human on this planet is the solar energy. The electromagnetic radiations given off by the sun are of three types. They are ultraviolet, visible, and infrared radiations. The wavelength of the ultraviolet electromagnetic wave (EMW) lies in between about 0.20–0.39, visible portion lies in between about 0.39–0.78, infrared lies in between about 0.78–4. Ultraviolet radiations are harmful to human beings and are being trapped by the atmosphere. The visible portion which is the biggest portion of the solar rays contains photons which could be converted directly into electricity by PV panels. Infrared portion carries heat along with it and could be utilized by solar thermal collectors.

India is one of the solar-rich countries among the 121 countries falling between the tropic of cancer and the tropic of Capricorn, which are full of sun rays at least 300 days a year. India is a naturally blessed country with a large portion of infrared radiations coming onto the surface. Country gets around 5000 trillion kWh of solar energy per year. However, many initiatives have been taken to harness this energy; it also demands a lot of research attention on new technologies for performance improvements of solar thermal collectors [20].

7 Methods for Harnessing Solar Energy

Various methods involve in the harnessing of solar energy is given in Fig. 3. The solar energy collected from the sun can be used either directly or indirectly. The direct utilization of solar energy includes the use of either thermal or photovoltaic systems for conversion. On the other hand, the indirect system involves the usage of either biomass, water power, the wind, and the difference in the temperature of the ocean.

The solar energy can be utilized using following processes as,

(a) Helio-chemical process: In this process, every plant on earth can make their food and maintaining the life on earth or converts the carbon dioxide into oxygen.

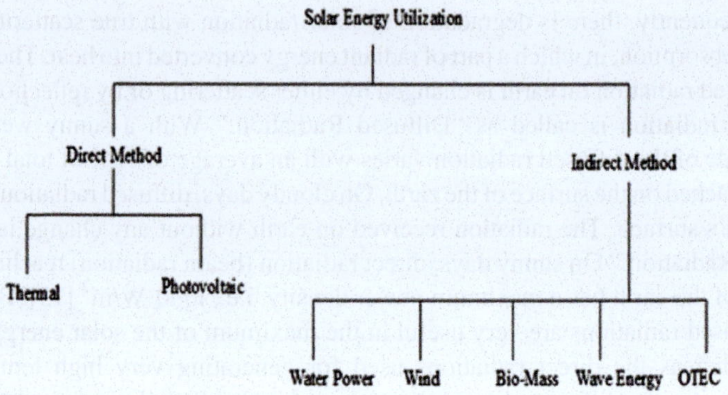

Fig. 3 Various methods of solar energy harvesting

(b) Helio-electrical process: It includes the photovoltaic converters which convert the solar radiation into electricity and these photovoltaic panels provide power to the spacecraft, satellite and can be easily utilized in the domestic and industrial applications.

(c) Helio-thermal process: It is utilized in the production of thermal energy, e.g., in space heating, solar drying, building heating, and water heating.

The main basic principle of the helio-thermal process is converting the solar radiation into thermal energy. Properly designed flat-plate collector is heated collector which is operating in the lower temperature range (i.e., atmospheric temperature to 100 °C) or can be used for medium temperature range (i.e., atmospheric temperature to 140 °C). Flat-plate collector is same as that of heat exchanger that transfers the radiating heat, obtaining from the solar radiation to sensible heat of fluid that may be air or liquid [21]. Flat-plate collector contains different types of combination which include grooved, flat or various types of shapes of absorbing surface with heat removal devices such as channels and tubes.

8 Different Utilizations of Solar Energy

Following are various applications of solar energy: solar furnace, photovoltaic conversion, food refrigeration, solar cooker, solar engines for pumping of water, salt production by evaporation, solar drying of agricultural products, solar air and water heating, heating and cooling of the building, high-temperature industrial processes, solar thermal power generation, etc. Broadly, one may classify electrical and thermal applications as two major categories of solar energy [22].

(a) **Direct Thermal Utilization**

Direct thermal utilization is one the best and easiest way to harvest the solar energy. Solar collector has to be placed in the direct solar radiation to harness the solar energy. These collecting devices are of mainly two types, i.e., absorbing type and concentrating type. The absorbing type solar collectors are usually dark in color (i.e., black color) in order to collect the maximum solar radiation. The energy collected is then transmitted to heat transfer fluid which may be either air or liquid. The concentrating type solar collector concentrates the solar radiation to its focal point and subsequently, the energy is transferred to the liquid. Following few devices which are being utilized for the direct thermal applications are given.

Flat-plate collector is having an absorber plate which receives the solar radiation, as the solar radiation passes through one or more translucent covers. The energy transfer from the collector to heat transfer fluid is known as an useful heat energy gain. The collectors are south faced if located in the northern hemisphere and north faced if located in the southern hemisphere. Further, the collectors have a tilt up to certain angle to the ground plane. If the fluid used in the flat-plate collector is water, then these collectors are known as "liquid flat-plate collector," and if air is used in these collectors are known as "solar air heater" [23].

It is a device in which the solar radiation are reflecting to a particular point or a path and which are, respectively, known as a focal point or a focal path of the solar collector [24–26]. In the concentrating collectors, the reflection can be obtained with the help of refracting and reflecting elements which result in an increase in the solar heat flux on the absorbing surface. In order to attain higher concentration, the sun-tracking system must be installed in these collectors. Focusing and concentrating collectors have three main parts: receiver system, solar tracking system, and the reflective or focusing device. Temperature more than that of 2000 °C may be easily achieved. So that this collector can be used for both electrical as well as for thermal applications.

A solar pond is a device which collects as well as stores the solar radiations. These devices have a pond of about 1 or 2 m depth containing water with different concentrations of salt (magnesium chloride or sodium chloride). Due to the different concentration of salt in the different layers of pond, the bottom layer of the solar pond has high temperature due to greater concentration of salt. Hence, the solar radiations get trapped in the bottom layers of the solar pond and upper layer behaves like a thermal protection for preventing the heat losses [20].

Space heating is important application performed using the solar collectors known as the air heaters. Flat-plate collector and concentrating or focusing type collectors may be employed as the air heaters. For space heating, the temperature to be raised is not more than 50–60 °C with either water or air as heat transfer fluid. Space heating is mostly useful in such places or countries which having low temperature and allots of energy utilized for space heating [27].

Generating the electricity from the solar radiation is another use of using thermal cycles of heat transfer fluid. The solar thermal cycle is divided into three parts: low temperature (80–100 °C), medium temperature (100–400 °C), and high temperature (>400 °C). The low-temperature cycle includes flat-plate collector and solar pond to collect the solar energy. Medium-temperature cycle includes focusing or concentrating collectors, whereas the high-temperature cycles include a central receiver and parabolic dish collectors in which receiver is present at the top of the tower [28–30].

Solar distillation is one of the efficient ways to obtain drinking water. Airtight shallow basin black in color made of impervious material contains saline water [31, 32]. As the solar radiation falls on this airtight shallow basin, the black color rapidly absorbs it. The temperature of saline water in the shallow basin increased by 20–30 °C and causing evaporation of water [33, 34]. As the evaporation of water starts, vapors rise upwards and get condensed in the collection channels. A well-designed solar distillation system has an output of 3 L/m^2 with efficiency of around 30%.

Drying of agricultural products is one of the convenient ways of using the solar energy. Drying is the method of removal of moisture from the agricultural products and preserves them for a much longer time. Traditionally, drying is done on the open ground, however, has disadvantages such as drying in open ground is very slow, dust, or insect contaminates the products, improper drying of products. By using solar dryer, these disadvantages can be eliminated [35]. Cabinet-type and box-type solar dryers are much preferable for the small-scale purpose. A simple dryer is having a translucent cover, perforated trays, and an absorber plate. The absorber plate inside the dryer absorbs the solar radiations and passes heat to the air, which flows along absorber plate. The heated air then passed along the perforated trays on which materials placed for drying. The hot air passes through the perforated trays remove the water or moisture inside the material. A small opening must be provided at the base or top of the dryer for the natural circulation. A normal temperature of around 60–80 °C is easily obtained inside the solar dryer. The drying time of material reduced from to 2–3 days depending upon the quantity [36]. For the large-scale application of solar dryer, forced circulation used instead of natural circulation.

Lastly, the most important thermal utilization of solar energy is solar cooking. A lot of solar cookers currently available with a different design such as (i) the box-type solar cooker or (ii) the concentrating type. The temperature about 100 °C is generally attainable in the box-type in the uncloudy days [37–39]. Concentrating-type solar cooker concentrates the solar radiation on its focal point, which is reflected by the parabolic surface. Cooking vessels are located at the focal point of the parabolic surface. These cookers may be equipped with a sun-tracking system. Temperatures above 400 °C may be achieved by concentrating type [40].

(b) **Electrical Utilization**

In the photovoltaic transformation, the devices utilized are known as solar cell. Solar radiation imparted on the surface of the semiconductor devices changes it into direct current (DC). The working principle of the solar cell includes two important steps: (i) creating a couple of positive and negative charges in the solar cell by absorbing the solar radiation and (ii) the converting part of both the negative and positive charges by a potential gradient which lies inside solar cell.

To carry out initial step, solar cell should have made of such material that can rapidly absorb the energy accomplice with the photons of solar radiation. The materials which are best suited for captivating the energy of photons associated with solar radiation are semiconductors, e.g., copper indium, cadmium sulfide, silicon, gallium arsenide [41–44]. A significant advantage incorporated with the utilization of solar cell is that it contains no moving parts so that it requires very less maintenance and works efficiently for both direct as well as for diffused radiation. There is one big disadvantage of solar cell is that these are having very high initial cost. But huge improvements have been made in a couple of years in the field of the solar cell. Various types of solar cells have created by using the innovative manufacturing processes. The conversing efficiencies have been increased significantly from a couple of years. So, in near future, the solar cell for harness the solar energy has become one of the best ways to generate electricity.

9 Solar Air Heater

The solar air heater is one the best examples of thermal application. The solar air heater is a system, which collects solar radiation and exchanges heat gathered with the moving air. It consists a conductive material usually of metal (i.e., copper), which acts as a heat exchanger. There are numerous designs of solar air heater, which may or may not be incorporated with the fan to enhance the flow rate of air enhancing the performance and efficiency of the solar air heater. Cloudy weather may affect the system's output so that heat storage system may be attached to solar air heaters. In order to obtain best results, the system should be employed in the unshaded and confront the heading toward the sun [45].

9.1 General Description of Solar Air Heater

The solar air heater is an example of the solar thermal application used for the space heating. Solar air heater is made of a flat-plate collector incorporated with a translucent cover which usually glass, an absorber plate, and insulation, which helps in reducing the heat losses from the various sides of solar air heater [46]. Materials used for fabricating solar air heater are identical to that of liquid flat-plate collector. As the solar radiation passes through the translucent cover of solar air heater, it gets rapidly absorbed on the absorbing plate. This absorbing plate transfers this heat to the air.

9.2 Various Components of the Solar Air Heater

(a) **Absorber Plate**

Absorber plate is a flat-plate made of materials, which have high thermal conductivity, better tensile strength, and non-corrosive (e.g., copper). The solar

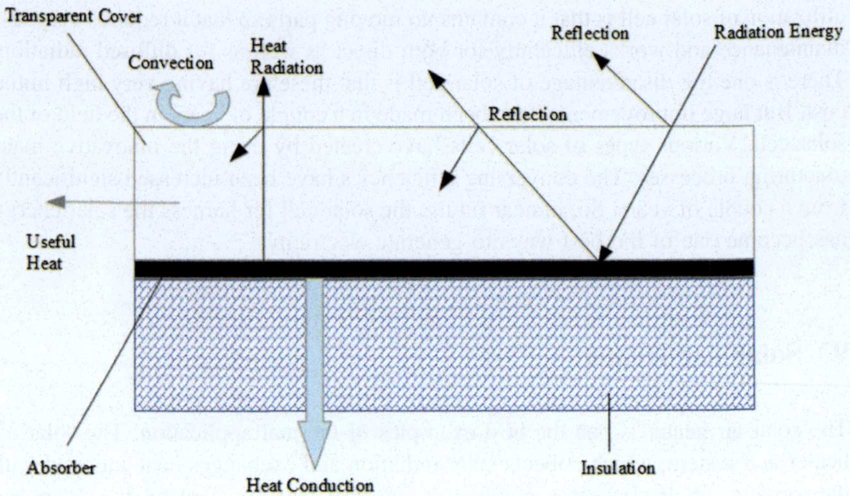

Fig. 4 Solar collector with absorber plate [46]

collector is made of steel or galvanized iron, aluminum, and types of thermoplastics. For fabrication of absorber plate, a thin sheet of metal (Cu or Al) may be taken with insulations, i.e., to non-flow sides. Radiations from the sun get rapidly absorbed by the thin sheet of aluminum or copper, and this heat is then transferred to air to increase its temperature (Fig. 4).

(b) **Cover Plate**

Most important part of solar air heater is a cover plate, which transmits the maximum amount of solar radiation to the absorbing plate. There are three purposes of the cover plate as (i) to transmit the solar radiation as much as possible to absorbing plate, (ii) to reduce heat losses to the atmosphere from the absorbing plate, and (iii) to serve as a protective shield (Fig. 5).

Important characteristics required for employing a cover plate are durable, energy transmission, non-degradability, and strength (e.g., glass). Before selecting the glass as a cover plate, its strength must be measured in order to ensure that it should stand against the harshest environment (e.g., snow loads in the winter season). The strength of the glass is directly proportional to the square of the thickness of the glass (about 0.33 cm thick).

Thermal impact on glass must be additionally considered due to day variations in the solar intensity of the sun, variation in the ambient temperature during the morning and evening time. Again, in the cloudy weather, glass temperature varies with a different rate. Mainly, somewhere at the center of solar collector subjected to more heat rather than the side of the collector. This results in thermal shock, which can lead to the breakage of glass.

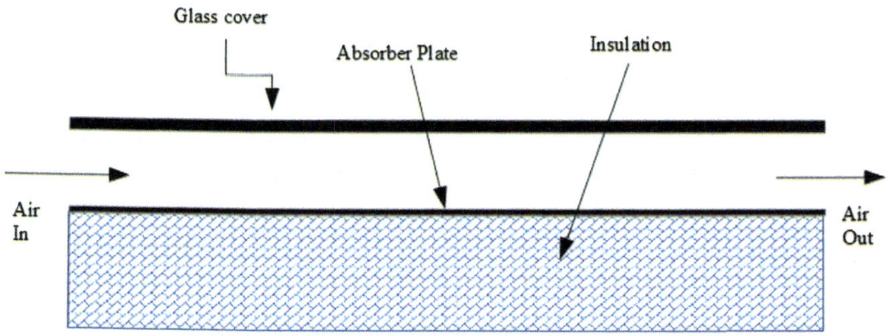

Fig. 5 Type-1 solar air heater [50]

One of another important characteristics must consider during the selection of glass is its rigidity. The rigidity of glass is directly proportional to the cube of the thickness of the glass. For better rigidity, plastic materials can be used. There are various types of plastics available which can be used in place of glass such as plastics coated with Tedlar and Mylar, Lexan, acrylic polycarbonate plastics, etc. But the plastics having one big disadvantage (i.e., reduction in the lifespan due to the impact of ultraviolet rays thus reducing in the transmissivity of plastics). Lightweight, resistance to breaking, and minimal cost are some benefits of plastics.

(c) **Insulation**

Insulation prevents heat losses by absorbing plates either in the form of conduction or convection. The most commonly used insulations nowadays is glass wool. Glass wool filled under the absorbing plate, and sides depends on the type of solar air heater is employed. The main use of insulation is to provide the best heat resistance [47, 48].

9.3 *Types of Solar Air Heater*

A distinct variety of solar air heater is available nowadays. The solar air heaters are divided into different types according to designs and arrangements. In some of solar air heater, absorber plate is placed under the glazing including smooth metal sheets, corrugated metal sheets, stacked screen or mesh, clear and black glass plate, etc. On the other hand, in some solar air heaters, air will pass through the absorber plate reducing the heat loss (i.e., downwards heat loss). One or two glass covers are provided for the proper resistance for upward convective heat losses. Depending upon the placement of the absorber plate inside the collector the solar air heater divided into two types [49] as (i) porous type solar collector and (ii) non-porous type solar collector.

9.3.1 Non-porous Type Solar Air Heater

(a) **Type-1**

In this solar air heater, air does not flow along with absorber plate, and it either flows above the absorber plate or below the absorber plate [50]; hence, it does not require any separate entry for the air. Air simply flows among translucent cover and absorber plate. However, as air flow above absorber plate, translucent cover gets the maximum part portion of heat and thus higher heat losses to the atmosphere.

(b) **Type-2**

In the type-2 non-porous solar air heater, the air vent lies underneath of the absorber plate. A separate plate is placed in between the insulation and absorber plate, making a passage and heat losses is minimized [51, 52] (Fig. 6).

(c) **Type-3**

In type-3, air flows over and below the absorber plate. This air heater has less heat transfer rate and low efficiency. It is observed that with the use of corrugated plate as an absorber, the efficiency may be increased [53, 54]. The increased flow turbulence results in improving the heat transfer rate. Further, implementing a suitable coating on the surface of the absorber plate may increase the overall efficiency. It is noted that utilizations of various types of fins on the absorber plate may result in the pressure drop. Hence, there are limitations in the use of non-porous type solar air heater (Fig. 7).

9.3.2 Porous Type Solar Air Heaters

A porous plate is utilized in this type of solar air heater and has few advantages: (i) radiation gets penetrated deep inside and gets absorbed along its path,

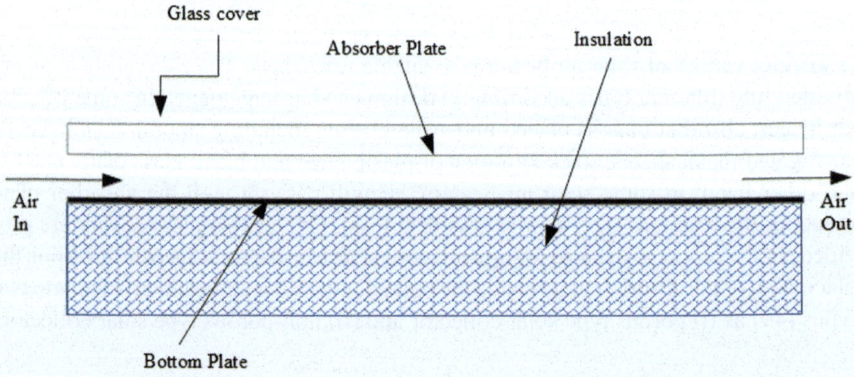

Fig. 6 Type-2 solar air heater [51]

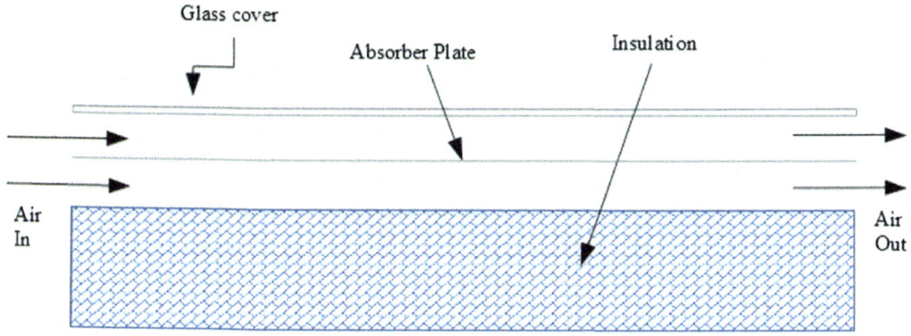

Fig. 7 Type-3 solar air heater [53]

(ii) minimum radiation losses, (iii) the drop-in pressure inside the solar collector is lower as compared to non-porous type. Despite having many advantages, they also having some disadvantages, e.g., improper choice of porosity or thickness in the matrix leads to the drop of efficiency.

(a) **Matrix-type solar air heater**

In the matrix-type solar air heater, solar radiation from the sun incidents on the air flowing over a porous matrix gets directly absorbed by the matrix. Cold air enters at the top of the matrix and gets warmed as it flows through the porous matrix in a reverse- or inverted-directions. The air entrance at the top, that is in between the translucent cover and top surface of matrix, minimizes the heat losses from the top [55–58].

(b) **Honeycomb porous-bed solar air heater**

Honeycomb porous-bed heater is slightly different from the matrix-type solar air heater, as it contains a honeycomb-like structure as matrix placed between translucent cover and absorber plate. Hence, at top, the convective heat losses are minimized. As the air is entering at the top, it gets heated by the incident radiation and as the air passes through the honeycomb matrix maximum temperature can be attained. Hexagonal and rectangular-shaped structures are mainly used in for the porous-bed [59] (Fig. 8).

(c) **Overlapped glass plate air heater**

This air heater has a series of parallel overlapped glass plates placed parallel to each other. The lower plates of such air heaters are black painted and air flows in between these parallel plates. Honeycomb structure is placed near the inlet for directing the air and to maintain the uniform velocity. Insulation is provided in the bottom of such air heaters [60]. Large collector face area and minimum pressure drop are the two main advantages of the overlapped glass plate solar air heater (Fig. 9).

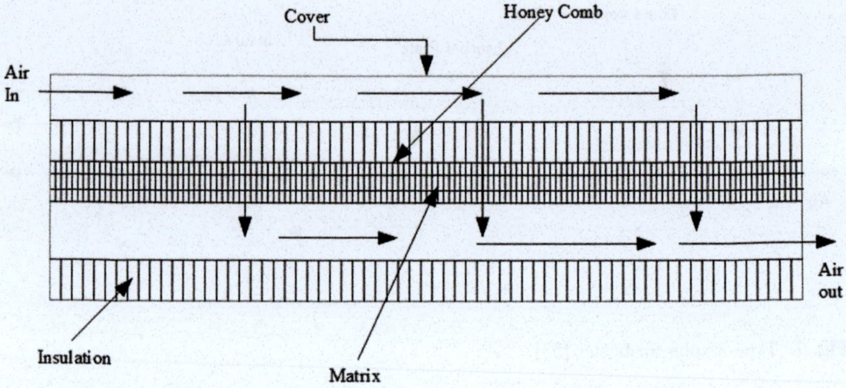

Fig. 8 Honeycomb porous-bed air heater [59]

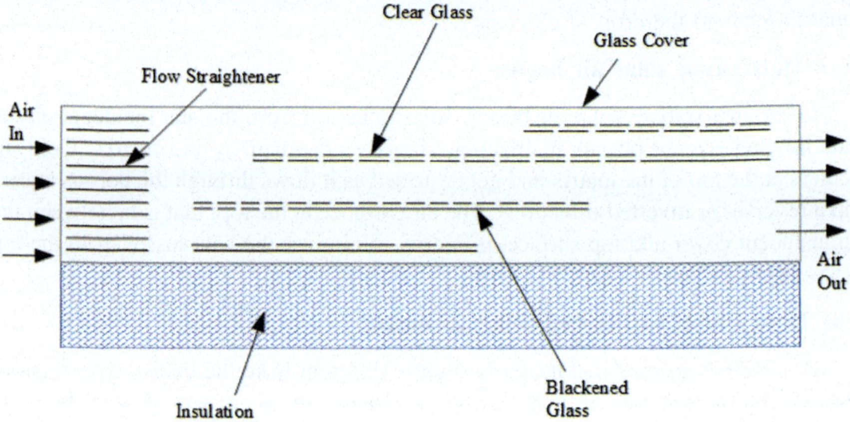

Fig. 9 Overlapped-glass plate solar air heater [60]

(d) **Jet-plate solar air heater**

This type of heater consists of extra translucent cover known as jet-plate placed between the bottom plate and absorber plate. This extra glass plate has equally spaced drilled holes. As air enters the jet-plate solar air heater, it flows between the jet-plate and the absorbing plate as well as between the bottom plate and jet-plate. This will increase the convection heat transfer coefficient, and there is an increase in the collector efficiency and useful heat energy gain [61] (Fig. 10).

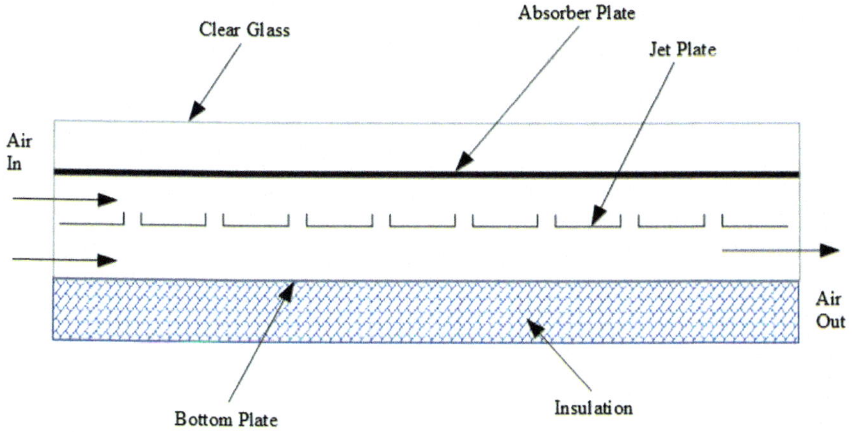

Fig. 10 Jet-plate solar air heater [61]

9.4 Parameters for Manufacturing of Solar Air Heater

Here, important parameters for manufacturing of the solar air heaters are listed: (i) **Heater configuration**: It is the aspect ratio of the length of air pass and the duct, (ii) **Air flow path**: Depending upon the air flow velocity, the efficiency of solar collectors varies, i.e., by increasing the air velocity efficiency of solar collector increases and vice versa, (iii) **Transmittance properties of the cover**: Spectral transmittance of translucent cover must be analyzed before applying it to the solar air heater. As the temperature required inside the solar air heater is high, more number of translucent covers are needed. As the numbers of translucent cover increase, the reflective losses would also increase. Translucent covers of low reflectivity and high transmissivity are required to minimize the absorbed and reflected radiation low, (iv) **Absorber plate material**: By using different types of materials with required properties enhance the efficiency of solar air heater by enhancing the overall efficiency of the solar collector. Dark color-coated absorber absorbs a greater amount of solar radiation, (v) **Natural convection barrier**: Stagnant air interposes greater impedance among ambient air and absorber plate. Using multiple covers or honeycomb structures, losses can be reduced, (vi) **Plate-to-air heat transfer coefficient**: For enhanced heat transfer coefficient, absorber plate must be coated with black and must be roughened. Roughness enhances the turbulence in flowing stream, (vii) **Insulation**: To minimize various losses from the various parts of solar collector (i.e., base and side), insulation is provided, (viii) **Solar radiation data**: For complete performance analysis of solar collector, solar radiation data must require [48, 62].

10 Thermal Energy Storage System

There are various types of storage technologies present, which are applicable for storing various kinds of energies (e.g., thermal energy, electrical energy, and mechanical energy) [63]. Thermal energy can easily be stored in a rightly insulated container. There is a change in internal energy of liquids or solid materials as a sensible heat or latent heat or thermo-chemical and by combination of all of these. The below figure represents major techniques for storing the solar thermal energy [64] (Fig. 11).

10.1 Sensible Heat Storage

In this type of storage, thermal energy is stored in fluid or solid by raising the temperature more than that of its nominal temperature without phase-change. The total amount of heat can be stored in fluid or solid is directly proportional to the medium change in temperature and specific heat of amount of storage materials. Table 3 represents the various materials with their properties used to find out the

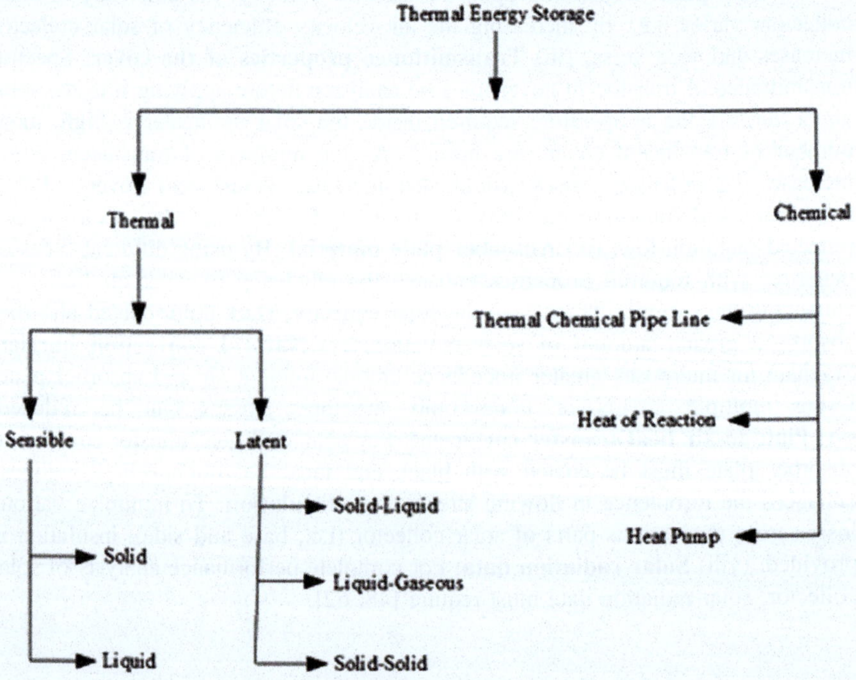

Fig. 11 Types of energy storage system [64]

Table 3 Various properties of materials for the use of sensible heat storage [65]

Fluid	Temp. variation (°C)	C_p(J/kgk)	Medium	Density(kg/m^3)
—	20	879	Rock	2560
—	20	840	Brick	1600
—	20	880	Concrete	1900–2300
—	0–100	4190	Water	1000
Oil	12–260	2200	Caloriea HT43	867
Oil	Up to 160	1880	Engine oil	888
Organic fluid	Up to 78	2400	Propanol	790
Organic fluid	Up to 97	2500	Butanol	800
Organic fluid	Up to 118	2400	Ethanol	809
Organic fluid	Up to 100	3000	Isobutanol	808
Organic fluid	Up to 148	2200	Octane	831
Organic fluid	Up to 126	2400	Isobutanol	704

sensible heat stored. As seen from the table, water is one of the best fluids having a good sensible heat storage capacity and being least expensive with a higher value of specific heat. Even, so-called the heat transfer fluids, liquid metals and molten salts, are utilized in applications above 100 °C. For air warming applications, rock bed-type storage material is generally utilized [65].

10.2 Latent Heat Storage (Using Phase-Change Materials)

It is the technique of absorbing and discharging of heat when the storage materials change its phase at a constant temperature. As given in Ref. [66], use of paraffin wax as latent heat storage is advantageous (Fig. 12).

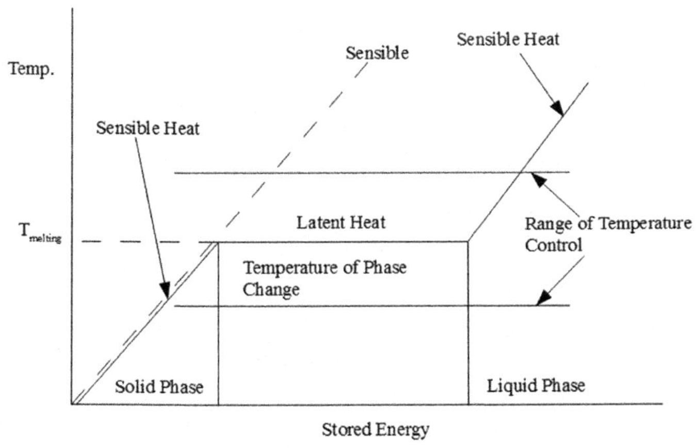

Fig. 12 Phase-change diagram for latent heat storage [66]

11 Integration of Solar Air Heater and Thermal Energy Storage

Solar air heaters along with the phase-change material(PCM)-based thermal energy storage system collect solar radiations during the daytime. For the PCM as solid phase then it melts and stores the energy by changing its phase. The energy stored in the liquid after melting can recover as the liquid change its phase from liquid to solid. In the thermal energy storage systems, solar energy stored during the daytime can be recovered during night or in cloudy weather. There are various types of phase-change material available for the different temperature ranges. And some of them are paraffin, non-paraffin, fatty acids, and hydrated salts (i.e., Glauber's salt, calcium chloride hexahydrate, etc.) [67–70]. Most suitable phase-change material for solar air heater is shown in Table 4.

The optimum properties of PCM materials in the solar air heater may be determined using simulation techniques, and the performance at various space heating loads for entire heating seasons may be estimated [71]. Transient behavior of storage system unit has been described by Jurinak et al. [72]. Paraffin wax and sodium sulfate decahydrate are the two PCM, which are used in solar air heater. Lacroix et al. [73] developed a hybrid energy storage system, which handles both thermal energy from the sun as well as electrical energy from the grid. This system stores solar heat in the clear weather and liberates at night. The electric energy in this system stored during the peak off period and discharged later in the peak period. They found that the electrical consumption decreased by 32% in the space heating. Kaygusuz et al. [74] examined the experimental and theoretical performance of solar air heater incorporated along the heat pump system which contains phase-change material. Various experimental studies show that the heat pump in parallel saves more energy as compared to the heat pump in series. It is due to the reason that parallel heat pump includes both solar and air as a heat source in the evaporator, whereas in parallel heat pump includes only solar energy. $CaCl_2 \cdot 6H_2O$ was used as phase-change material. Efficiency about 70% was obtained from this system during experiments. Mettawee et al. [75] integrated the solar collector with the latent heat energy storage system containing PCM (paraffin wax) to absorb maximum solar energy. The result of this experiment showed that as the molten layer thickness increases during the charging process, average heat transfer coefficient enhances as the natural convection increases. Zheo et al. [76] investigated the solar heating system which was built for obtaining warm water throughout the year in the northern region of China and keep the building warm for the winter season.

For such a system, volume/height of the tank, water storage tank, heat transfer coefficient of the insulation layer, and volume of the sensible heat storage material bed, installation angle of solar collector, solar collector area, and mass flow rate through the system can be analyzed with the help of TRNSYS software.

Table 4 Potential phase-change materials [67]

Materials	Latent heat (kJ/kg)	Melting point (°C)
Organic materials		
Capric acid	152	36
Lauric acid	178	49
Elaidic acid	218	47
Tristearin	191	56
Myristic acid	199	58
Pentadecanoic acid	178	52.5
Palmitic acid	163	55
Acetamide	241	81
Stearic acid	199	69.4
Inorganic materials		
$Mn(NO_3)_2 \cdot 4H_2O$	115	37
$FeCl_3 \cdot 6H_2O$	223	37
$Zn(NO_3)_2 \cdot 6H_2O$	134	36
$CoSO_4 \cdot 7H_2O$	170	40.7
$Na_2HPO_4 \cdot 12H_2O$	279	40
$Zn(NO_3)2 \cdot 4H_2O$	110	45
$K_2HPO_4 \cdot 7H_2O$	145	45
$CaI_2 \cdot 6H_2O$	162	42
$MgI_2 \cdot 8H_2O$	133	42
$Na_2SiO_3 \cdot 4H_2O$	168	48
$Fe(NO_3)_3 \cdot 9H_2O$	155	47
$Ca(NO_3) \cdot 4H_2O$	153	47
$Mg(NO_3) \cdot 4H_2O$	142	47
$Ca(NO_3)_2 \cdot 3H_2O$	104	51
$MgSO_4 \cdot 7H_2O$	202	48.5
$Na_2S_2O_3 \cdot 5H_2O$	210	48.5
$K_2HPO_4 \cdot 3H_2O$	99	48
$MnCl_2 \cdot 4H_2O$	151	58
$Ni(NO_3)_2 \cdot 6H_2O$	169	57
$FeCl_3 \cdot 2H_2O$	90	56
$Zn(NO_3)_2 \cdot 2H_2O$	68	55
$NaAl(SO_4)_2 \cdot 10H_2O$	181	61
$Fe(NO_3)_2 \cdot 6H_2O$	126	60.5
$MgCl_2 \cdot 4H_2O$	178	58
$Ba(OH)_2 \cdot 8H_2O$	265	78
$Al(NO_3)_2 \cdot 9H_2O$	155	72
$LiCH_3COO \cdot 2H_2O$	150	70
$Na_3PO_4 \cdot 12H_2O$	190	65
$NaOH \cdot H_2O$	273	64.3

References

1. WEC (2016) World energy resources. World Energy Council report, vol 1, p 468
2. IDB (2011) Chaglla hydropower project
3. Hekkert MP, Hendriks FHJF, Faaij APC, Neelis ML (2005) Natural gas as an alternative to crude oil in automotive fuel chains well-to-wheel analysis and transition strategy development. Energy Policy 33(5):579–594
4. Holmberg K, Andersson P, Erdemir A (2012) Global energy consumption due to friction in passenger cars. Tribol Int 47:221–234
5. Umbach F (2010) Global energy security and the implications for the EU. Energy Policy 38 (3):1229–1240
6. Burroughs WJ (2003) Book reviews. Prometheus 21(1):120–139
7. Cox PM, Betts RA, Jones CD, Spall SA, Totterdell IJ (2000) Acceleration of global warming due to carbon-cycle feedbacks in a coupled climate model. Nature 408(6809):184–187
8. Meinshausen M et al (2009) Greenhouse-gas emission targets for limiting global warming to 2 °C. Nature 458(7242):1158–1162
9. Young OR (2017) A. Society, I. Law, T. A. Journal, and I. Law, Review reviewed work (s): The collapse of the kyoto protocol and the struggle to slow global warming by David G. Victor. Am J Int Law 96(3) (Jul 2002):736–741
10. Harb A (2011) Energy harvesting: state-of-the-art. Renew. Energy 36(10):2641–2654
11. Galik CS, Abt RC, Latta G, Meley A, Henderson JD (2016) Meeting renewable energy and land use objectives through public-private biomass supply partnerships. Appl Energy 172:264–274
12. Sen Z (2008) Solar energy fundamentals and modeling techniques: atmosphere, environment, climate change and renewable energy
13. Kalogirou SA, Karellas S, Braimakis K, Stanciu C, Badescu V (2016) Exergy analysis of solar thermal collectors and processes. Prog Energy Combust Sci 56:106–137
14. Karanasios K, Parker P (2016) Recent developments in renewable energy in remote aboriginal
15. (2016) Graph, See Government, The Ministry, The Pumps, Solar Ministry, The Energy, Renewable Ministry
16. Meisen P, Quéneudec E (2006) Overview of renewable energy potential of India, October, pp 1–20
17. Bhawan SP, Marg S (2016–17) Annual report of contribution of different sectors to gross value added in 2015–16. Government of india, Ministry of statistics and programme implementation
18. Duffie JA, Beckman WA, McGowan J (1985) Solar engineering of thermal processes. Am J Phys 53(4):382
19. Liu BYH, Jordan RC (1960) The interrelationship and characteristic distribution of direct, diffuse and total solar radiation. Sol Energy 4(3):1–19
20. Samimi A, Zarinabadi S, Samimi M (2012) Solar energy application on environmental protection, vol 1, no 8, pp 21–24
21. Lodhi MAK (2004) Helio-hydro and helio-thermal production of hydrogen. Int J Hydrogen Energy 29(11):1099–1113
22. Buchberg H, Catton I, Edwards DK (1976) Natural convection in enclosed spaces—a review of application to solar energy collection. J Heat Transf 98(2):182
23. Klein SA (1978) Calculation of flat-plate collector utilizability. Sol Energy 21(5):393–402
24. Singh PL, Sarviya RM, Bhagoria JL (2010) Heat loss study of trapezoidal cavity absorbers for linear solar concentrating collector. Energy Convers Manag 51(2):329–337
25. Coventry JS (2005) Performance of a concentrating photovoltaic/thermal solar collector. Sol Energy 78(2):211–222
26. Sultana T, Morrison GL, Rosengarten G (2012) Thermal performance of a novel rooftop solar micro-concentrating collector. Sol Energy 86(7):1992–2000

27. Esen M (2000) Thermal performance of a solar-aided latent heat store used for space heating by heat pump. Sol Energy 69(1):15–25
28. Hu E, Yang Y, Nishimura A, Yilmaz F, Kouzani A (2010) Solar thermal aided power generation. Appl Energy 87(9):2881–2885
29. Huang BJ, Ding WL, Huang YC (2011) Long-term field test of solar PV power generation using one-axis 3-position sun tracker. Sol Energy 85(9):1935–1944
30. Zhang XR, Yamaguchi H, Uneno D, Fujima K, Enomoto M, Sawada N (2006) Analysis of a novel solar energy-powered Rankine cycle for combined power and heat generation using supercritical carbon dioxide. Renew Energy 31(12):1839–1854
31. Tiwari GN, Dimri V, Chel A (2009) Parametric study of an active and passive solar distillation system: energy and exergy analysis. Desalination 242(1–3):1–18
32. Khas H (1996) Pergamon PII: s0360-5442(%)ooo15-1, vol 21, no 9, pp 805–808
33. Kumar S, Tiwari GN, Singh HN (2000) Annual performance of an active solar distillation system. Desalination 127(1):79–88
34. Tiwari GN, Singh HN, Tripathi R (2003) Present status of solar distillation. Sol Energy 75 (5):367–373
35. Aberle AG (2009) Thin-film solar cells. Thin Solid Films 517(17):4706–4710
36. Guha S, Yang J (1999) Science and technology of amorphous silicon alloy photovoltaics. IEEE Trans Electron Devices 46(10):2080–2085
37. Thirugnanasambandam M, Iniyan S, Goic R (2010) A review of solar thermal technologies. Renew Sustain Energy Rev 14(1):312–322
38. Nandwani SS (1996) Solar cookers—cheap technology with high ecological benefits. Ecol Econ 17(2):73–81
39. Telkes M (1959) Solar cooking ovens. Sol Energy 3(1):1–11
40. Sharma SD, Iwata T, Kitano H, Sagara K (2005) Thermal performance of a solar cooker based on an evacuated tube solar collector with a PCM storage unit. Sol Energy 78(3):416–426
41. Savin H et al (2015) Black silicon solar cells with interdigitated back-contacts achieve 22.1% efficiency. Nat Nanotechnol 10(7):624–628
42. Shah AV et al (2004) Thin-film silicon solar cell technology. Prog Photovolt Res Appl 12 (23):113–142
43. Shah AV, Platz R, Keppner H (1995) Thin-film silicon solar cells: a review and selected trends. Sol Energy Mater Sol Cells 38(1–4):501–520
44. Wenham SR, Green MA (1996) Silicon solar cells. Prog Photovolt 4(1):3–33
45. Mohamad AA (1997) High efficiency solar air heater. Sol Energy 60(2):71–76
46. Kumar A, Saini RP, Saini JS (2014) A review of thermohydraulic performance of artificially roughened solar air heaters. Renew Sustain Energy Rev 37:100–122
47. Close DJ (1963) Solar air heaters for low and moderate temperature applications. Sol Energy 7(3):117–124
48. Kumar A, Saini RP, Saini JS (2012) Heat and fluid flow characteristics of roughened solar air heater ducts—a review. Renew Energy 47:77–94
49. Gupta CL, Garg HP (1967) Performance studies on solar air heaters. Sol Energy 11(1):25–31
50. Bhargava AK, Garg HP, Sharma VK (1982) Evaluation of the performance of air heaters of conventional designs. Sol Energy 29(6):523–533
51. Biondi P, Cicala L, Farina G (1988) Performance analysis of solar air heaters of conventional design. Sol Energy 41(1):101–107
52. Loveday DL (1988) Thermal performance of air-heating solar collectors with thick, poorly conducting absorber plates. Sol Energy 41(6):593–602
53. Satcunanathan S, Deonarine S (1973) A two-pass solar air heater. Sol Energy 15(1):41–49
54. Garg HP, Sharma VK, Bhargava AK (1985) Theory of multiple-pass solar air heaters. Energy 10(5):589–599
55. Wijeysundera NE, Ah LL, Tjioe LE (1982) Thermal performance study of two-pass solar air heaters. Sol Energy 28(5):363–370
56. Science E (1966) An investigation on packed-bed collectors

57. Lansing FL, Clarke V, Reynolds R (1979) A high performance porous flat-plate solar collector. Energy 4(4):685–694
58. Parker BF, Lindley MR, Colliver DG, Murphy WE (1993) Thermal performance of three solar air heaters. Sol Energy 51(6):467–479
59. Lalude O, Buchberg H (1971) Design and application of honeycomb porous-bed solar-air heaters. Sol Energy 13(2):223–242
60. Selçuk K (1971) Thermal and economic analysis of the overlapped-glass plate solar-air heater. Sol Energy 13(2):165–191
61. Choudhury C, Garg HP (1991) Evaluation of a jet plate solar air heater. Sol Energy 46 (4):199–209
62. Klein SA, Beckman WA, Duffie JA (1976) A design procedure for solar heating systems. Sol Energy 18(2):113–127
63. Bal LM, Satya S, Naik SN (2010) Solar dryer with thermal energy storage systems for drying agricultural food products: a review. Renew Sustain Energy Rev 14(8):2298–2314
64. Bal LM, Satya S, Naik SN, Meda V (2011) Review of solar dryers with latent heat storage systems for agricultural products. Renew Sustain Energy Rev 15(1):876–880
65. Schröder J, Gawron K (1981) Latent heat storage. Int J Energy 5(March 1980):103–109
66. Salunkhe PB, Krishna DJ (2017) Investigations on latent heat storage materials for solar water and space heating applications. J Energy Storage 12:243–260
67. Rabin Y, Bar-Niv I, Korin E, Mikic B (1995) Integrated solar collector storage system based on a salt-hydrate phase-change material. Sol Energy 55(6):435–444
68. Hasan A (1994) Phase change material energy storage system employing palmitic acid. Sol Energy 52(2):143–154
69. Tyagi VV, Buddhi D (2007) PCM thermal storage in buildings: a state of art. Renew Sustain Energy Rev 11(6):1146–1166
70. Naphon P, Kangtragool B (2003) Theoretical study on heat transfer characteristics and performance of the plat-plate solar air heaters. Int Commun Heat Mass Transf 30(3):1125–1136
71. Morrison DJ, Abdel-Khalik SI (1978) Effects of phase-change energy storage on the performance of air-based and liquid-based solar heating systems. Sol Energy 20(1):57–67
72. Jurinak JJ, Abdel-Khalik SI (1978) Properties optimization for phase-change energy storage in air-based solar heating systems. Sol Energy 21(5):377–383
73. Hammou ZA, Lacroix M (2006) A new PCM storage system for managing simultaneously solar and electric energy, vol 38, pp 258–265
74. Energy R (2000) Experimental and theoretical investigation of a solar heating system with heat pump, vol 21
75. Mettawee ES, Assassa GMR (2006) Experimental study of a compact PCM solar collector, vol 31, pp 2958–296
76. Zhao DL, Li Y, Dai YJ, Wang RZ (2011) Optimal study of a solar air heating system with pebble bed energy storage, vol 52, pp 2392–2400

Solar Thermal Energy Storage Using Graphene Nanoplatelets-Added Phase Change Materials

S. Suresh and Srikanth Salyan

Abstract Thermal energy storage (TES) is a key system to reduce the gap between energy supply and energy demands. Energy storage materials play a very important role in the design of TES. Energy storage materials store energy in form of sensible heat, latent heat and thermochemical energy storage. Compared to the various forms of energy storage, latent heat-based energy storage system can store a lot of energy at isothermal temperature during melting and can release the stored heat during solidification. Phase change material (PCM) includes organic, inorganic and eutectic materials. One of the primary disadvantages of PCMs is their very low thermal conductivity. This challenge could be overcome by the addition of high thermal conductive additives to form a composite. Thermo-physical property determination of these composites is very important to determine the feasibility of using such composites as energy storage materials. Characterisation techniques like differential scanning calorimetry (DSC), Fourier-transform infrared (FTIR), thermo-gravimetric analysis (TGA) and laser flash apparatus (LFA) are very effective and useful methods to determine the potential use of PCM in TES applications. This work offers the characteristics of pentaerythritol (PE) and D-Mannitol (DM) and the effect of adding high conductive graphene nanoplatelets (GnPs) to form a phase change composites. Thermal cycling is done to evaluate the thermal reliability of the composite after repeated melt/freeze cycles. Finally, the composite will be evaluated for its suitability of being used as a PCM in LHES systems.

Keywords Phase change materials · Nanoparticles · Solar thermal energy storage · Latent heat

S. Suresh (✉) · S. Salyan
Department of Mechanical Engineering, National Institute of Technology, Tiruchirappalli 620015, Tamil Nadu, India
e-mail: ssuresh@nitt.edu

H. Tyagi et al. (eds.), *Applications of Solar Energy*, Energy, Environment, and Sustainability, https://doi.org/10.1007/978-981-10-7206-2_10

Nomenclature

DM	D-Mannitol
DM-GnP	D-Mannitol with 1 wt% graphene nanoplatelets composite
DSC	Differential scanning calorimetry
FTIR	Fourier-transform infrared
GnPs	Graphene nanoplatelets
LHES	Latent heat energy storage
PCM	Phase change material
PE	Pentaerythritol
PE-GnP	Pentaerythritol with 1 wt% graphene nanoplatelets composite
PO	Polyalcohols
TES	Thermal energy storage
TGA	Thermo-gravimetric analysis
ρ	Density (kg/m^3)
d	Sample thickness (cm)
C_p	Specific heat (J/kg K)
t	Time(s)
k	Thermal conductivity (W/m K)
α	Thermal diffusivity of sample (cm^2/s)

Subscripts

½ Half maximum

1 Introduction

Effective use of renewable energy is important due to increased levels of greenhouse gases and steep increase in fuel prices in recent years. In this view, the most abundant form of renewable energy is solar energy. The challenge exists in the storage of this energy as solar energy, which is basically intermittent in nature and cannot supply energy during the night-time. Hence, developing energy storage devices to negate this disadvantage is very important for scientists all-over the world. Energy storage devices can store the energy in form of sensible heat, latent heat and thermochemical or any of the combination. In sensible heat storage system, energy is stored in the form of raised temperature, increasing from one temperature to another. The amount of energy stored depends on the specific heat of the storage medium. Latent heat storage depends on the heat absorbed or released during phase transition of the material. A huge amount of energy is transferred while the material changes its phase at a constant temperature. Thermochemical systems depend on the strength of molecular bonds between molecules of the materials. The energy is stored and released in the form of a complete reversible chemical reaction. The advantages of using phase change materials (PCMs) in energy storage devices help in bridging the gap between energy supply and demand and meeting the energy requirement throughout the day and night. Most of the PCMs are easily available, possess good thermal, physical, chemical properties and are processed in industrial scale. PCMs are broadly classified into organic, inorganic and eutectic materials. Organic materials are further classified into paraffins and non-paraffins [1–3]. Paraffins mostly consist of straight-chain alkanes. Mostly paraffins are safe, stable, corrosion resistive and chemically inert. Non-paraffin organic materials include esters, fatty acids, alcohols and glycols. Alcohols (also called as polyalcohols) are evaluated as one particular category of PCM material interest, owing to their moderate phase change temperatures for the range of 0–200 °C and higher mass-specific and volume-specific enthalpies than other organic PCMs [4]. Furthermore, many polyols are non-toxic and safe. D-Mannitol ($C_6H_{14}O_6$) is a solid–liquid phase change material with a melting temperature of 163–168 °C with an enthalpy of 270–275 J/g. Pentaerythritol is a solid–solid phase change material, which undergoes a solid mesophase with high orientation disorder during phase transition. It has phase transition temperature of around 182–188 °C with the latent heat of transition of 240–280 J/g. Many studies have been reported in the literature regarding the suitability of polyalcohols for thermal energy storage medium. Aran et al. [5] studied experimentally the chemical and thermal stability of D-Mannitol, myo-inositol and galactitol using DSC and FTIR before and after repeated thermal cycles. They have reported a polymorphic change of myo-inositol in the temperature range of 50–200 °C. They analysed the effect of thermal cycling of PCM using different crucible in DSC and their study showed poor thermal cycling stability of galactitol. Kumaresan et al. [6] in their study reported a large temperature difference of more than 130 °C between the melting point and decomposition temperature of DM, making it thermally stable for use in medium temperature applications. Abhijit et al. [7] have done an experimental investigation on the eutectic mixture of galactitol and mannitol with a 30:70 molar ratio, and the studies found that the eutectic mixture exhibited high latent heat value of 293 J/g with good thermal cycling stability in the working temperature range of 50–200 °C. Solid–solid PCMs undergo solid/solid phase transition where these materials change their crystalline structure from one lattice configuration to another at fixed temperature and absorb and release large amount of heat which is comparable to the fusion enthalpies of solid–liquid PCMs [8]. At these phase change temperatures, solid/solid polyalcohols change their molecular structure from a body-centred tetrahedral (α-phase) to a homogenous face-centred cubic crystalline structure (γ-phase) accompanied with a huge amount of absorption of the hydrogen bond energy [9]. Various studies have been reported on solid/solid phase change materials and their associated phase transition enthalpies in the literature [10–13]. However, the major disadvantage of using PCMs in TES is their very low thermal conductivity. This limits the rate of heat absorbed/released and affects the thermal response of the PCMs. To address this issue, several additives having a high thermal conductivity are added to the base PCM to develop a high thermal conductive composite. A complete review on the effect of thermo-physical properties by the addition of nanoparticles in PCMs was presented by Kibira et al. [14]. Some of the most commonly used additives are expanded graphite, graphene aerogel, metal oxides, metals, carbon nanotubes, graphene nanoplatelets and graphene foam.

In this work, the addition of graphene nanoplatelets in the mass fraction of 1 wt% on solid–liquid PCM (D-Mannitol) and solid–solid PCM (Pentaerythritol) has been reported. Thermal and chemical characterisation has been done by differential scanning calorimetry (DSC), thermo-gravimetric analysis (TGA), Fourier-transform infrared (FTIR), and the thermal conductivity enhancement study is analysed by laser flash apparatus (LFA). The effect of addition of additives before and after thermal cycling has been presented. The prepared composites provide a novel approach for a new phase change material for the application in thermal energy storage system like solar water heating, solar desalination units and mainly in medium temperature solar-concentrated solar thermal power plants working in the temperature range of 100–250 °C.

2 Methods

2.1 *Materials*

Pentaerythritol with purity of 98.0% (analytical reagent grade) in powder form was purchased from Alfa Aeser, USA. Similarly, 99% pure DM was obtained from Spectrochem Pvt. Ltd, Mumbai, India. Graphene nanoplatelets with sub-micron particles having a surface area of 500 m^2/g were procured from Alfa Aesar, USA.

2.2 *Sample Preparation and Thermal Cycling*

Composite of PE with 1 wt% GnP (PE-GnP) and DM with 1 wt% GnP (DM-GnP) was prepared by addition of GnP into base PCM and mixed in a low energy ball mill. Milling was done for 2.5 h at 200 rpm. Three stainless steel balls were used to provide centrifugal force during rotation to produce uniform mixing of GnP with PCM. The repeating thermal cycling tests, which consisted of heating the PCM composite up to its melting point and then solidifying it to room temperature, were done in a custom fabricated hot plate. The hot plate was connected to an autotransformer for regulating power supply. The temperature of the PCM during thermal cycling was obtained by K-type thermocouples calibrated in a constant temperature oil bath with an accuracy of ±0.5 °C. Sample weighing 50 g each was used for thermal cycling, and the temperature of the sample is recorded using a data acquisition system. The melting and freezing cycles were repeated for 100 times using the hot plate thermal cycler, and the samples were tested for their thermo-physical properties before and after the thermal cycles, respectively.

3 Characterisation Techniques

3.1 *Differential Scanning Calorimetry (DSC)*

Differential scanning calorimetry (DSC) is a widely used method to characterise the thermal properties of PCMs among other methods like the differential thermal analysis (DTA). In DSC analysis, the system measures the difference in the amount of heat required to increase the temperature of a material sample (sample pan) and an empty sample reference pan as a function of temperature. The DSC measurements were carried out using Mettler Toledo DSC 822e. The instrument has a temperature range of −130 to 450 °C, measurement range of ±350 mW at RT, measurement resolution of 0.04 mW at RT, temperature accuracy of ±0.2 °C and a temperature reproducibility of ±0.1 °C. The PCM samples were subjected to heating and cooling cycles between the temperature range of 30–280 °C at 10 °C/min.

3.2 *Thermo-gravimetric Analysis (TGA)*

TGA is an analysis method in which physical and chemical properties of materials are measured as a function of increased temperature with constant heat rate or as a function of time with fixed constant temperature. TGA is commonly used to characterise the material for their mass loss with an increase of temperature. In PCM applications, TGA helps in determining the thermal stability of samples within the working range temperature of LHES. TGA analysis was done by NETZSCH STA 449 Jupiter. Measurements were conducted with an airflow rate of 60 ml/min with a temperature range from room temperature to 500 °C. The sensitivity of the instrument is 0.2 mg from room temperature to 1200 °C.

3.3 *Fourier-Transform Infrared Analysis (FTIR)*

FTIR technique is used to study the infrared spectrum of the sample and their individual bonding. FTIR is used to determine the chemical stability of the prepared samples before and after thermal cycles. The FTIR spectra of all the PCM samples were obtained using infrared spectrophotometer (Frontier FTIR/FIR, PerkinElmer) in the wave number range of 400–4000 cm^{-1} with the minimum resolution of 4 cm^{-1}.

3.4 *Laser Flash Apparatus (LFA)*

LFA is a thermal analysis instrument for indirect measurement of thermal conductivity. The thermal conductivity is found by measuring thermal diffusivity using NETZSCH LFA 467 HyperFlash apparatus. The diffusivity is measured by applying an energy pulse on one side of the disc, which heats the plane parallel to the sample, and the resulting time-dependent temperature rise on the rear side is measured using a liquid nitrogen cooled detector.

The thermal diffusivity of the sample is given by the Eq. (1)

$$\alpha = 0.1388\ \frac{d^2}{t_{1/2}} \tag{1}$$

Finally, the thermal conductivity is found by the Eq. (2)

$$k = \alpha \cdot \rho \cdot C_p \tag{2}$$

4 Results and Discussion

4.1 *Thermal Stability Analysis*

Figure 1 shows the TGA curve of pure pentaerythritol before and after 100 thermal cycles. The sample is heated from room temperature to 230 °C. It can be seen from the Fig. 1 that the samples are very stable in the working temperature range of 50–200 °C. There was a negligible weight loss (less than 2%) up to the temperature of 240 °C. PE after 100 thermal cycles also showed a similar trend. Figure 2 shows the thermo-gravimetric curves of pentaerythritol added with 1% GnP before and after thermal cycles. The TGA curve of the sample indicated negligible weight loss at a temperature below 220 °C and a faster rate of loss was observed beyond 240 °C. Figure 3 shows the TGA curves for DM before and after 100 cycles. The samples are tested up to a temperature of 500 °C for DM samples. The melting temperature of DM is around 164–168 °C. From the graphs, it is clear that the DM and its composite are very stable in the working temperature range. A weight loss of less than 1% was found up to a temperature of 280 °C for DM before thermal cycling. DM observed a single-step degradation at a temperature of 320 °C. A similar trend was observed with cycled DM samples. With 1 wt% GnP sample (Fig. 4), the decomposition temperature decreased, yet there was a good thermal

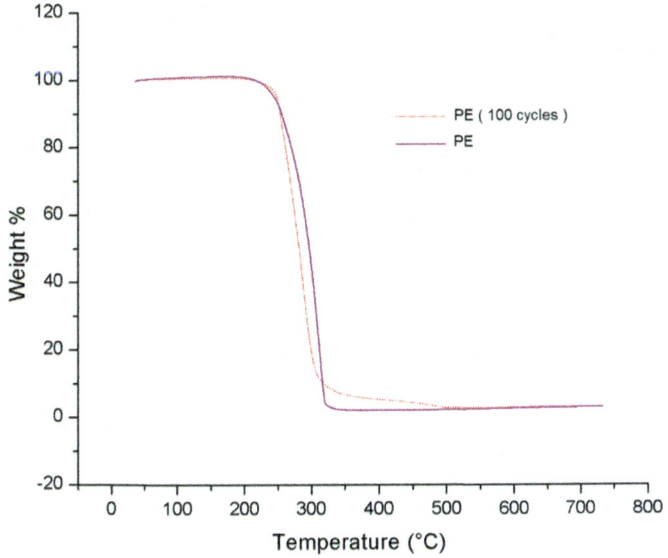

Fig. 1 TGA graphs of PE before and after 100 cycles

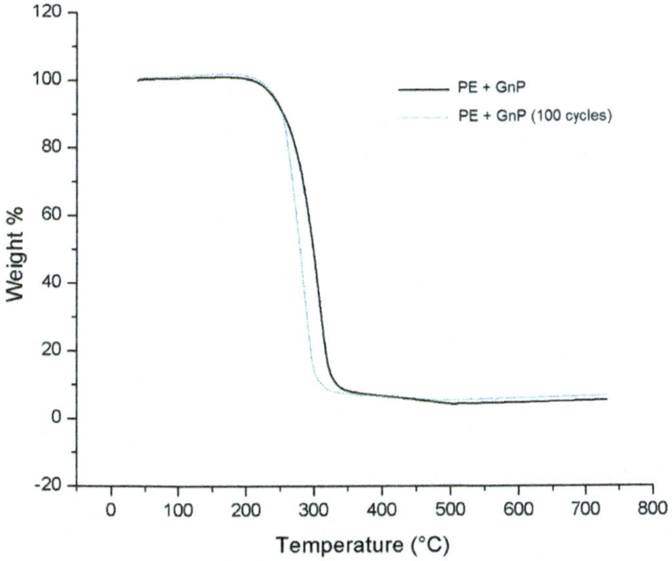

Fig. 2 TGA graphs of PE + GnP before and after 100 cycles

stability up to a temperature of 240 °C. From the TGA curve, we can conclude that both PE and DM are very stable even after thermal cycling and are suitable for application in LHES.

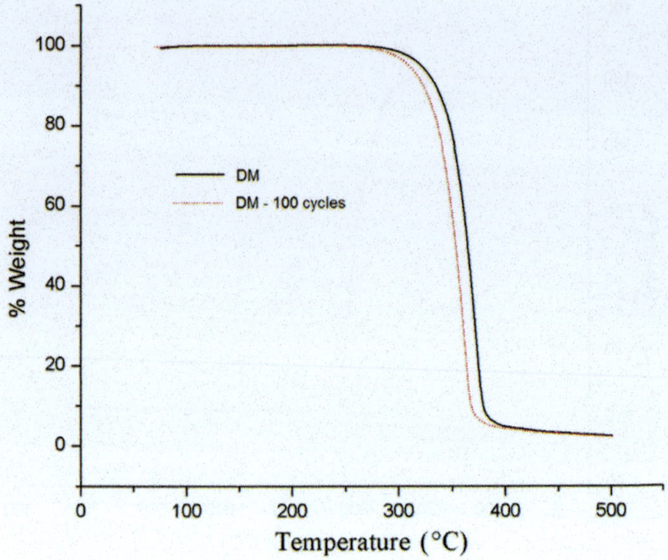

Fig. 3 TGA graphs of pure DM before and after 100 cycles

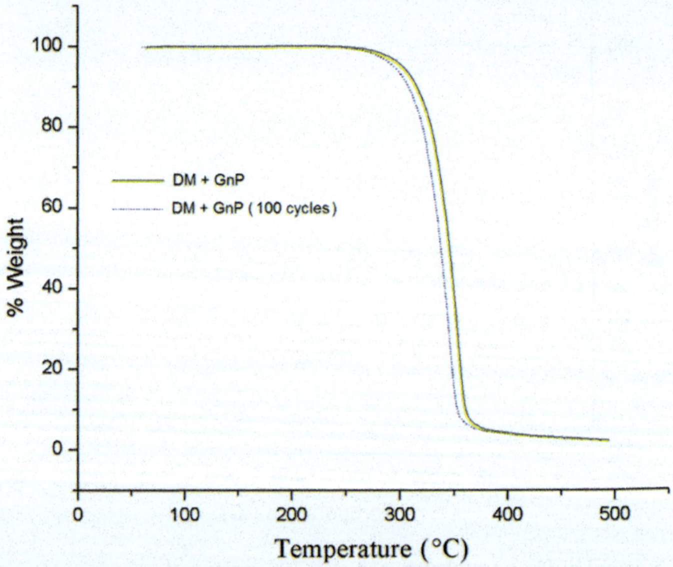

Fig. 4 TGA graphs of DM-GnP before and after 100 cycles

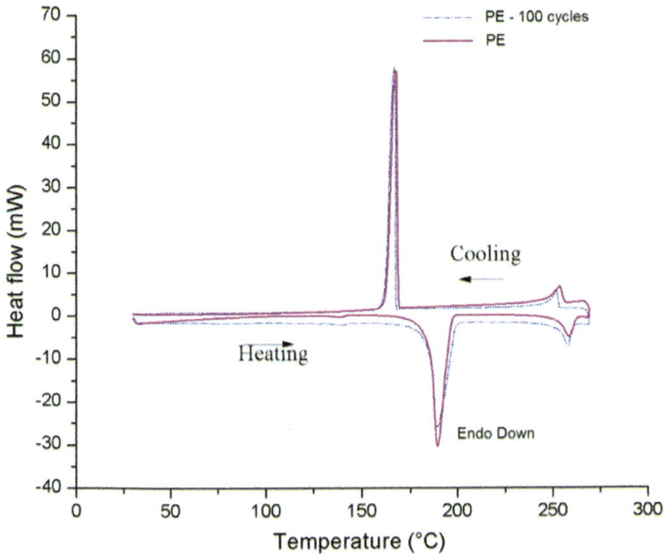

Fig. 5 DSC graphs of PE before and after 100 cycles

4.2 Phase Change Temperature and Enthalpy

Differential scanning calorimetry (DSC) is a widely used method to determine the phase change temperatures and enthalpies of PCMs. Figure 5 shows the DSC curve plotted for pure PE before and after thermal cycles. During heating, the onset of solid–solid transition temperature is observed at 181.35 °C with a peak at 187.82 °C. The enthalpy of transition was calculated to be 250.98 J/g.

This is attributed to the change of crystalline structure of PE from a body-centred tetrahedral molecular structure (α-phase) to a homogenous face-centred cubic crystalline structure (γ-phase) accompanied with hydrogen bond absorption energy [9]. The onset of melting was obtained at 258.13 °C with a peak melting temperature at 261.11 °C. The latent heat of melting was found to be 42.7 J/g.

During the cooling cycle, the onset temperature for solidification was at 247.51 °C with a peak temperature at 246.16 °C. The onset temperature for solid–solid transition during cooling was seen at 169.12 °C with peak value found at 168.56 °C. The enthalpy change during solid–solid transition during the discharge cycle was found at 238.92 J/g. After 100 thermal cycles, the onset and peak temperatures for the solid–solid transition were found to be at 186.31 and 190.96 °C with an enthalpy change of 235.98 J/g. Solid–liquid phase change was observed to begin at 253.61 °C with a peak temperature at 258.93 °C. The latent heat of fusion was obtained to be 39.08 J/g. During the discharging cycle, the onset of solid–solid transition took place at 168.72 °C with a peak temperature of 167.27 °C and a release of latent heat enthalpy of 226.6 J/g.

Figure 6 represents the DSC graphs of PE + 1 wt% GnP. During the heating cycle, the peak of solid to solid phase transition was found to be having an onset at 182.12 °C and a peak temperature of 183.69 °C. The enthalpy change during this transition was estimated as 242.31 J/g. The solid–liquid phase change occurred at 257.45 °C (onset) and 259.23 °C (peak) temperatures with a phase change enthalpy of 37.55 J/g. During solidification, the onset occurred at 249.22 °C and peak at 247.88 °C with the release of 21.22 J/g. The peak corresponding to solid–solid transition was observed at 169.33 °C with an onset temperature of 169.82 °C and phase change enthalpy of 221.35 J/g. DSC results obtained by testing the PE + 1 wt% GnP subjected to 100 thermal cycles are also shown in Fig. 6. During heating cycles, the onset and peak temperatures were found at 186.45 °C, respectively, and 189.56 °C with solid–solid phase change enthalpy at 229.16 J/g. The solid to liquid phase change temperature occurred at 255.32 °C with a peak temperature of 260.32 °C with the latent heat of solidification of 14.36 J/g. The solid to solid transition during the discharging process started at 168.78 °C with peak temperature value of 169.77 °C. The heat released during the cooling cycle was estimated at 215.35 J/g.

Figure 7 shows the DSC curves for DM before and after 100 cycles. During the heating cycle, the onset and peak temperatures were found to be at 166.32 °C and 168.38 °C, respectively, with the latent heat of fusion of 281.89 J/g. During the cooling cycle, the solidification started at 120.32 °C with the peak of solidification at 118.55 °C, and release of enthalpy of solidification is 219.52 J/g. DM after 100 thermal cycles is also shown in Fig. 7. The results show that during the heating

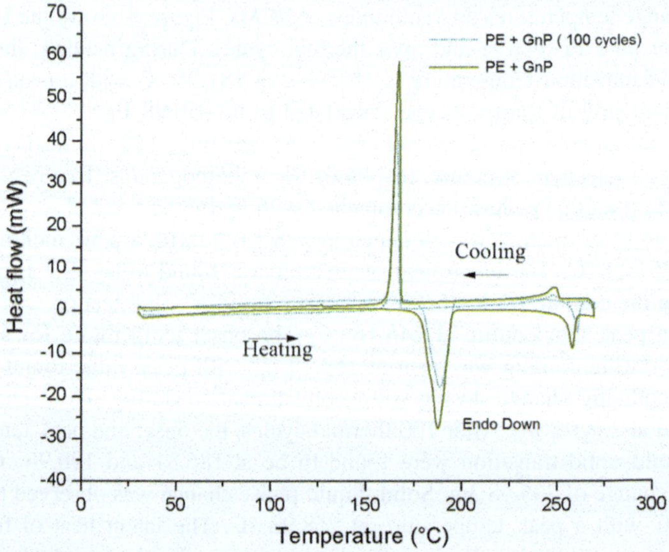

Fig. 6 DSC graphs of PE + GnP (1 wt%) before and after 100 cycles

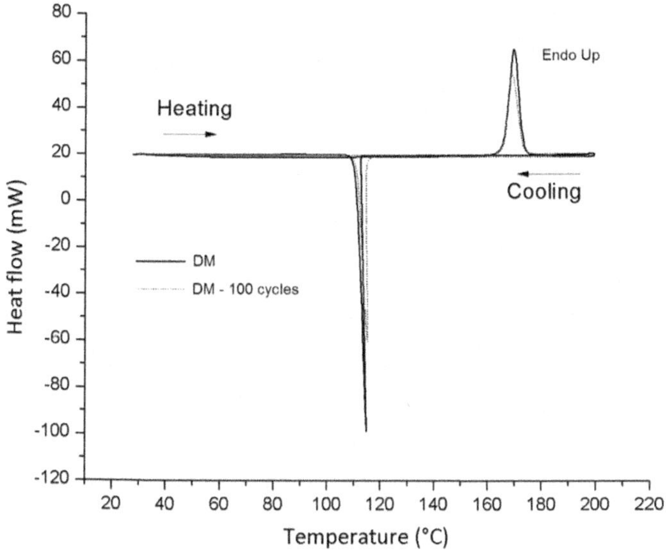

Fig. 7 DSC graphs of DM before and after 100 cycles

cycle, the onset and peak temperatures were found to be at 166.28 °C and 167.87 °C, respectively, with a latent heat of fusion of 240.56 J/g. During the cooling cycle, the peak temperature was found to be at 118.78 °C with a release of heat of 206.88 J/g.

Figure 8 shows the DSC curves of DM + 1 wt% GnP. During the heating cycle, the onset and peak temperatures were found to be 166.72 °C and 167.55 °C, respectively, with a change of enthalpy of 238.43 J/g. During the solidification cycle, the onset and peak temperatures were found to be at 118.7 and 117.89 °C with the release of 208.34 J/g of the heat of solidification. After 100 cycles, DM + GnP composite showed the onset and peak temperatures during heating at 166.45 and 167.32 °C with the latent heat of fusion at 195.33 J/g. During the solidification cycle, the onset and peak temperatures were found to be at 116.67 and 115.76 °C with the latent heat of solidification released found to be 158.33 J/g. Table 1 summarises the DSC analysis results.

From the DSC results summarised, it can be seen that after the thermal cycles, both PE and DM have resulted in a decrease in enthalpies, i.e. enthalpy of solid–solid transition in case of PE and enthalpy of solid–liquid phase change in case of DM. Addition of GnP particles showed a slight decrease in the latent heat value, and latent heat further decreased after the thermal cycles. The decrease in latent heat is due to non-participation of additives in transition processes. Even with a decrease in the latent heat value, both PE and DM composite showed high latent heat capacity of more than 170 J/g which indicates that both PCMs could be used for potential use in latent heat storage system.

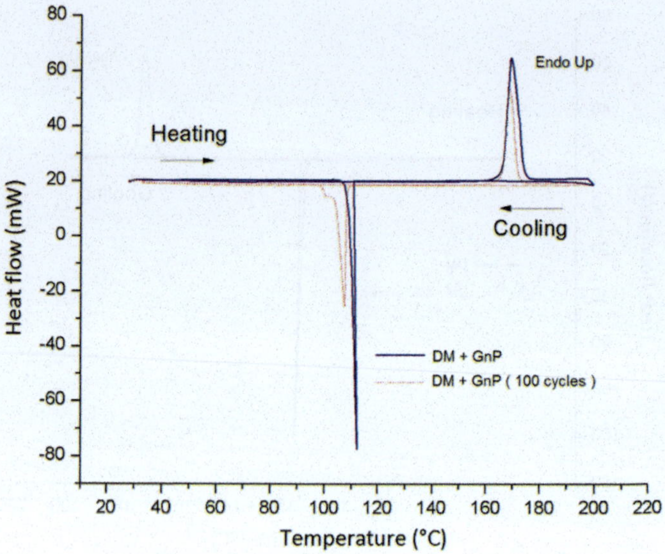

Fig. 8 DSC graphs of DM + GnP before and after 100 cycles

Table 1 DSC data for PE, PE + GnP, DM and DM + GnP before and after thermal cycles

Sample	Cycles	Heating			Cooling		
		Onset (°C)	Transition (°C)	Enthalpy change (J/g)	Onset (°C)	Transition (°C)	Enthalpy change (J/g)
PE	0	181.35	187.82	250.98	169.12	168.58	238.92
	100	186.31	190.96	235.98	168.72	167.27	226.96
PE + GnP	0	182.12	187.69	242.31	169.92	169.35	221.35
	100	186.45	189.56	229.16	168.78	169.77	215.35
DM	0	166.32	168.38	281.89	120.32	118.55	219.52
	100	166.28	167.87	240.56	118.78	117.30	206.88
DM + GnP	0	166.72	167.55	238.43	118.71	117.89	208.34
	100	166.45	167.32	195.33	116.67	115.76	158.33

4.3 Chemical Compatibility Studies

The chemical stability of the composite is analysed using FTIR techniques. Figure 9 shows the FTIR spectra of PE before and after thermal cycles. In the case of pure PE, a broad peak due to O–H bond stretching vibration is found between the wavenumber 3600 and 2800 cm^{-1}. The C–H anti-symmetric stretching vibrations are observed at the wavenumber of 2940 cm^{-1}. The 2884 cm^{-1} peak represents the –CH2 symmetric stretching vibration and a double peak at 1400 and 1500 cm^{-1} are due to the –CH2 and –CH3 bending vibrations. The peaks due to the C–O bond

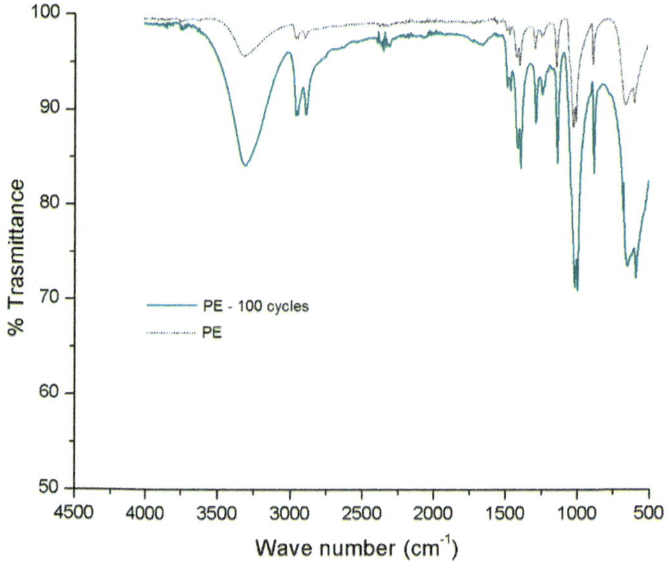

Fig. 9 FTIR graphs of PE before and after 100 cycles

stretching vibration appear at 1301 and 1011 cm^{-1}. FTIR samples of PE after thermal cycles are shown in Fig. 9. It is evident from the figure that no new peaks have been found as compared to the spectra of pure PE. Figure 10 shows the FTIR spectra of GnP particles. From the spectra, it is clear that GnP does not exhibit major peaks since it has a hexagonal lattice structure on the atomic scale and is only made of carbon atoms. However, minor peaks are observed in the range of wavenumber of 1500–1750 cm^{-1}. This can be attributed to C=C (aromatics) bond vibrations observed at 1611 cm^{-1} (weak and multiple bands). Multiple weak absorption bands in the range of 2000–2500 cm^{-1}, which represents the conjugate (duplet) C=C bonds. Figure 11 shows the FTIR spectra of PE + 1 wt% GnP before and after 100 thermal cycles. As seen with pure PE samples, similar peaks are seen even in spectra of PE + GnP with a slight change of transmittance value. This change could be attributed to the presence of GnP in PE having only physical interaction between them. This means that there was no chemical reaction between added GnP particles and the base PE.

Figures 12 and 13 show the FTIR spectra of DM and DM + 1 wt% GnP before and after 100 cycles. For pure DM samples, O–H stretching vibration peaks were found at 3300 cm^{-1}. Peaks obtained at 2939 cm^{-1} were assigned to a stretching frequency of C–H functional group. The peaks obtained due to the stretching vibration of C–O of alcohols and phenols functional class were found at 1463 and 1382 cm^{-1}. Peaks observed at 942, 876 and 723 cm^{-1} correspond to O–H bond bending, C–H bond bending and O–H (out of plane for alcohols). FTIR spectra of thermally cycled DM also showed similar peaks with a change in percentage

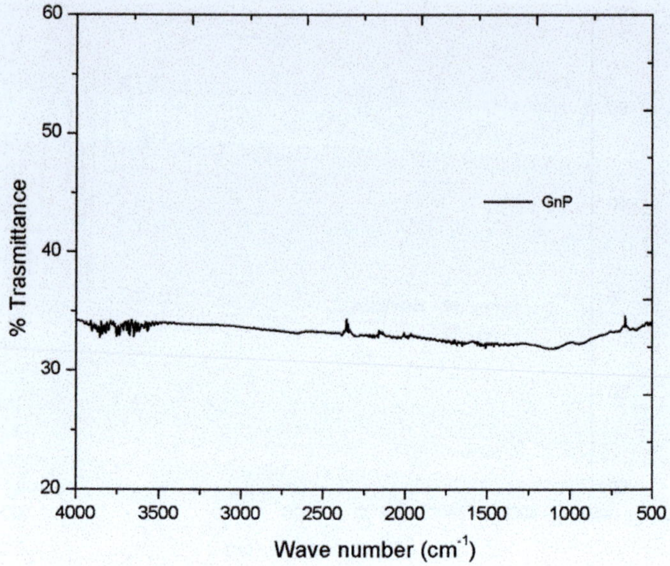

Fig. 10 FTIR graphs of graphene nanoplatelets (GnPs)

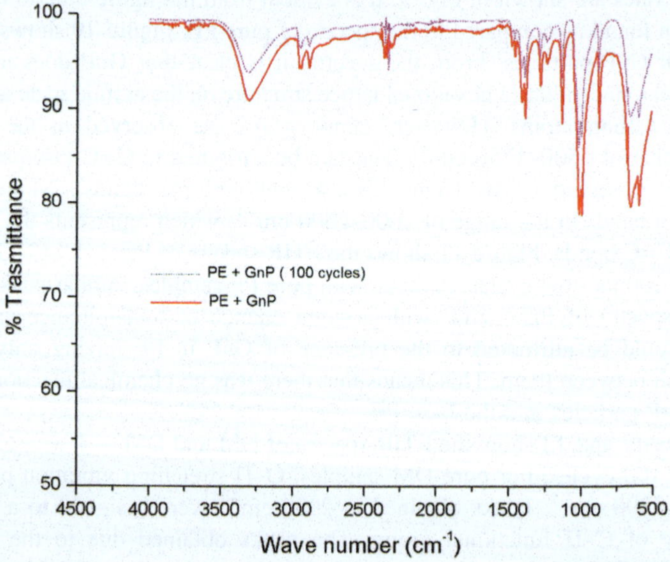

Fig. 11 FTIR graphs of PE + GnP before and after 100 cycles

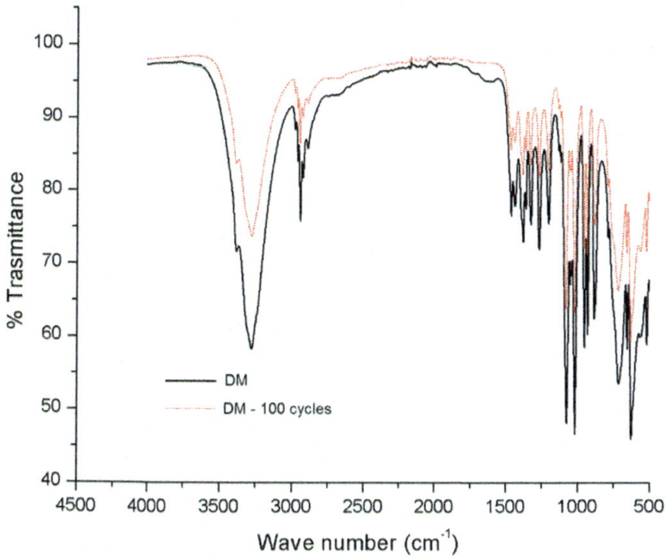

Fig. 12 FTIR graphs of DM before and after 100 cycles

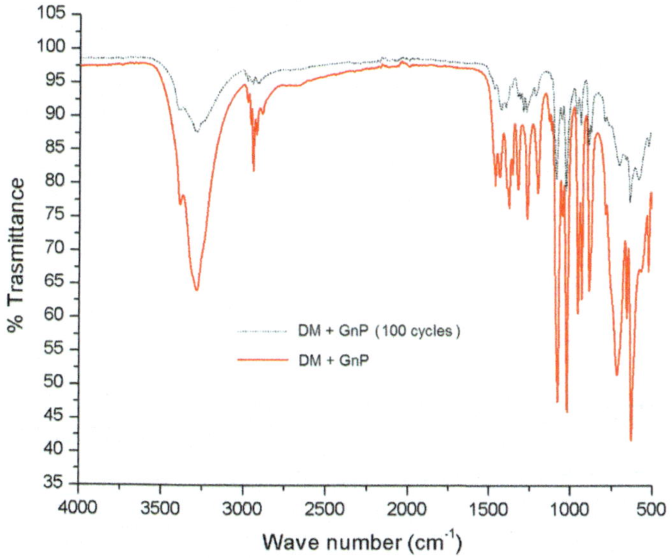

Fig. 13 FTIR graphs of DM + GnP before and after 100 cycles

transmittance value. Figure 13 shows the FTIR spectra of DM-1 wt% GnP before and after thermal cycles. As compared to spectra of pure DM, DM-GnP showed similar peaks with a change in transmittance values. This suggests that the prepared

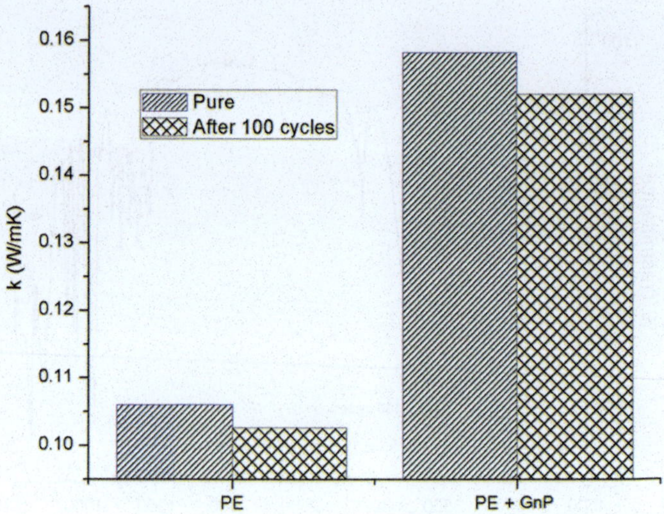

Fig. 14 Thermal conductivity value of PE and PE-GnP before and after 100 cycles

composite was chemically stable and there was no chemical reaction between the added GnP particles and DM. The composites exhibited high chemical stability even after 100 thermal cycles. The change in transmittance values suggests the possible dehydration (evaporation of moisture) of PCM samples after repeated heating and cooling cycles. Both composites of PE and DM with GnP exhibited good chemical stability before and after thermal cycles. This means that there is no bond valence between PE, DM and GnP particles.

Thus, it can be concluded that the prepared composite is very stable chemically and could be used as an energy storage material in LHES.

4.4 Thermal Conductivity Measurements

Figure 14 shows the thermal conductivity of PE and PE + 1 wt% GnP before and after thermal cycles. It is evident from the figure that the thermal conductivity of PE increased with an increase in the weight fraction of the GnP particles. The thermal conductivity value of pure PE was obtained as 0.106 W/m K. The thermal conductivity increased to 0.158 W/m K with the addition of 1 wt% GnP particles. After 100 thermal cycles, the thermal conductivity of PE and PE + 1 wt% GnP was obtained as 0.1026 and 0.1521 W/m K.

Figure 15 shows the thermal conductivity value of DM and DM-GnP before and after 100 cycles. The thermal conductivity of pure DM was obtained as 1.303 W/m K, whereas the thermal conductivity increased to 2.13 W/m K for DM-GnP composite with 1 wt% composite. After 100 cycles, the thermal conductivities of DM

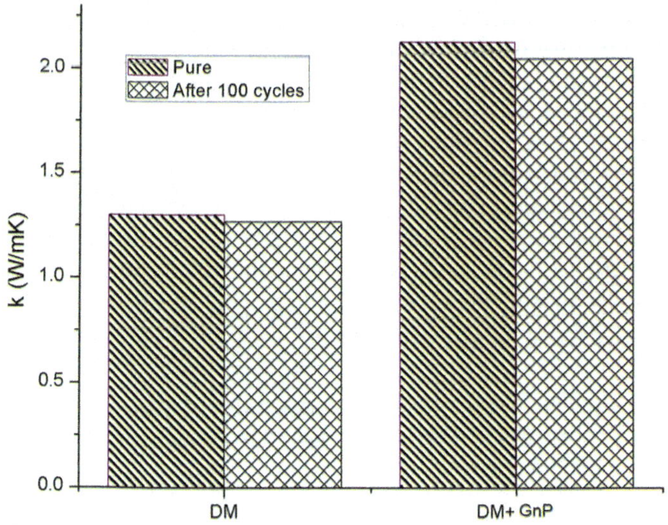

Fig. 15 Thermal conductivity value of DM and DM-GnP before and after 100 cycles

and DM-GnP slightly reduced to 1.267 W/m K and 2.046 W/m K, respectively. The reasons for the increase in thermal conductivity with increase in the mass fraction of nanoadditives could be attributed to the increase in inter-particulate diffusion, heat exchange potential and clustering of nanocomposites.

5 Conclusion

The characterisation of pentaerythritol (PE) and D-Mannitol (DM) added with graphene nanoplatelets (GnPs) has been carried out to determine the thermal and chemical stability of the prepared composites. GnP with 1 wt% is added to both PE and DM, and its thermo-physical properties were determined. A thermal cycling study was done to test the durability of the composites, and the following observations were made from the results:

(a) TGA analysis of DM and PE with GnP showed a noticeable change in material thermal behaviour due to thermal cycling at elevated temperature. Both DM and PE showed a negligible weight loss of less than 2% at the temperature of around 220 °C even after thermal cycling which shows that the composite is very stable in the temperature range of 50–200 °C.
(b) The FTIR analysis of DM and PE with GnP composite before and after thermal cycling showed that addition of GnP did not produce any new peaks in the IR spectra. This means that there was no chemical reaction between the base PCM

and added GnP particles. After thermal cycling, the composites were stable, which shows that there is no chemical degradation of the prepared composites.

(c) DSC results showed that the addition of GnP in DM and PE decreases the latent heat of transition and phase change slightly. Addition of GnP particles had no effect on the phase transition temperature of the PCM. Even after 100 thermal cycles, the composite exhibited very high enthalpies above 170 J/g, which makes it a very good potential as energy storage materials.

(d) The thermal conductivity of PE and DM increased by 49.1 and 63.4% with the addition of 1 wt% GnP in the base PCM. After 100 thermal cycles, the thermal conductivity of PE, PE + GnP, DM and DM + GnP decreased by 3.21%, 3.91%, 2.72% and 3.94%, respectively.

Acknowledgements The authors wish to thank Department of Science and Technology (DST Sanction letter No. DST/TM/SERI/DSS/275(G) dated 9 September 2015), Government of India, for its financial support to this work.

References

1. Abhat A (1983) Low temperature latent heat thermal energy storage: heat storage materials. Sol Energy 30(4):313–332
2. Sharma A, Chen CR (2009) Solar water heating system with phase change materials. Int Rev Chem Eng 1(4):297–307
3. Kenisarin M, Mahkamov K (2007) Solar energy storage using phase change materials. Renew Sustain Energy Rev 11(9):1913–1965
4. Gunasekara SN, Pan R, Chiu JN, Martin V (2016) Polyols as phase change materials for surplus thermal energy storage. Appl Energy 15(162):1439–1452
5. Aran Sole A, Neumann H, Niedermaier S, Martorell I, Schossig P, Cabeza LF (2014) Stability of sugar alcohols as PCM for thermal energy storage. Sol Energy Mate Sol Cells 126:125–134
6. Kumaresan G, Velraj R, Iniyan S (2011) Thermal analysis of d-mannitol for use as phase change material for latent heat storage. J Appl Sci 11:3044–3048
7. Paul A, Shi L, Bielawski CW (2015) A eutectic mixture of galactitol and mannitol as a phase change material for latent heat storage. Energy Convers Manag 31(103):139–146
8. Benson DK, Webb JD, Burrows RW, McFadden JDO, Christensen C (1985) Materials research for passive solar systems: solid-state phase change materials. Sol Energy Res Inst SERI/TR-255-1828
9. Wang XW, Lu ER, Lin WX, Liu T, Shi ZS, Tang RS, Wang CZ (2000) Heat storage performance of the binary systems neopentyl glycol/pentaerythritol and neopentyl glycol/trihydroxy methyl-aminomethane as solid-solid phase change materials. Energy Convers Manag 41:129–134
10. Xing DQ, Chi GS, Ruan DS, He LF, Li DH, Zhang TP, Zhang DS (1995) Solid state phase transition in binary systems of polydric alcohols. Sol Energy 16:133–137
11. Benson DK, Burrows RW, Webb JD (1986) Solid state phase transitions in pentaerythritol and related polyhydric alcohols. Sol. Energy Mater 13:133–152
12. Gao WF, Lin WX, Liu T, Xia CF (2007) An experimental study on the heat storage performances of polyalcohols NPG, TAM, PE, and AMPD and their mixtures as solid–solid phase-change materials for solar energy applications. Int J Green Energy 4:301–311

13. Barrio M, Font J, Muntasell J, Navarro J, Tamarit JL (1988) Applicability for heat storage of binary systems of neopentylglycol, pentaglycerine and pentaerythritol: a comparative analysis. Sol Energy Mater 18
14. Kibria MA, Anisur MR, Mahfuz MH, Saidur R, Metselaar IH (2015) A review on thermophysical properties of nanoparticle dispersed phase change materials. Energy Convers Manag 1(95):69–89

Part V
Various Applications of Solar Energy: Cooling, Cooking, Efficient Buildings

Water–Lithium Bromide Absorption Chillers for Solar Cooling

Ashok Verma, Satish and Prodyut R. Chakraborty

Abstract Solar thermal resources can be effectively utilized to meet the refrigeration and air-conditioning demands for both household and industrial purposes. Considerable fraction of total available electricity is consumed by the conventional vapor compression refrigeration systems (VCRS) during the summer season in countries with tropical climate. The leakage of VCRS refrigerants in the atmosphere has also been identified as one of the major contributors toward ozone layer depletion and hence global warming. The utilization of solar thermal energy for obtaining refrigeration and air-conditioning is the key to address these issues concerning high electricity demand as well as the environmental pollution. Solar thermal energy, being one of the leading resources of green energy, can reduce the carbon footprint considerably, when used for sorption cooling process. The advantages of using sorption cooling systems powered by solar thermal energy over VCRS are twofold when we consider environmental issues. Sorption-based solar thermal cooling reduces the electricity demand for cooling to a large extent, which in turn reduces usage of fossil fuels to produce this electricity, and thus leads to low-carbon footprint. Also, the refrigerants used for sorption cooling are less prone to cause ozone layer depletion. Although sorption-based refrigeration systems driven by solar thermal energy are mature technologies, wide acceptability of such cooling system is yet to be achieved. Two major limitations of sorption-based solar thermal cooling are relatively low coefficient of performance (COP), and large volume requirement. Other than these two limitations, the intermittent nature of solar thermal resource and heat exchanger and control mechanism design complicacies also pose considerable challenge. Sorption cooling technology can be broadly classified base on absorption and adsorption processes. Absorption is a volumetric phenomenon where a substance of one state gets absorbed in another substance in a different state with or without having chemical reaction, such as liquid being absorbed by solid or gas being absorbed by liquid. On the other hand, adsorption is a surface phenomenon due to physical bonding forces such as Van der walls forces between a solid surface and adjacent fluid or due to chemical bonding

A. Verma · Satish · P. R. Chakraborty (✉)
Indian Institute of Technology Jodhpur, Jodhpur, India
e-mail: pchakraborty@iitj.ac.in

H. Tyagi et al. (eds.), *Applications of Solar Energy*, Energy, Environment, and Sustainability, https://doi.org/10.1007/978-981-10-7206-2_11

between the two. The discussion in this chapter is attributed to Water–Lithium Bromide-based absorption cooling systems. The discussion emphasizes on fundamental concepts of absorption refrigeration cycle, starting with simplest intermittent vapor absorption refrigeration system and gradually elaborating toward the operating principles of commercially used chillers at the end. Cycle analysis of most commonly used single-effect absorption chillers is discussed in a detailed manner along with the background knowledge of how to determine pertinent thermodynamic properties at the inlet and outlet of individual components. Finally, methods and design criteria that can improve the system performance are discussed.

1 Absorption Cooling

Vapor absorption refrigeration systems which are also commercially known as VARSs, are a vapor cycle-based refrigeration system quite similar to more popular vapor compression refrigeration system (VCRS) in operating terms. However, the major difference between these two systems is demarcated by the nature of obtaining high-pressure vapor for the condenser. While VCRS uses a mechanical compressor requiring electrical energy input, the VARS utilizes thermal energy input to achieve the same effect of pressurizing the refrigerant vapor to high condensation pressure. Although not as popular as its counterpart VCRS, VARS has also been commercialized for many applications pertaining to refrigeration and air-conditioning. From the economic aspect, absorption refrigeration is particularly suitable to deploy when there is a source of inexpensive heat energy at a temperature range of 100–200 °C. Such low-grade thermal energy can be easily harnessed from waste heat or solar energy. Also, the usage of environment-friendly refrigerants such as water and ammonia makes VARS very attractive from the environmental point of view.

The most commonly used absorption cooling systems use water (H_2O)-Lithium Bromide (LiBr) and ammonia (NH_3)-water (H_2O) refrigerant–absorbent pair. For H_2O–LiBr-based absorption cooling systems, H_2O is used as refrigerant and LiBr plays the role of absorbent. As water acts as refrigerant, the major limitation of H_2O–LiBr-based absorption cooling system is incapability of such system to provide refrigeration at sub-zero temperature. However, the generation of pure water vapor from the heated H_2O–LiBr renders the system analysis and design much more simpler as compared to NH_3-H_2O (NH_3 as refrigerant and H_2O as absorbent) solution-based cooling systems. The vapor obtained from heated NH_3-H_2O solution contains mixture of gaseous ammonia and water vapor and additional rectifier is required to separate water vapor from the refrigerant (Ammonia). The limitation of H_2O–LiBr absorption cooling system below 0 °C restricts its utilization only for the purpose of air-conditioning (above 0 °C). On the other hand, NH_3-H_2O-based cooling systems can be considered as more versatile as compared to H_2O–LiBr-based systems, as they can be utilized for both sub-zero (refrigeration) as well as above 0 °C (air-conditioning) applications.

Since H_2O–LiBr absorption cooling systems are easy to analyze, the first few subsections on absorption cooling will be devoted to H_2O–LiBr systems before moving on to more complicated NH_3-H_2O-based cooling systems.

1.1 Working Principles of Basic Water–Lithium Bromide-Based Absorption Cooling

Dissolving Lithium Bromide salt (solute) in water (solvent) causes reduction of the vapor pressure of water in the solution below that of the saturation pressure corresponding to pure water at that same temperature. Since the difference between boiling point between LiBr and H_2O is very large [1, 2] (> 1000 °C at atmospheric pressure), the vapor obtained upon heating the H_2O–LiBr solution at the temperature range of 60 – 200 °C (typical generation temperature range) contains pure water vapor and the vapor pressure can be considered to be almost equal to the vapor pressure of water.

The simplest absorption refrigeration system is designed on the basis of this decrement of vapor pressure of the solvent in the solution as compared to the pure solvent. Figure 1 shows such a simple design of a refrigeration system, where two vessels are connected through a valve with one vessel (vessel 1) containing pure water and the other (vessel 2) containing a rich solution of H_2O–LiBr. Initially, the valve is closed (Fig. 1a) and both the vessels are in thermal equilibrium with the ambient (i.e., temperature at both the vessels are equal). However, the vapor pressure of water in the vessel 2 containing rich solution of H_2O–LiBr is much lower than the vapor pressure $p_s(T_\propto)$ of pure water in vessel 1. If the valve is now opened (Fig. 1b), the water vapor from the vessel 1 containing pure water will flow to the vessel 2 containing rich H_2O–LiBr solution due to the pressure difference and get absorbed in the solution. The flow of vapor from vessel 1 will cause gradual reduction of vapor pressure in vessel 1, causing farther evaporation to proceed, while temperature of the pure water keeps decreasing and the heat of evaporation is absorbed from the surrounding of vessel 1. The dotted envelop surrounding vessel 1 represents the domain from where heat is removed and cooling effect is obtained. On the other hand, absorption of water vapor by the H_2O–LiBr solution in vessel 2 is an exothermic process (details of this exothermic absorption will be discussed in later sections). If the heat generated from this exothermic process is not removed from vessel 2, the solution temperature will rise rapidly, causing rapid elevation of vapor pressure of water in the H_2O–LiBr solution. As a result, the vapor pressure of the two vessels will equilibrate very fast and the cooling effect will cease to proceed farther. That is why heat needs to be removed continuously to the atmosphere from vessel 2 during the process in order to sustain cooling effect. The cooling process of the surrounding of vessel 1 will continue till the vapor pressure in vessel 1 and 2 comes to equilibrium. During the process, vapor pressure in vessel 1 will gradually decrease while depleting the mass of water and bringing down the temperature of pure water in the vessel 1. On the other hand, the vapor pressure in vessel 2 will gradually increase and the absorption of water vapor by the H_2O–LiBr solution will make the solution leaner in LiBr composition. Once the vapor pressures in the two vessels come to equilibrium, no farther cooling can be attained. In order to start the cooling process once more, the system needs to be restored to its initial condition described by Fig. 1a. The process of restoring the system to its initial state is called the regeneration. The regeneration process (Fig. 1c) is the reverse of the cooling process involving heating of the weak H_2O–LiBr solution in vessel 2 to a high temperature (T_G) which allows desorption or liberation of water vapor from the solution making it richer in LiBr composition once again, while the liberated water vapor gets condensed in vessel 1 by rejecting heat to the surrounding. The regeneration process continues till the depleted amount of water is restored in vessel 1. Once the depleted water in vessel 1 is restored, the valve is closed and heat

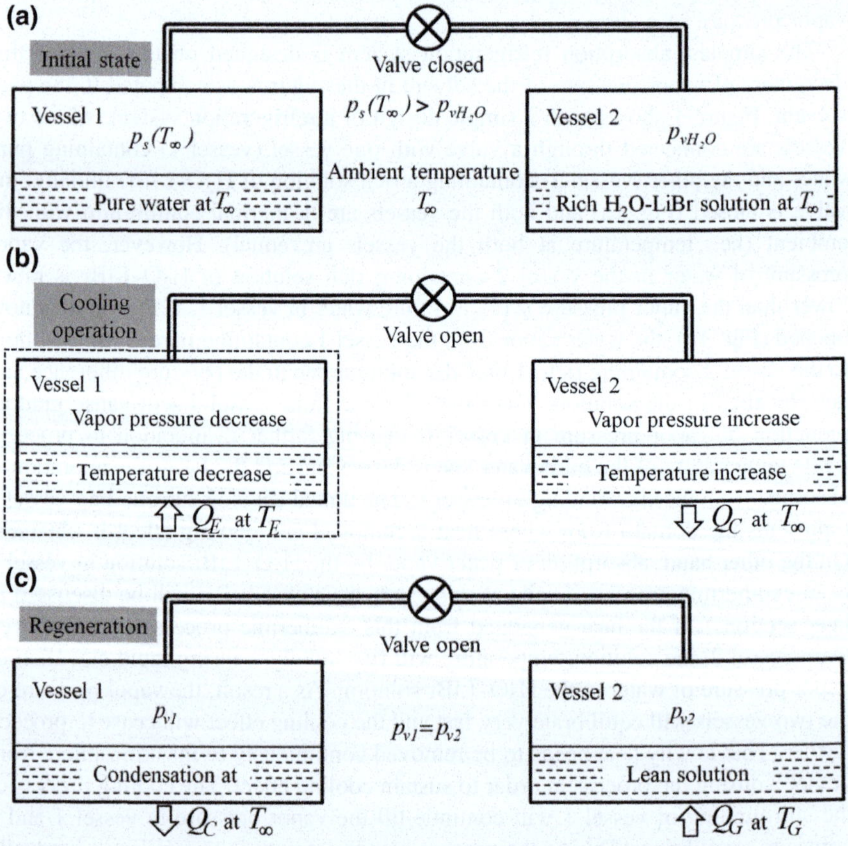

Fig. 1 Simple absorption cooling mechanism (**a**) initial state, (**b**) cooling process, (**c**) regeneration process

supply to vessel 2 is stopped, and both the vessels are allowed to attain thermal equilibrium with the ambient to restore the initial state described by Fig. 1a.

The described system is evidently an intermittent type absorption cooling system. This simple system although intermittent in nature is capable of providing nighttime cooling in remote rural locations devoid of sufficient electricity supply by using solar energy only. While the regeneration process can be attained during daytime by means of supplying regeneration heat from solar energy, the cooling effect can be obtained at night. The simple design of the system allows it to be inexpensive, user-friendly, easy to maintenance operation. The duration of cooling, however, depends on the size of the system; the larger the size of the system, the longer is the cooling duration. Although the described system has many advantages from the perspective of simplicity and economy, the inherent intermittency renders such system to be inadequate when the application demands continuous cooling. The design and operation of such continuous-type H_2O–LiBr cooling system is described in the next section.

1.2 Working Principles of Water–Lithium Bromide Absorption Cooling System for Continuous Operation

In order to obtain continuous operation with H_2O–LiBr absorption cooling system certain modifications of the simple system described in the previous section are required. The modification involves additional solution pump and expansion valves along with separate condenser and evaporators connected with two vessels containing rich and lean solution of H_2O–LiBr (Fig. 2). The two vessels containing H_2O–LiBr are called generator and absorber, respectively. As shown in Fig. 2, the

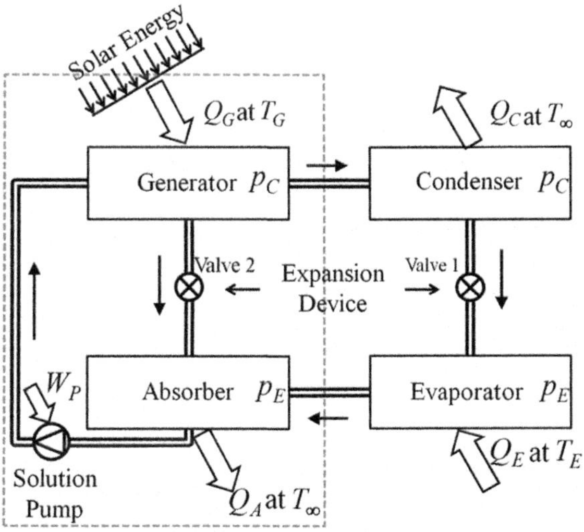

Fig. 2 H_2O–LiBr vapor absorption refrigeration system for continuous operation

generator subjected to heat input ($\mathbf{Q_G}$) at a higher temperature ($\mathbf{T_G}$) and the solution liberates water vapor at high pressure ($\mathbf{p_C}$). The input heat ($\mathbf{Q_G}$) source can be waste heat from some other device or heat obtained from solar energy. In the process of producing high-pressure vapor in the generator, the liquid solution becomes rich in LiBr composition. The high-pressure high-temperature water vapor moves to the condenser, liberates heat to the atmosphere at high condenser pressure ($\mathbf{p_C}$), and condenses to liquid water at high pressure. The high-pressure liquid water produced at the condenser is then expanded through an expansion valve or throttling device (valve 1) to bring down the pressure to evaporator pressure ($\mathbf{p_E}$), so that the water entering the evaporator is either in saturated liquid state or in wet vapor (with low vapor quality) state. In the evaporator, the saturated liquid water or the wet vapor evaporates completely by absorbing heat from the immediate surrounding designated for desired cooling effect and maintained at a low temperature ($\mathbf{T_E}$). The saturated or superheated water vapor generated at the evaporator enters the absorber maintained at same pressure as the evaporator. The LiBr-rich liquid solution produced in the generator is expanded through another throttling device (valve 2) to bring the solution to the absorber pressure ($\mathbf{p_E}$) which is equal to the evaporator pressure. In the absorber, LiBr-rich liquid solution coming from the generator absorbs the saturated or superheated water vapor coming out from the evaporator producing a LiBr lean liquid solution. Since this absorption process in the absorber is an exothermic process, the generated heat ($\mathbf{Q_A}$) is removed from the absorber to the ambient. Finally, the LiBr lean liquid solution produced at the absorber is sent back to the generator operating at condensation pressure ($\mathbf{p_C}$). It is pertinent to mention here that apart from the generation and absorption pair connected by the expansion valve 2 and the solution pump, rest of the refrigeration cycle resembles exactly with the conventional vapor compression refrigeration system. If the system inside the dotted envelop in Fig. 2 is replaced by a vapor compressor, we simply obtain the conventional vapor compression refrigeration system (VCRS). The major difference between the conventional VCRS with the absorption refrigeration system (VARS) described in this section is the different modes of power input to achieve refrigeration effect. While the VCRS requires electrical power input to operate the compressor in order to pressurize the refrigerant vapor from evaporator pressure to condenser pressure, the same is achieved in VARS majorly through the input thermal power to the generator. Although the solution pump requires certain amount of electrical power input ($\mathbf{W_P}$) to pressurize the solution from absorber pressure ($\mathbf{p_E}$) to generator pressure ($\mathbf{p_C}$), it is negligible as compared to the power requirement of the compressor ($\mathbf{W_C} \gg \mathbf{W_P}$) in a VCRS system. The ideal work input per unit mass of fluid flow through compressor and pump is defined by $\int_{\mathbf{p_E}}^{\mathbf{p_C}} -\mathbf{v}\mathbf{dp}$. Since the specific volume ($\mathbf{v_g}$) of gas or vapor is several order higher than the specific volume of liquid ($\mathbf{v_l}$), the required work input to compressor to raise the pressure of the gas or vapor is several order higher than the work input to a pump to pressurize liquid to the same level. Hence, the work input to VARS solution pump can be considered to be negligible as compared to the thermal energy input at the generator.

1.2.1 Coefficient of Performance (COP) and Maximum Obtainable COP

The coefficient of performance of the absorption cooling system discussed in the foregoing discussion can be obtained as the ratio of heat removal at the evaporator (Q_E) to the sum of net input energy ($Q_G + W_P$).

$$COP = \frac{Q_E}{Q_G + W_P} \quad (1.2.1)$$

Since the thermal energy supplied at the generator is much larger than that of the work input to the solution pump ($Q_G \gg W_P$), we can ignore the pump work while calculating the *COP* of absorption refrigeration system. The *COP* can be quite reasonably expressed as:

$$COP = \frac{Q_E}{Q_G} \quad (1.2.2)$$

It is pertinent to mention here that for prescribed values of condensation and evaporation temperatures, *COP* of a VCRS system is much larger than that of *COP* of a VARS system ($COP_{VCRS} \gg COP_{VARS}$). The reason being utilization of high-grade mechanical energy by the VCRS as compared to the low-grade thermal energy is used by the VARS. However, comparison of VARS and VCRS performances on basis of their *COP* s is not really an ideal way of doing so as thermal energy is comparatively less expensive as compared to the mechanical energy. An ideal way of comparing the two would be to compare their second law efficiency obtained from exergy analysis. The second law efficiencies obtained for these two systems are found out to be of the same order of magnitude.

In order to find the maximum obtainable *COP* for the VARS, we first identify from Fig. 2 that the exchange of thermal energy between the system and surrounding occurs at generator, condenser, evaporator, and absorber. At the generator and the evaporator, the net input to the system occurs from the surroundings maintained at T_G and T_E, respectively, while at the condenser and the absorber, heat is rejected to the surrounding maintained at ambient temperature T_∞. So it is evident that the VARS operates between three temperature levels of the surrounding, namely: T_G, T_∞, and T_E. We will apply first and second law of thermodynamics to find out maximum obtainable *COP* for the VARS operating between these three temperature levels. Applying first law of thermodynamics to the system shown in Fig. 2, we obtain

$$Q_G + Q_E - Q_C - Q_A + W_P = 0 \tag{1.2.3}$$

Applying second law of thermodynamics $\Delta S_{Uni} = \Delta S_{Sys} + \Delta S_{Sur} \geq 0$, and observing VARS operating in a closed cycle for which change in entropy for the system can be considered as $\Delta S_{Sys} = 0$, we can write:

$$\Delta S_{Uni} = \Delta S_{Sur} = -\frac{Q_G}{T_G} - \frac{Q_E}{T_E} + \frac{Q_C}{T_\infty} + \frac{Q_A}{T_\infty} \geq 0 \tag{1.2.4}$$

Substituting Eq. 1.2.3 in Eq. 1.2.4 and rearranging we obtain:

$$Q_G\left(\frac{T_G - T_\infty}{T_G}\right) \geq Q_E\left(\frac{T_\infty - T_E}{T_E}\right) - W_P \tag{1.2.5}$$

Neglecting solution pump work, W_P; the *COP* of VARS can be expressed as:

$$\frac{Q_E}{Q_G} \leq \left(\frac{T_G - T_\infty}{T_G}\right)\left(\frac{T_E}{T_\infty - T_E}\right) \tag{1.2.6}$$

For a completely reversible VARS without any internal or external irreversibility, the '$\leq$' sign in Eq. 1.2.6 can be replaced by an '$=$' sign. Hence, the maximum obtainable coefficient of performance for a VARS operating between three temperature levels T_G, T_∞, and T_E can aptly be represented as the *COP* of this ideal system which is given by:

$$COP_{Max} = COP_{Ideal} = \left(\frac{T_G - T_\infty}{T_G}\right)\left(\frac{T_E}{T_\infty - T_E}\right) = \eta_{Carnot} COP_{Carnot} \tag{1.2.7}$$

Since $(T_G - T_\infty)/T_G$ is efficiency of a Carnot engine operating between temperature levels T_G and T_∞, and $T_E/(T_\infty - T_E)$ is the *COP* of Carnot refrigerator operating between T_∞ and T_E, the ideal VARS can be described as a combination of a Carnot heat engine and a Carnot refrigerator (Fig. 3). And the maximum obtainable coefficient of performance (COP) of a VARS can be expressed as:

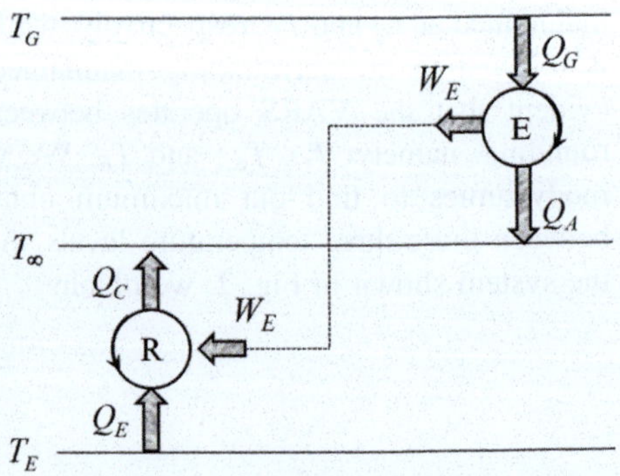

Fig. 3 Representation of ideal VARS as a combination of Carnot heat engine and Carnot refrigerator

$$COP_{Max} = COP_{Ideal} = [\eta_{Carnot}]_{T_\infty}^{T_G} [COP_{Carnot}]_{T_E}^{T_\infty} \tag{1.2.8}$$

The actual VARS, however, contains various external and internal irreversibility and the *COPs* of such systems are much less than that predicted by Eq. 1.2.8.

1.3 Steady Flow Analysis of Single-Effect Water–Lithium Bromide Systems

A single-effect H_2O–LiBr absorption refrigeration system is the simplest and most frequently utilized manifestation of absorption chilling technology. The only difference in the design from the basic design shown in Fig. 2 is being the addition of a solution heat exchanger between the absorber and the generator. A schematic of single-effect H_2O–LiBr absorption cooling system is presented in Fig. 4. The addition of the solution heat exchanger helps to decrease heat input to the generator. In the schematic shown in Fig. 4, all the major components of the single-effect VARS are shown along with the state points assigned to specific numbers at the inlet and outlet of these components. The energy transfers occurring at the system boundaries are shown with arrows pointing the directions of the transfer. Similar to the previously discussed absorption system in the previous section, there are four heat transfer ($\mathbf{Q_G}$, $\mathbf{Q_C}$, $\mathbf{Q_E}$, and $\mathbf{Q_A}$) and one work transfer ($\mathbf{W_P}$) term.

The schematic shown in Fig. 4 can be directly superimposed on a Dühring chart which shows solution temperature, pressure and composition variation at different components of the VARS (Fig. 5). Dühring chart is a very effective way of representing temperature versus the vapor pressure data of H_2O–LiBr solutions. In a Dühring chart, the solution temperature is linearly plotted along x-axis, the

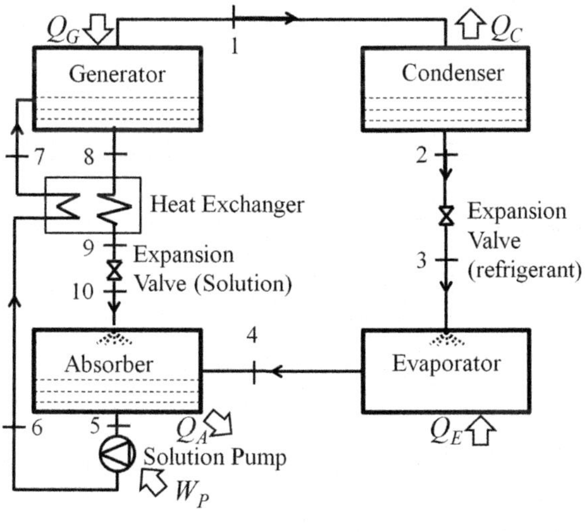

Fig. 4 Schematic of single-effect H_2O–LiBr absorption cooling system

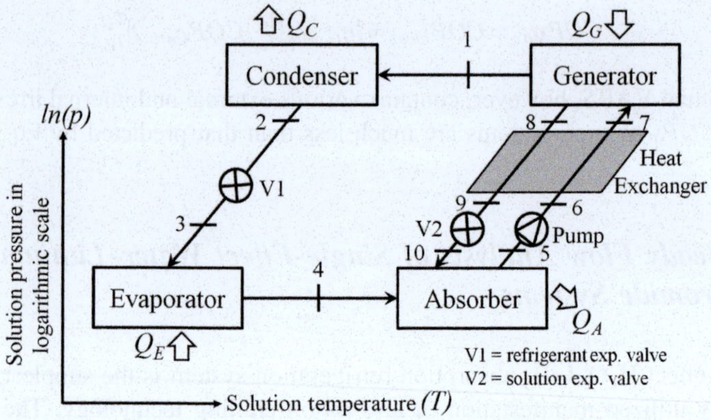

Fig. 5 Single-effect H_2O–LiBr absorption cooling system superimposed on Dühring chart

saturation temperature of pure water is plotted linearly along y-axis on the right-hand side and the equilibrium vapor pressure of the liquid solution is plotted along the left-hand side ordinate (y-direction) on a logarithmic scale. The vapor pressure and corresponding temperature valued are plotted along isosteric or constant solution composition lines. Figure 6 shows a typical single-effect H_2O–LiBr absorption system on a Dühring plot [3]. The relative position of the individual components involving liquid to vapor or vapor to liquid phase change in the representative schematic on Dühring chart (Fig. 5) indicates the relative temperature and pressure of the H_2O–LiBr solution inside those components [4]. It is pertinent to mention here that Dühring chart is not adequate to represent the subcooled liquid and superheated vapor states and are capable to display only saturated states.

In order to analyze the VARS cycle, thermodynamic states at the inlet and outlet of each device must be well comprehended. In order to obtain first-cut analysis before the design of the VARS, wherever possible, the liquid and vapor states are considered to be saturated. Such a summary of the state point description for the ten inlet and outlet conditions described in Figs. 4 and 5 are presented in Table 1. According to the listed table, three of the state points are saturated liquid (2, 5 and 8), two of them are saturated vapor (1 and 4), three of them are subcooled liquid (6, 7 and 9), and two of them are two-phase (liquid-vapor) state (3, 10). In real systems, the liquid streams may be in subcooled condition and the vapor steams may attain superheated condition. However, modeling them proves to be much more complicated as the superheated or subcooled conditions require more data. For the sake of simplicity, outlet conditions for each of the major components, i.e., generator, condenser, evaporator, and absorber are considered to be in saturated states.

Figure 6 represents the solution cycle on a representative Dühring plot. The saturated states (1–5, 8, 10) are shown on their state point location, and the subcooled states are located at their corresponding temperatures and composition. It is

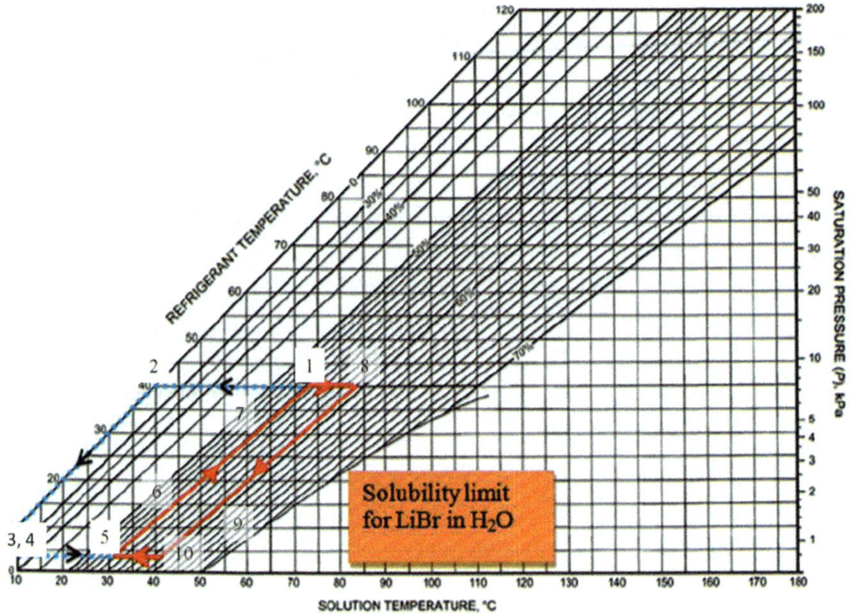

Fig. 6 Dühring plot for cycle solution [3, 4]

Table 1 State point description at the outlet of individual components

Point	Location	State	Remarks
1	Generator exit and Condenser inlet	Saturated vapor	Vapor quality 1
2	Condenser exit	Saturated liquid	Vapor quality 0
3	Exp. valve for refrigerant (V1) exit	Two-phase liquid vapor of refrigerant	State calculated from adiabatic expansion
4	Evaporator exit	Saturated vapor	Vapor quality 1
5	Absorber lean solution exit	Saturated liquid solution	Vapor quality 0
6	Solution pump exit	Subcooled liquid	State calculated from isentropic pump model
7	Heat exchange exit for lean solution	Subcooled liquid	State calculated from solution heat exchanger model
8	Generator exit for rich solution	Saturated liquid solution	Vapor quality 0
9	Heat exchanger exit for rich solution	Subcooled liquid	State calculated from solution heat exchanger model
10	Exp. valve solution (V2) exit	Two-phase liquid vapor solution	State calculated from adiabatic expansion

pertinent to mention here that the pressure coordinate is irrelevant for the depiction of subcooled states, as Dühring chart is not adequate to represent vapor pressure of the subcooled liquid.

The steady mass and energy conservations for individual components are car-ried out with the following assumptions:

i. Steady state and steady flow;
ii. Negligible changes in potential and kinetic energies across each component;
iii. Negligible pressure drop due to flow friction;
iv. Vapor liberated at the generator is pure water vapor.

Mass flow rate of refrigerant (water) and rich and lean solutions (rich or lean in LiBr) are denoted by: $\dot{m}$, $\dot{m}_{rs}$, and $\dot{m}_{ls}$, respectively. One commonly used parameter known as the circulation ratio (λ) is defined as the ratio of the rich solution flow rate to refrigerant flow rate so that

$$\lambda = \dot{m}_{rs}/\dot{m} \tag{1.3.1}$$

The mass conservation of all the components depicted in figure 1.3.1 and 1.3.2 are as follows:

Mass conservation across condenser:

$$\dot{m}_1 = \dot{m}_2 = \dot{m} \tag{1.3.2}$$

Mass conservation across refrigerant expansion valve (V1):

$$\dot{m}_2 = \dot{m}_3 = \dot{m} \tag{1.3.3}$$

Mass conservation across evaporator:

$$\dot{m}_3 = \dot{m}_4 = \dot{m} \tag{1.3.4}$$

Mass conservation across absorber:

$$\dot{m}_5 = \dot{m}_4 + \dot{m}_{10} \Rightarrow \dot{m}_{ls} = \dot{m} + \dot{m}_{rs} \Rightarrow \dot{m}_{ls} = \dot{m}(1 + \lambda) \tag{1.3.5}$$

Mass conservation across solution pump:

$$\dot{m}_5 = \dot{m}_6 = \dot{m}_{ls} = \dot{m}(1 + \lambda) \tag{1.3.6}$$

Mass conservation across solution heat exchanger:

$$\dot{m}_6 = \dot{m}_7 = \dot{m}_{ls} = \dot{m}(1 + \lambda) \tag{1.3.7}$$

$$\dot{m}_8 = \dot{m}_9 = \dot{m}_{rs} = \lambda\dot{m} \tag{1.3.8}$$

Mass conservation across generator:

$$\dot{m}_7 = \dot{m}_1 + \dot{m}_8 \Rightarrow \dot{m}_{ls} = \dot{m} + \dot{m}_{rs} \Rightarrow \dot{m}_{ls} = \dot{m}(1 + \lambda) \tag{1.3.9}$$

Finally, mass conservation across refrigerant expansion valve (V2):

$$\dot{m}_9 = \dot{m}_{10} = \dot{m}_{rs} = \lambda \dot{m} \tag{1.3.10}$$

An alternative formulation for circulation ratio (λ) can be obtained from mass balance in generator and absorber if the compositions of rich and lean mixtures in terms of mass fraction are known at the absorber and generator. If the composition is defined in terms of mass fraction of anhydrous LiBr in the solution, the mass fraction of LiBr can be expressed as $f^{LiBr} = m_{LiBr}/(m_{LiBr} + m_{H_2O})$, where m_{LiBr} and m_{H_2O} are the mass of pure anhydrous LiBr and pure water in the solution, respectively. Using this definition of the mass fraction and from mass balance of pure water in the absorber of generator, we obtain:

$$\dot{m} + \left(1 - f_{rs}^{LiBr}\right)\dot{m}_{rs} = \left(1 - f_{ls}^{LiBr}\right)\dot{m}_{ls} \tag{1.3.11}$$

Substituting expressions for $\dot{m}_{rs}$ and $\dot{m}_{ls}$ from Eqs. 1.3.1 and 1.3.5 in Eq. 1.3.11 and after little algebraic manipulation, we obtain the alternative formulation for circulation ratio (λ) as:

$$\lambda = \frac{f_{ls}^{LiBr}}{f_{rs}^{LiBr} - f_{ls}^{LiBr}} \tag{1.3.12}$$

Next we consider the energy balance across individual components.

Energy conservation across condenser:

$$Q_C = \dot{m}(h_1 - h_1) \tag{1.3.13}$$

Where, h_1 and h_2 are enthalpies of saturated vapor and saturated liquid of the refrigerant (water) at condenser pressure p_C and corresponding saturation temperature $T_{sat}(p_C)$. Values of h_1 and h_2 are easily obtainable from standard steam table.

Energy conservation across refrigerant expansion valve (V1):

$$h_2 = h_3 \tag{1.3.14}$$

The expansion process is considered to be an isenthalpic process. The pressure change occurs across the expansion valve is an adiabatic free expansion process. During this process, there can be change in the volume, if the fluid generates a small amount of vapor due to expansion at low pressure. The other thermodynamic change is the possibility of a temperature drop at the end of the expansion, as some amount of energy must be supplied by the liquid phase itself in order to generate vapor.

Energy conservation across evaporator:

$$Q_E = \dot{m}(h_4 - h_3) \tag{1.3.15}$$

Where, h_4 is the enthalpy of saturated vapor of the refrigerant water at evaporator pressure p_E and corresponding saturation temperature $T_{sat}(p_E)$. Values of h_4 is again obtainable from standard steam table.

Energy conservation across absorber:

$$Q_A = \dot{m}h_4 + \dot{m}_{rs}h_{10} - \dot{m}_{ls}h_5 \tag{1.3.16}$$

Substituting relation given by Eqs. 1.3.1 and 1.3.5 in Eq. 1.3.16, one can get the following expression for energy conservation:

$$Q_A = \dot{m}[(h_4 - h_5) + \lambda(h_{10} - h_5)] \tag{1.3.17}$$

The first term in LHS of Eq. 1.3.17 $\dot{m}(h_4 - h_5)$ is the enthalpy change of water as it changes its state from saturated vapor at state point 4 to liquid at state 5. The second term $\lambda\dot{m}(h_{10} - h_5)$, on the other hand, represents the sensible heat transferred from the solution as the solution at state 10 is cooled to solution at state 5. The procedure to obtain solution enthalpy will be described in details in subsequent section.

Energy conservation across solution pump:

$$W_P = \dot{m}_{ls}(h_6 - h_5) = \dot{m}(1 + \lambda)(h_6 - h_5) \tag{1.3.18}$$

As the pump handles liquid solution, the flow across the pump can be aptly considered to be incompressible, and the alternative expression for energy conservation across the pump can be written as:

$$W_P = \dot{m}(1 + \lambda)v_{sol}(h_6 - h_5) \tag{1.3.19}$$

where v_{sol} is the specific volume of the liquid H_2O–LiBr solution which can be taken to be close to 0 0.00055 m3/kg [2].

Energy conservation across solution heat exchanger:

$$\dot{m}(1 + \lambda)(h_7 - h_6) = \lambda\dot{m}(h_8 - h_9) \tag{1.3.20}$$

The foremost role of solution heat exchanger is to preheat the lean H_2O–LiBr solution before it enters the generator so that the input heat required at the generator is reduced. It also removes unwanted heat from the rich solution sent to the absorber and thus reduces the amount of rejected heat from the absorber.

One of the important parameters to analyze the system performance is the effectiveness of solution heat exchanger ε_{HX} which is defined as [2, 4]:

$$\varepsilon_{HX} = \frac{T_7 - T_6}{T_8 - T_6} \tag{1.3.21}$$

It is pertinent to mention here that this definition of ε_{HX} is based on the assumption $c_{pls} \leq c_{prs}$ [5], where c_{pls} and c_{prs} represents the specific heats of lean and rich solution. Temperature of the weak solution entering the generator (T_7) can be determined by using Eq. 1.3.21 as $T_6 \approx T_5$. Since point 5 and 8 are well defined as liquid solutions in equilibrium their temperature–pressure–compositions are well defined and if the effectiveness of the solution heat exchanger is assumed Eq. 1.3.21 provides reasonable estimation of property values at state point 7.

Energy conservation across generator:

$$Q_G = \dot{m}h_1 + \dot{m}_{rs}h_8 - \dot{m}_{ls}h_7 \tag{1.3.22}$$

Substituting relation given by Eqs. 1.3.1 and 1.3.5 in Eq. 1.3.22, one can get the following expression for energy conservation across the generator:

$$Q_G = \dot{m}[(h_1 - h_7) + \lambda(h_8 - h_7)] \tag{1.3.23}$$

The first term in LHS of Eq. 1.3.23 $\dot{m}(h_1 - h_7)$ is the heat required to change the state of water to saturated water vapor at state 1 from the solution at state 7, while the second term $\lambda\dot{m}(h_8 - h_7)$ is the amount of sensible heat added to the solution to change the state of the solution from state 7 to state 8.

Finally, energy conservation across refrigerant expansion valve (V2) is given by:

$$h_9 = h_{10} \tag{1.3.24}$$

The coefficient of performance of the VARS is given by Eq. 1.2.1 or 1.2.2 and the second law efficiency can be obtained as:

$$\eta_{II} = \frac{COP}{COP_{Max}} = \frac{Q_E}{Q_G}\left(\frac{T_G}{T_G - T_\infty}\right)\left(\frac{T_\infty - T_E}{T_E}\right) \tag{1.3.25}$$

The thermodynamic properties are of utmost importance to analyze the steady-state performance of the VARS by Eqs. 1.3.1–1.3.24 defined in the foregoing section. The knowledge of temperatures, pressures, composition of lean and rich solution, effectiveness of solution heat exchanger, and the refrigeration load are some of the important parameters that define the first-cut design. With the general assumption of the solution to be at equilibrium state at the outlet of generator and absorber enables us to utilize the equilibrium pressure–temperature–composition and enthalpy–temperature–composition charts for evaluating solution properties.

1.4 Thermodynamic Properties of H_2O–LiBr Solutions

Thermodynamic properties such as vapor pressure at a given temperature, enthalpy, specific volume, and entropy for H_2O–LiBr solution are available in the empirical forms [1, 3, 6–13] and charts [2, 3, 6]. Equations 1.4.1–1.4.3 shows some typical empirical formulas between the temperature of the refrigerant (H_2O) $\mathbf{T_r}$, aqueous-LiBr solution temperature $\mathbf{T_{sol}}$ and its LiBr composition $\mathbf{f_{LiBr}}$ with prescribed values of constants $\mathbf{A_n}$ and $\mathbf{B_n}$ [15]. The assumption for these formulations is considering the refrigerant to be pure water without.

$$T_r = \frac{\left(T_{sol} - \sum_{n=0}^{3} B_n f_{LiBr}^n\right)}{\sum_{n=0}^{3} A_n f_{LiBr}^n} \quad (1.4.1)$$

$$T_{sol} = \sum_{n=0}^{3} B_n f_{LiBr}^n + T_r \sum_{n=0}^{3} A_n f_{LiBr}^n \quad (1.4.2)$$

Similarly enthalpy relation for H_2O–LiBr solutions at equilibrium conditions is given by.

$$h = \sum_{n=0}^{4} A_n f_{LiBr}^n + T_{sol} \sum_{n=0}^{4} B_n f_{LiBr}^n + T_{sol}^2 \sum_{n=0}^{4} C_n f_{LiBr}^n \quad (1.4.3)$$

The values of constants A_n, B_n, and C_n are listed in Table 2.

Among the charts, we have already discussed about Dühring chart (Fig. 6). Another version of similar chart showing vapor pressure data for H_2O–LiBr systems are also found in literature [7]. Figure 7 shows such a chart with the mass fraction of LiBr being plotted on abscissa, while saturation temperature of pure water and vapor pressure are plotted on y-axis. Solution isotherms are also shown in this chart.

Table 2 Values of constants A_n, B_n, and C_n for calculation of T_r, T_{sol}, and h

T_r, T_{sol}					
	$n=0$	$n=1$	$n=2$	$n=3$	$n=4$
A_n	−2.008	−0.17	-3.133×10^{-3}	1.977×10^{-5}	–
B_n	124.937	−7.716	0.152	-7.951×10^{-4}	–
h					
A_n	−2024.33	163.309	−4.881	6.303×10^{-2}	-2.914×10^{-4}
B_n	18.283	−1.169	3.248×10^{-2}	-4.034×10^{-4}	1.852×10^{-6}
C_n	-3.701×10^{-2}	2.888×10^{-3}	-8.131×10^{-5}	9.912×10^{-7}	-4.444×10^{-9}

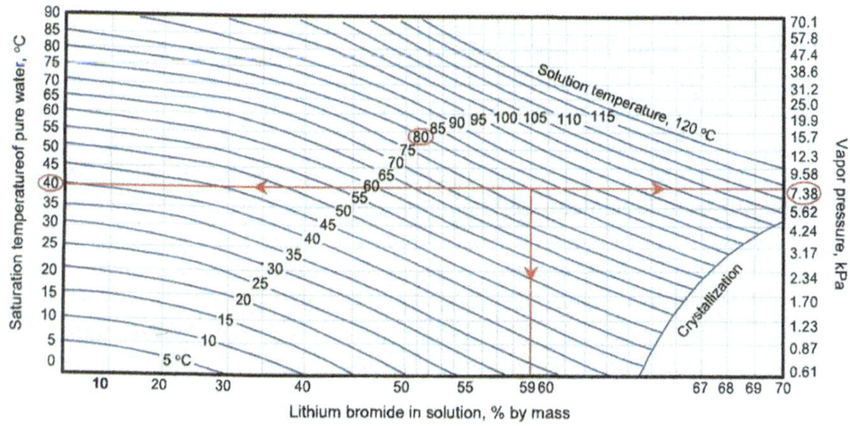

Fig. 7 Pressure–Temperature–Concentration diagram for H_2O–LiBr solution [2, 7]

As an example, the vapor pressure and saturation temperature of water corresponding to solution composition of 59% LiBr mass fraction at 80 °C is shown in the chart depicted in Fig. 7.

Figure 8 shows specific enthalpy–temperature–mass fraction plot for water–lithium bromide solutions [2, 7]. The chart is prepared by considering reference enthalpy of 0 kJ/kg for liquid water at 0 °C and solid anhydrous LiBr at 25 °C.

The enthalpy data of refrigerant water for H_2O–LiBr are obtained from pure water property table. For subcooled liquid water enthalpy, h_W, at any temperature T can be expressed as:

$$h_{W,liquid} = 4.2\left(T - T_{ref}\right) \tag{1.4.4}$$

Here, reference temperature T_{ref} is chosen to be 0 °C, unit of enthalpy is kJ/kg.

The water vapor obtained from the generator is often in superheated state as the generator temperature can be much higher than the saturation water temperature at that pressure. In order to calculate the enthalpy of the superheated water vapor liberated at the generator, $h_{W,sup}$ at low pressure and temperature can be approximated as:

$$h_{W,sup} = 2501 + 1.88\left(T - T_{ref}\right) \tag{1.4.5}$$

Again the reference temperature T_{ref} is considered to be 0 °C, and unit of enthalpy obtained from Eq. 1.4.3 is kJ/kg. In fact, the number 2501 appearing on the LHS of Eq. 1.4.5 is nothing but the latent heat of vaporization for water at 0 °C, and average specific heat for water vapor $c_{p,g}$ is 1.88 kJ/kg-K.

It should be noted from all different property charts that the entire H_2O–LiBr absorption cooling cycle operates under vacuum condition.

1.4.1 Crystallization

The pressure–temperature–composition charts are shown by Figs. 6 and 7 and enthalpy–temperature–composition chart is shown by Fig. 8 show distinct zone marked as crystallization zone (bonded by isosteric and isothermal lines) in the lower right section of each of those plots. If the solution composition and temperature fall within this region formation of solid crystal occurs and the solution is no longer liquid but a two-phase solid liquid mixture also known as slush. In the slush, H_2O–LiBr solution and pure LiBr crystals exist in equilibrium. Since the solid crystal in the solution can cause blocking of the pipes and valves, care should be taken to stay away from this region. The Dühring plot representation is of paramount importance as it helps to visualize the occurrence of such situation. That

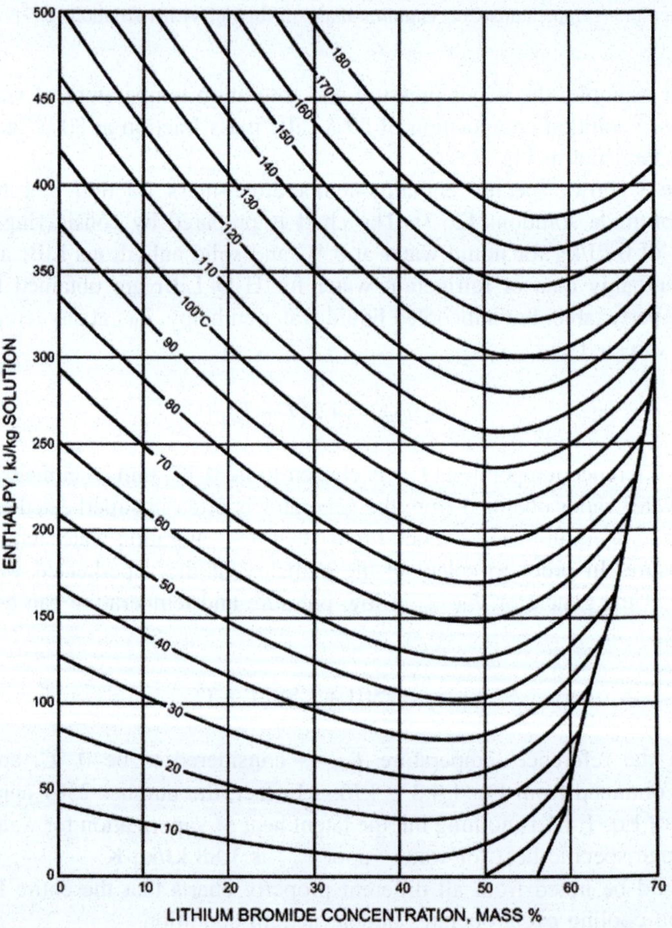

Fig. 8 Enthalpy–Temperature–Concentration diagram for H_2O–LiBr solution [3, 7]

is why, such a plot is strongly recommended to check any solution cycle in order to make sure that the cycle operates comfortably away from the crystallization region. In order to avoid crystallization combination of low temperature, low vapor pressure, and high concentration of LiBr in the solution must be avoided.

Other than the crystallization, the common problems in the operation of H_2O–LiBr absorption cooling systems are associated with air leakage and pressure drops. The operating condition for the entire cycle being vacuum, chances of air leakage are very large. That is why air purging system is must for all the practical applications. Also the pressure drop due to friction must be minimized as the operating pressures are very small and volume flow rate is high due to large specific volume of vapor. The recent developments of single-effect H_2O–LiBr absorption cooling systems can be found form the references [14–19].

1.5 Commercial H_2O–LiBr Absorption Cooling Systems and Multi-Effect Absorption Cooling Systems

Commercially used single effect systems are of two types namely: (i) Twin drum type (more common), and (ii) Single drum type.

As the operating pressure of evaporator–absorber pair (p_E) and generator–condenser pair ($p_C > p_E$) are same, each pair is located in two drums. Thus, a twin drum system contains two separate vessels operating at low- and high-pressure conditions. Figure 9 shows the schematic of a commercial single-effect, twin drum system.

As shown in the Fig. 9, the same cooling water (heat sink) is used for heat extraction from the condenser after it extracts heat from the absorber in series. That is why, this particular arrangement is called series arrangement. The series arrangement of heat sink is particularly advantageous as the requirement of cooling water flow rate is less. By routing the cooling water first through the absorber also ensures that the condensation can occur at a higher pressure to prevent crystallization. Parallel flow arrangement of cooling water is also common; however, the requirement of cooling water is much higher than the series arrangement. The spraying of refrigerant water in the evaporator tubes by means of a refrigerant pump also ensures good heat and mass transfer at the evaporator. Heater tubes carrying heat transfer fluids like steam, hot water, or hot oil are directly immersed in the rich solution pool of the generator in order to generate vapor. Locating the evaporator–absorber pair and generator–condenser pair in the same housing reduces the pressure drop as long vapor lines are eliminated. Air leakages also reduce drastically due to less number of joints.

Figure 10 shows the schematic of a single-effect single-drum arrangement in which all four components are located in the same vessel. The vessel is, however, divided into high- and low-pressure sides by placing a diaphragm between these two domains.

For multi-effect systems, [2, 4, 22, 23] consist of a series of generators operating at progressively reducing pressures. While the heat is supplied to the generator

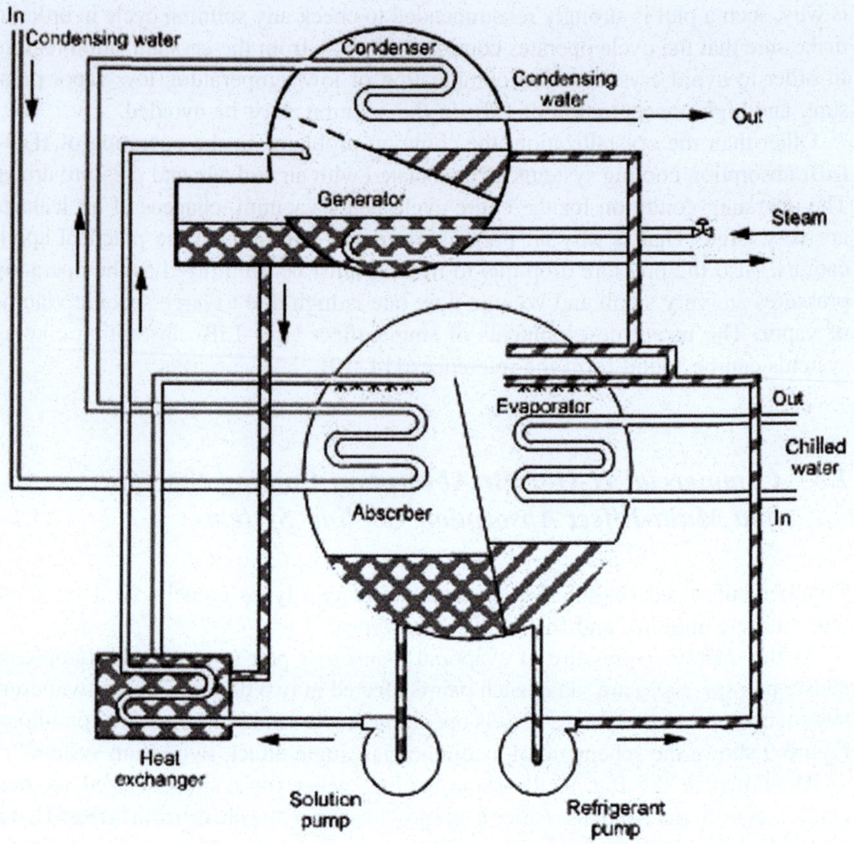

Fig. 9 Single-effect two-drum H_2O–LiBr water chiller (ASHRAE, 1983)

operating at the highest pressure, the later stages generate vapor by absorbing the energy (enthalpy) from the superheated steam generated at the highest pressure and temperature at the first stage. The subsequent generators produce more steam by absorbing the energy (enthalpy) from the superheated steam generated at previous stages. In this manner, the input heat is utilized more economically and efficiently and large amount of refrigerant vapor is produced leading to higher coefficient of performance (*COP*). Although heat utilization occurs very efficiently in such systems, the complicacies arising in constructing such device is paramount and initial installation cost is very high. Also such systems require much higher heat source temperatures in the highest stage generator. Figure 11 shows the schematic of a solar-assisted double-effect H_2O–LiBr vapor absorption chilling system [21], and Fig. 12 shows the schematic of commercial double-effect systems [2].

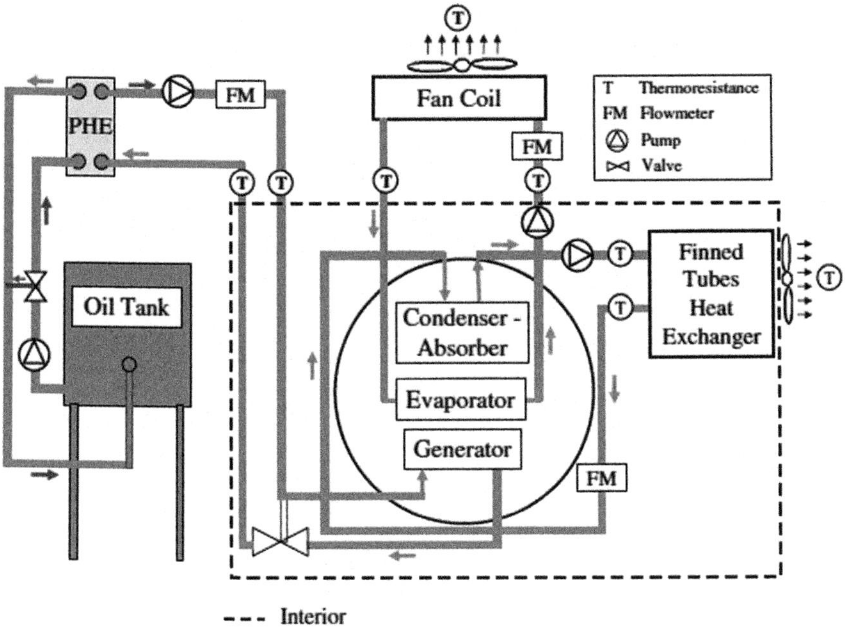

Fig. 10 A commercial, single-drum type, water–lithium bromide system [20]

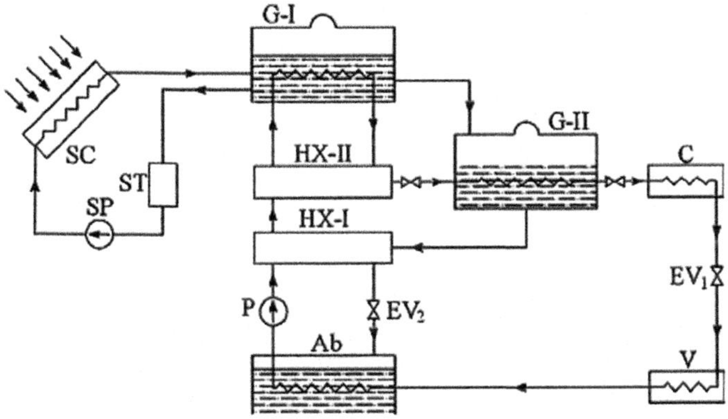

Fig. 11 Schematic of solar-assisted double-effect H_2O–LiBr absorption system

1.6 Closure

A wide variety of heat sources can be utilized for driving H_2O–LiBr absorption cooling systems. Since these systems are large in volumetric aspect such systems can only be appreciated for addressing large cooling requirement. Large capacity

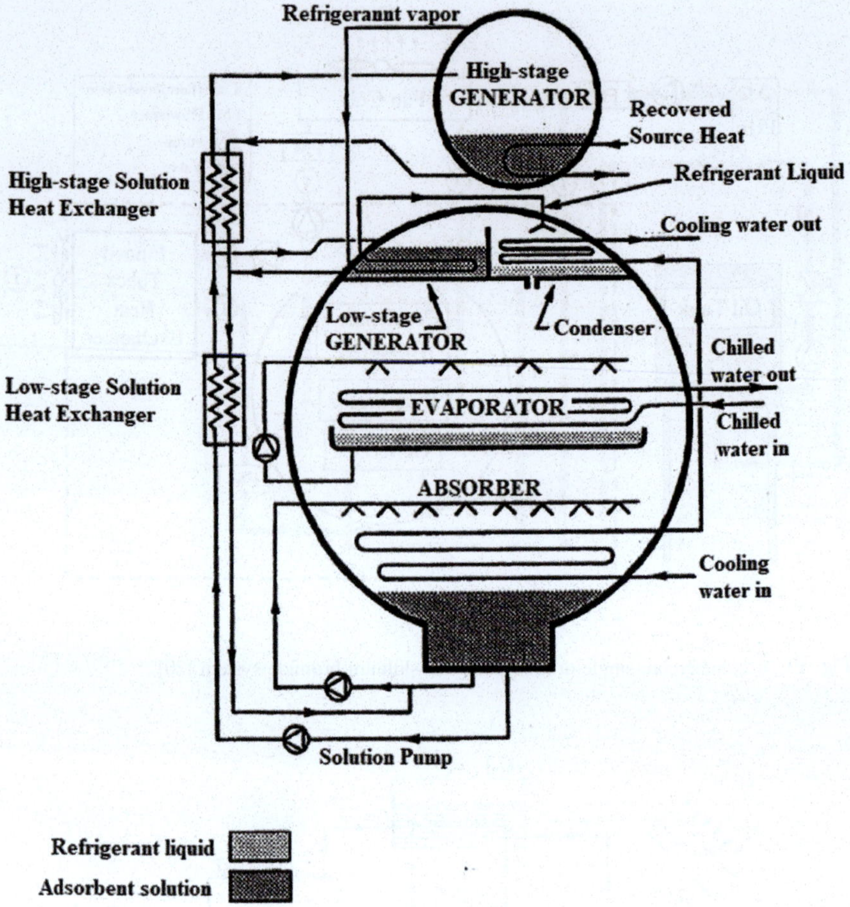

Fig. 12 A commercial, double-effect, water–lithium bromide system

systems are usually driven by steam or hot water obtained through solar heating or waste heat from industry. Small capacity systems on the other hand are usually driven by oil or gas. The heat source temperature requirement for a typical single-effect system to produce chilled water at 7 °C is about 120 °C while condenser and absorber operate at about 46 °C and 40 °C, respectively. The **COP** of single-effect system varies within the range of 0.6–0.8, while the range is almost double for multi-effect systems.

The major obstacle to successfully implement solar energy technologies for cooling application is the intermittent availability of solar energy. Significant economic and environmental benefits can be achieved if such problem can be addressed by effective solar thermal energy storage [24]. The recent trend in developing cascaded storage system offers vast potential for the improvement of a solar cooling system performance [24].

References

1. Kaita Y (2001) Thermodynamic properties of lithium bromide–water solutions at high temperatures. Int J Refrig 24(5):374–390
2. https://www.nptel.ac.in/courses/112105129/pdf/RAC%20Lecture%2014.pdf
3. Muhumuza R, Strachan P (2010) Modelling, implementation and simulation of a single-effect absorption chiller in MERIT. Master of Science in Renewable Energy Systems and the Environment. University of Strathclyde Engineering, Scotland
4. Herold KE, Radermacher R, Klein SA (2016) Absorption chillers and heat pumps. CRC Press (2016)
5. Bergman TL, Frank PI (2011) Fundamentals of heat and mass transfer. Wiley
6. Iyoki S, Uemura T (1989) Vapour pressure of the water—lithium bromide system and the water—lithium bromide—zinc bromide—lithium chloride system at high temperatures. Int J Refrig 12(5):278–282
7. ASHRAE Handbooks. Fundamentals (2009) American Society of Heating, Refrigerating and Air-Conditioning Engineers, Inc, Atlanta
8. Florides GA, Kalogirou SA, Tassou SA, Wrobel LC (2003) Design and construction of a LiBr-water absorption machine. Energy Convers Manag 44:2483–2508
9. Jeter SM, Lenard JLY (1992) A.S. TejaProperties of lithium–bromide–water solutions at high temperatures and concentrations—part IV: vapor pressure. ASHRAE Trans 98(1):167–172
10. Feuerecker G, Scharfe J, Greiter I, Frank C, Alfeld G (1993) Measurement of thermophysical properties of LiBr–solutions at high temperatures and concentrations. In: International absorption heat pump conference ASME, AES, vol 31, pp 493–99
11. Uemura T, Hasaba S (1964) Studies on the lithium bromide-water absorption refrigeration machine. Technology Reports of Kansai University, vol 6, pp 31–55
12. McNeely LA (1979) Thermodynamic properties of aqueous solutions of lithium bromide. ASHRAE Trans 85(Part 1):413–434
13. Patterson MR (1988) H. Perez-BrancoNumerical fits of the properties of lithium–bromide water solutions. ASHRAE Trans 94(part2):2379–2388
14. Ali AHH, Noeres P, Pollerberg C (2008) Performance assessment of an integrated free cooling and solar powered single-effect lithium bromide-water absorption chiller. Sol Energy 82.11:1021–1030
15. Kaushik SC, Arora Akhilesh (2009) Energy and exergy analysis of single effect and series flow double effect water–lithium bromide absorption refrigeration systems. Int J Refrig 32 (6):1247–1258
16. Kaushik SC, Arora Akhilesh (2009) Energy and exergy analysis of single effect and series flow double effect water–lithium bromide absorption refrigeration systems. Int J Refrig 32 (6):1247–1258
17. Şencan Arzu, Yakut Kemal A, Kalogirou Soteris A (2005) Exergy analysis of lithium bromide/water absorption systems. Renew Energy 30(5):645–657
18. Srikhirin Pongsid, Aphornratana Satha, Chungpaibulpatana Supachart (2001) A review of absorption refrigeration technologies. Renew Sustain Energy Rev 5(4):343–372
19. Kilic Muhsin, Kaynakli Omer (2007) Second law-based thermodynamic analysis of water-lithium bromide absorption refrigeration system. Energy 32(8):1505–1512
20. Izquierdo M, Lizarte R, Marcos JD, Gutiérrez G (2008) Air conditioning using an air-cooled single effect lithium bromide absorption chiller: results of a trial conducted in Madrid in August 2005. Appl Therm Eng 28(8):1074–1081
21. Sarbu I, Sebarchievici C (2013) Review of solar refrigeration and cooling systems. Energy Build 67:286–297
22. Puig-Arnavat M, López-Villada J, Bruno JC, Coronas A (2010) Analysis and parameter identification for characteristic equations of single-and double-effect absorption chillers by means of multivariable regression. Int J Refrig 33(1):70–78

23. Yin H, Qu M, Archer DH (2010) Model based experimental performance analysis of a microscale LiBr–H 2 O steam-driven double-effect absorption Chiller. Appl Therm Eng 30 (13):1741–1750
24. Chidambaram LA, Ramana AS, Kamaraj G, Velraj R (2011) Review of solar cooling methods and thermal storage options. Renew Sustain Energy Rev 15(6):3220–3228

Solar Assisted Solid Desiccant—Vapor Compression Hybrid Air-Conditioning System

D. B. Jani, Manish Mishra and P. K. Sahoo

Abstract Solar assisted solid desiccant—vapor compression based hybrid space cooling system for building thermal comfort is made by integrating desiccant based dehumidification system as well as by use of solar heating system along with conventional VCR system. This is because in humid climates, excessive humidity during summer leading to inefficiencies of conventional cooling devices. Humidity increment in ambient air as well as ventilation requirement suddenly rises the latent load of the space to be cooled. Conventional VCR systems are not effective in handling both temperature and relative humidity of cooling air independently. The application of solid desiccant based hybrid cooling systems significantly ameliorates the humidity control irrespective of temperature of supply air. The application of renewable free energy like as use of solar energy for regenerating the desiccant used in desiccant wheel helps us to alleviate the major requirement of electric energy needed by conventional VCR air-conditioning system for hot sunny days. So, it ameliorates overall energy efficiency and reduces energy costs. Performance of solid desiccant—vapor compression based hybrid building space comfort cooling system has been evaluated during hot and humid period from April to September for the ambient conditions of the Roorkee for various important parameters such as temperatures of supply air, room air and regeneration air etc. Regeneration temperature is one of the most important parameters having the key role in changing the performance of desiccant based cooling system.

Keywords Desiccant dehumidifier · Hybrid cooling · Regeneration
Solar

D. B. Jani (✉)
Gujarat Technological University, G.T.U., Ahmedabad, Gujarat, India
e-mail: dbjani@rediffmail.com

M. Mishra · P. K. Sahoo
Department of Mechanical & Industrial Engineering, I.I.T., Roorkee, Uttarakhand, India
e-mail: mmishfme@iitr.ac.in

P. K. Sahoo
e-mail: sahoofme@iitr.ac.in

H. Tyagi et al. (eds.), *Applications of Solar Energy*, Energy, Environment, and Sustainability, https://doi.org/10.1007/978-981-10-7206-2_12

Nomenclature

COP	Coefficient of performance
E	Electrical power consumption (kW)
h	Enthalpy (kJ/kg)
m	Mass flow rate of air stream (kg/s)
MRR	Moisture removal rate (kg/h)
Q_{cc}	Cooling capacity (kW)
Q_{reg}	Reactivation heat addition (kW)
RH	Air relative humidity (%)
T	Dry bulb temperature (K)
TRNSYS	Transient system simulation
VCR	Vapor compression refrigeration

Greek letters

ε	Dehumidifier effectiveness
ω	Air humidity ratio (g/kg)

Subscripts

a	Dry air
dw	Rotary dehumidifier
hrw	Enthalpy wheel
p	Process air
r	Regeneration air
1,2, etc.	Reference state points

1 Introduction

Energy consumption has increased since last decades with the development of worldwide economy. Requirement for the energy needed for the cooling and air-conditioning is estimated between 30–40% of total building energy use [20]. Moreover, because of increased living standards and occupants demands, cooling energy demand will further increase in near future. The peak load on the electricity grid increases in hot summer days, because of high cooling requirement [7]. This could cause blackouts and grid failure. Conventional vapour compression systems increase greenhouse gases in the environment responsible for depleting the ozone layer. Solar assisted desiccant cooling can help to solve this matter. The major cooling requirement in summer is associated with peak solar radiation availability gives a good opportunity to exploit solar assisted desiccant cooling technology [18]. The optimal use of solar energy is based on the operational strategy that consumes the least electrical energy and uses thermal energy efficiently while maintaining the indoor comfort level at optimum operating cost. Thus, desiccant cooling can be a perfect supplement to the traditional vapour compression system by controlling temperature and humidity independently [27].

Solid desiccant based comfort space cooling systems are heat driven cooling units in which heat energy is needed to regenerate the rotary desiccant dehumidifier. Desiccant wheel is the main part of the system which removes humidity from the supply process air. Heated and dehumidified process air is first cooled sensibly in sensible heat recovery wheel and then in conventional cooling coil of vapor compression refrigeration air-conditioner. Latent heat load is removed in rotary desiccant dehumidifier while sensible heat load is removed later in heat wheel and in conventional cooling coil [21]. Thus, solid desiccant cooling systems can take care of both sensible and latent load of air conditioned space separately as well as of fresh ventilated outside ambient air requirements. Hybrid comfort space cooling systems also have the merit of being regenerated by waste heat from plant etc. or renewable solar energy which can easily be obtained from low grade or renewable free energy sources. It also improves overall energy efficiency and reduces energy costs for the comfort space cooling. Besides, the desiccant cooling systems provides higher percentage of fresh air to achieve better air quality at lower energy cost. Thus, these hybrid systems can achieve the augmentation in comfort, energy and cost saving for the comfort space cooling [19].

Reactivating the desiccant materials used in solid desiccant based rotary dehumidifiers with the help of freely available solar energy not only reduce the electricity consumption but also ameliorates the substantial fossil energy saving [22]. The use of solar air heating for reactivation of desiccant wheel is an interesting option to meet the cooling demand of conditioned space as the peak intensity of solar energy and demand for cooling are the highest during the same period. Joudi and Madhi [28] conducted experimental tests on solar assisted desiccant cooling system under hot and humid weather conditions. The desiccant dehumidifier is regenerated at 70 °C using solar energy. Enhancement in the system performance was further examined by using higher regeneration temperature and larger mass flow rate of the air to produce the comfort conditions. Halliday et al. [15] also carried out an experimental analysis to check the feasibility of hybrid space desiccant cooling system to evaluate the application of renewable solar energy to regenerate the desiccant dehumidifier. The maximum saving in system energy was found to be 39%.

Optimization of rotary dehumidifier reactivated by solar energy is carried out by Ahmed et al. [1]. Influence of different operating parameters such as air stream flow rate, regeneration temperature, wheel speed, wheel thickness, and porosity on the performance is evaluated by numerical simulation. Solid desiccant cooling system regenerated by solar energy having direct flow vacuum tube collector, is simulated and validated with experimental results by Bourdoukan et al. [11]. It was found that raising the reactivation temperature by 12 °C decreases the overall system performance by 45%. Kim and Ferreira [29] reviewed different space cooling systems that use solar energy for air conditioning comfort space cooling. Brief comparisons have been carried out among various systems on the basis of energy efficiency and economic suitability. Simulation and optimization of solid desiccant based comfort space cooling system regenerated by use of solar heat was carried out by Fong et al. [14] to evaluate performance of the system. It was found that the requirement of ventilated increases to augment quality of the indoor air.

Koronaki et al. [30] experimented to determine overall performance of a solid desiccant cooling system regenerated by solar energy in Mediterranean areas. It was found that the thermal COP of overall system increases when the humidity ratio of process air at inlet of dehumidifier is high. Lower temperature and humidity ratio of reactivation air at dehumidifier inlet can ameliorates system performance. A solid desiccant cooling system regenerated by solar energy was experimentally investigated by Baniyounes et al. [7]. Performance of the system, economic feasibility, carbon credit and energy savings are experimentally evaluated based on parameters such as coefficient of performance (COP), availability of peak solar fraction, life cycle calculation, payback period determination and present worth factor. Results showed that system COP can be enhanced from 0.7 to 1.2 by the use of solar energy for desiccant regeneration.

An experimental setup for evaluating the reactivation and adsorption capability of various desiccant material like as silica gel, activated alumina and activated charcoal, was proposed by Yadav and Bajpai [37]. Different desiccants were regenerated by using solar energy in the range of 54.3–68.3 °C. It is found that the regeneration performance was mostly influenced by the reactivation temperature and also depends on relative humidity of process air, temperature of the desiccants and flow rate of reactivation air. A brief comparison between different desiccants showed that silica gel was observed to perform better both at high as well as at low flow rates for reactivation and adsorption than the activated alumina and activated charcoal.

Commonly used evaporative type solid desiccant cooling system in the field of comfort space cooling can handles sensible heat load via over-drying the process air to permit further cooling by direct or indirect evaporative type desiccant space cooling. For hot and humid climates, the performance of this systems would be degraded due to possible dehumidification may not be satisfactory to assist evaporative cooling of supply air. To enhance the output or quality of supply air, the other space cooling technologies such as traditional vapour-compression air-conditioning unit should be provided to the present configuration to formulate a hybrid system. Many researchers previously studied performance of hybrid cooling system using various simulation tools such as Yadav [39] had carried out simulations of a hybrid desiccant cooling system comprise of a conventional vapour-compression refrigeration air-conditioning system integrated to desiccant wheel which is regenerated by solar heat. It is shown that when the latent load constitutes the major share of the total cooling load, the hybrid desiccant cooling system saves substantial energy needed for the cooling. Dhar and Singh [13] evaluated the performances of different configurations of hybrid desiccant cooling systems for hot-dry and hot-humid weather by using analogy method. It is found that the use of hybrid desiccant cooling cycle saves substantial energy as compared to the conventional vapour-compression air-conditioners. The influence of parameters such as sensible heat factor of test room, process and regeneration air stream mixing ratio and reactivation temperature on overall cycle performance has also been carried out.

Desiccant dehumidifier integrated VCR air-conditioning system was further experimentally evaluated by Subramanyam et al. [35] for comfort space cooling and dehumidification. The influence of various parameters like as air flow rate, compressor speed, and desiccant wheel speed on system performance has been conducted. It is found that for the range of parameters selected, wheel speeds of the order of 17.5 rph is the optimum speed for overall system performance. Yong et al. [40] conducted various experiments on solid desiccant VCR hybrid comfort space cooling system with rotary dehumidifier made up of LiCl. The reactivation temperature and air stream flow rate have found significant influence over cycle performance. It can also be seen that the hybrid desiccant comfort space conditioning system achieves a better part load performance with a capacity to work at various operational modes such as ventilation, recirculation, mixed etc. and assures satisfactory performance in high humid regions. Experimental tests on hybrid desiccant cooling system was also carried out by Jia et al. [27] to examine the role of various parameters such as sensible heat factor, refrigerant mass flow rate and length of evaporator tubes on overall system performance. They also suggested that the hybrid comfort space cooling system can make significant energy savings as compared to traditional vapor compression based air-conditioning system.

Energy consumption model of a hybrid comfort space cooling system that is integration of a desiccant based dehumidifier and a vapour-compression refrigeration system was established by Liu et al. [33]. Significance of various parameters such as regeneration temperature, temperature and ambient air relative humidity, R/P ratio etc. over the system performance has been studied. Al-Alili et al. [2] carried out an experimental study to get the performance of a hybrid system consisting of desiccant and enthalpy wheel with a traditional air-conditioner. The effect of different parameters such as temperature and relative humidity of process air, ventilation flow rate, on the performance of hybrid air-conditioning system is investigated. The experimental results show that the hybrid air conditioning system is more effective than the traditional system in maintaining the indoor comfort conditions for hot and humid climate. Angrisani et al. [3] have conducted experiments on a hybrid desiccant cooling system integrated with small cogenerator. Performance of the hybrid cooling system was analyzed by varying different parameters such as regeneration temperature, volume air flow rates, outdoor temperature and humidity ratio, and desiccant wheel speed. The obtained result indicates that the overall performance of hybrid cooling system reduces with the rise in regeneration temperature and regeneration flow rate. Furthermore, optimal operation of the system is found for a narrow range of desiccant wheel speed ranging 2–7.5 rph.

Solar assisted vapor compression based hybrid solid desiccant cooling system substantially saves high grade electrical energy as compared to the conventional vapor compression space cooling system that uses lots of electricity for operating the reactivation heater. Yadav and Kaushik [38] carried out studies on solid desiccant based dehumidification system and VCR hybrid air-conditioning system reactivated by solar heat. It is found that a hybrid system with desiccant cycle is more effective in higher ambient humidity conditions with high latent heat load and significant energy saving can be achieved as compared to the conventional VCR system. Maalouf et al. [34] studied the performance of a LiCl dehumidifier integrated solar collectors to optimize the reactivation heat energy. The effect of desiccant wheel speed, regeneration temperature, and air stream flow rate has been studied for decreasing the primary energy consumption and increasing the overall system performance. The decrease in air change hours during occupation period lowers the required reactivation heater operation time up to 40%. Bourdoukan et al. [10] conducted experimental tests on the desiccant air handling unit integrated vacuum-tube solar collectors. The overall system performance has been evaluated for different ambient conditions. COP of the system for a moderate humid day is calculated as 0.4–4.3 on the basis on primary energy consumption and electrical energy consumption respectively.

La et al. [32] reviewed the current status of rotary desiccant dehumidification and comfort space cooling applications in aspects of development of advanced desiccant materials, the optimization of system configuration and the utilization of solar heat and similar other low grade energy sources, such as district heating, waste heat and bio-energy. It is found that as compared to the conventional VCR system, desiccant cooling has the merits being environment-friendly, energy saving, healthy and comfortable. Dai et al. [12] theoretically analyzed a solar assisted two-stage desiccant cooling system. The system achieves average cooling capacity of 70 kW with the corresponding thermal COP and electrical COP of 1.1–12.5, respectively. Moreover, it removes about 45.3% moisture of the total cooling load and hence the electric power consumption is reduced by 39.5% in comparison to the conventional configuration.

To the best of authors' reference, none of the previous study of solid desiccant based comfort space cooling system, have a detailed experimental tests on solar assisted hybrid comfort space cooling system which integrates rotary desiccant dehumidifier and enthalpy wheel to the traditional vapor compression space conditioner for typical hot and humid North Indian Himalayan climate (Roorkee). In the current study, numbers of tests have been carried out to determine the overall performance of the hybrid air conditioning system during a period of March to September mid for hot and humid climatic conditions. Moreover, the effect of variations in outdoor climate on overall system performance has been discussed in detail by use of TRNSYS prediction tool. The performance tests covers the great influence of operating parameters such as flow rate, pressure drop etc. on system performance. Results obtained by simulations using TRNSYS software predict that the system performance is highly influenced by the changing outdoor conditions.

The change in regeneration temperature based on ambient condition has also been depicted. The impact of variations in ambient conditions on dehumidifier performance was also been evaluated in detailed.

2 Experimental Investigation

A room with capacity of 5 people having dimensions $4\,m \times 3\,m \times 3\,m$ constructed for the present study. Sensible and latent cooling cooling loads for the same are 1.37 kW and 0.39 kW respectively [31]. The room sensible heat ratio (RSHR) is 0.78. The inside room comfort is obtained as 50% RH and 26 °C DBT [6] while Outdoor conditions is taken as 35 °C DBT and 70% RH [23]. The mass flow rates of process and regeneration air streams are 0.12–0.08 kg/s respectively.

A schematic layout of solar assisted solid-desiccant vapour-compression hybrid comfort space cooling system has been illustrated in Fig. 1. In case of recirculation mode, the return room air stream from room at state 1 passes through the rotary desiccant dehumidifier. Its relative humidity is partly but significantly adsorbed by the desiccant wheel material and the heat of adsorption makes it dry as well as elevates its temperature. Desiccant wheel adsorbs moisture from the air because of the difference between partial pressure of moisture in air and that in desiccant material. Thus, a warm and dry air stream exits the dehumidifier at state 2. The dry and hot air is cooled down sensibly up to the state 3 by fresh outdoor air in heat recovery wheel. The VCR sensible cooling coil reduces the temperature of supply air to state 4 without affecting its humidity ratio.

At the regeneration side, outdoor air at state 6 initially passes though enthalpy wheel to recover heat from warm and dried air (state 7). The preheated air is further heated by a liquid to air heating coil to a required regeneration temperature (state 8). Regeneration heat provided by the solar heating system which consists of solar collector, back up heater, storage tank, circulating pump and liquid to air heating coil. When insufficient thermal energy is available from solar collector due to rainy weather, the backup auxiliary heater is used to obtain the required heat necessary for dehumidifier reactivation. At last, hot reactivation air removes humidity from desiccant wheel and exits to ambient at state point 9.

Experimental tests were conducted by simultaneous measurements of dry bulb temperature, relative humidity, pressure and flow rate of air steams by use of data acquisition system. Measurement of dry bulb temperature and relative humidity are electronically recorded by use of digital temperature and hygrometer transmitter. Pressure variation across desiccant wheel, enthalpy wheel and sensible cooling coil is measure by use of pressure transmitter. The air steam flow rate of process and regeneration air is measured by use of digital velocity transmitter. Energy meter measures the electricity consumption of the system. A humidistat is incorporated to control the dehumidifier operation according to the room humidity change. A temperature controller incorporating a relay is also used to control the operation of compressor of used in the hybrid system as per changes in the room temperature.

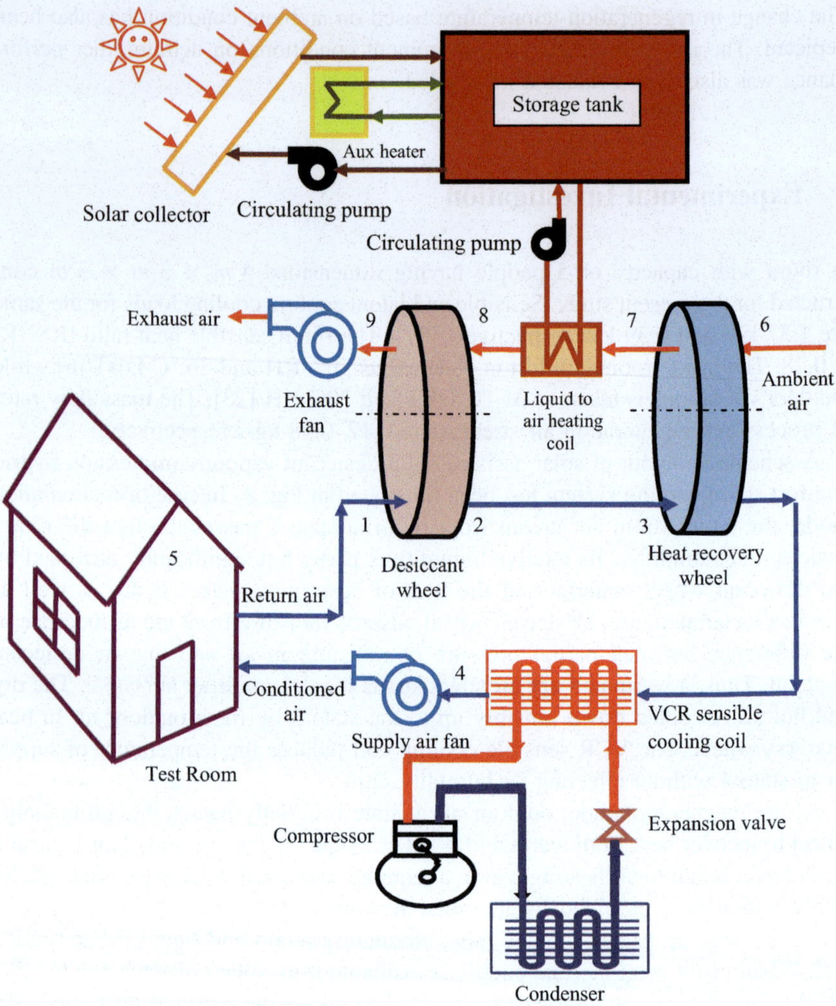

Fig. 1 Schematic layout of solar assisted hybrid desiccant cooling

The overall performance of solid-desiccant vapour-compression based hybrid comfort space cooling system used for the space cooling greatly depends upon a group of parameters and conditions, implicating the systems operational parameters such as air stream flow rate, regeneration temperature etc. and variable space cooling requirements owing to the variations in the environmental conditions and occupancy. The following calculations are performed to determine overall system performance and performances of individual components used in the system.

The performance of rotary desiccant dehumidifier is evaluated by calculating the moisture removal rate (MRR) and effectiveness (ε_{dw}). The moisture removal rate or moisture transfer rate [4] from air to the desiccant surface is defined as follows

$$MRR = m_{pa}(\omega_1 - \omega_2) \quad (1)$$

where, m_{pa} is the mass flow rate of process air stream at inlet of the dehumidifier, ω_1 and ω_2 are the humidity ratios of process air at the dehumidifier inlet and outlet respectively.

The effectiveness of rotary desiccant dehumidifier is evaluated as the ratio of the change in actual humidity ratio of the air to the maximum possible change in humidity ratio. The effectiveness of the rotary desiccant dehumidifier [8] can be expressed by the following relation

$$\varepsilon_{dw} = \frac{\omega_1 - \omega_2}{\omega_1 - \omega_{2,\,ideal}} \quad (2)$$

where, $\omega_{2,ideal}$ is the ideal specific humidity of the process air stream at the exit of the desiccant dehumidifier. Assuming that the air is completely dehumidified at this state point, the value of $\omega_{2,ideal}$ can be taken as zero.

Similarly, the effectiveness [9] of heat recovery wheel (ε_{hrw}) is given as

$$\varepsilon_{hrw} = \frac{T_2 - T_3}{T_2 - T_6} \quad (3)$$

The coefficient of performance [5] of the overall system as per electricity usage is defined as the ratio of the cooling capacity to the total power consumption (E_{total}) by the system. It is given by

$$COP = \frac{Q_{cc}}{E_{total}} \quad (4)$$

In the above equation, Q_{cc} is the cooling capacity [16, 24] and it is defined as

$$Q_{cc} = m_{pa}(h_1 - h_4) \quad (5)$$

where, m_{pa} is the mass flow rate of process air at the inlet of dehumidifier. While E_{total} represents the electricity usage to drive the system. Hence, E_{total} is calculated [25] as

$$E_{total} = E_{compressor} + E_{fan} + E_{heater} + E_{others} \quad (6)$$

where, $E_{compressor}$ and E_{fan} depict the electricity usage to run the VCR compressor and fans respectively. Fans are incorporated to circulate the regeneration and the process air streams and also for the traditional VCR unit. E_{heater} is electricity usage by the regeneration heater used in the dehumidifier. E_{others} depict electricity usage of other equipments such as desiccant wheel motor and enthalpy wheel driving motor. E_{total} is the total electrical power consumption of the system measured by using energy meter. The heat supplied for regeneration (Q_{reg}) can be given [17, 26] by

$$Q_{reg} = m_{ra}(h_8 - h_7) \tag{7}$$

where, m_{ra} is the mass flow rate of regeneration air stream, while h_7 and h_8 are the enthalpies of the regeneration air at inlet and outlet of the electric heater, respectively.

The inaccuracies in measurement of DBT, RH, air steam flow rate, pressure drop and electricity usage are $\pm$ 0.3 K @ 296 K, $\pm$ 2.0%, $\pm$ 3.0%, $\pm$ 1.5% and $\pm$ 2.0% respectively. The maximum uncertainty observed in case of coefficient of performance is found to be $\pm$ 14.63%.

Figure 2 depicts the impact of variation in ambient temperature on humidity ratio of process air at its outlet of rotary desiccant dehumidifier for various regeneration temperatures. It has been seen that the increase in ambient temperature at constant regeneration temperature leads to augment in humidity ratio of process air outlet of the desiccant wheel. This is attributed to the fact that the diffusion of the water vapor from process air to the desiccant material surface available in wheel matrix is mainly governed by the difference in partial pressure of air in process air and in hot desiccant surface. Thus, rise in ambient temperature makes to increment in process air temperature which in turn lowers the partial vapor pressure of process air at inlet of the desiccant wheel. This further leads to lower vapor pressure

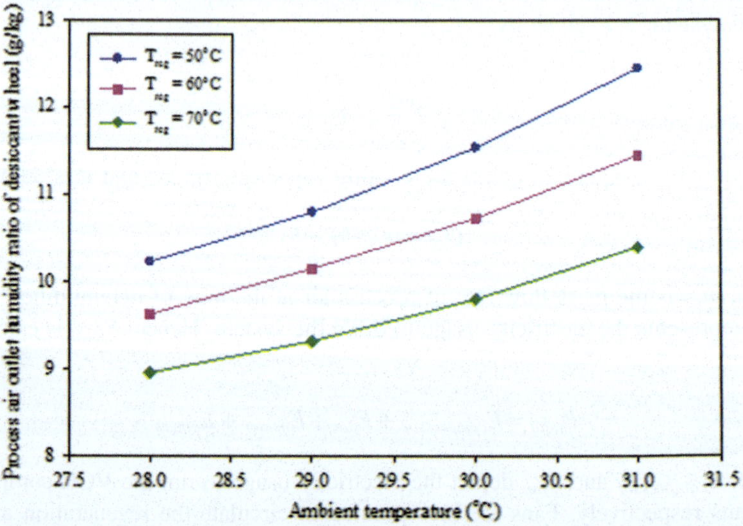

Fig. 2 Effect of variation in ambient temperature on humidity ratio of process air at desiccant wheel outlet

difference between the air and the desiccant along the channel. Since the moisture attraction by the desiccant material from the process air is based on the difference in vapor pressure between desiccant material surface in channel and moist air flowing through it, the moisture removal rate of the dehumidifier gets lowered. Thus, humidity ratio of process air at the dehumidifier outlet ameliorates. It is observed that rate of rise in humidity ratio of process air at desiccant wheel outlet is ameliorates when the ambient temperature increases above 30 °C.

Figure 3 illustrates the impact of variation in outdoor temperature on dehumidifier effectiveness at various regeneration temperatures. It is seen that increase in outdoor temperature leads to lower desiccant wheel effectiveness at particular regeneration temperature. An increase in process air temperature at the inlet of desiccant wheel due to increase in ambient temperature, the difference of vapor partial pressure between process air and the desiccant material surface lowers leading to lowers diffusion process of water vapor droplets from process air to desiccant felt. Thus, rise in temperature of process air at desiccant wheel inlet due to ameliorated ambient temperature makes decrement in moisture removed and thus reduces the dehumidifier effectiveness. Since the adsorption process inside the dehumidifier is exothermic hence it is progressed by low temperature of moist process air for better effectiveness of dehumidifier. At a fixed ambient temperature, the rise in regeneration temperature increases the vapor partial pressure difference between desiccant matrix and moist process air due to large temperature difference between them which ameliorates adsorption and ultimately leads to higher effectiveness.

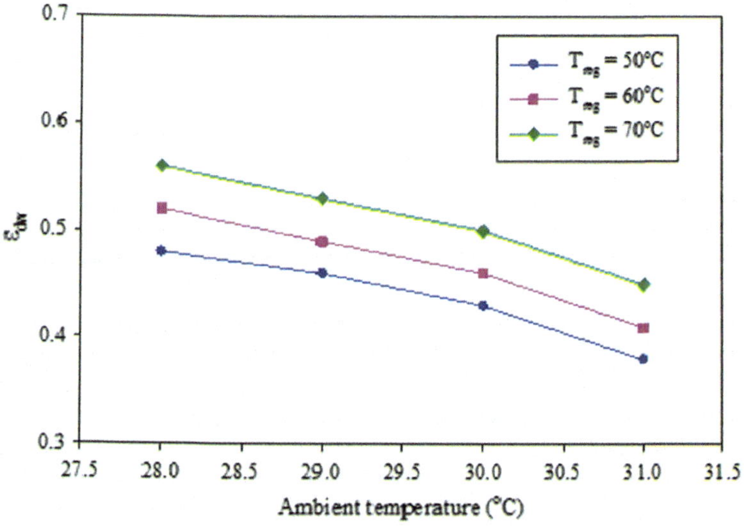

Fig. 3 Effect of variation in ambient temperature on dehumidifier effectiveness at different regeneration temperatures

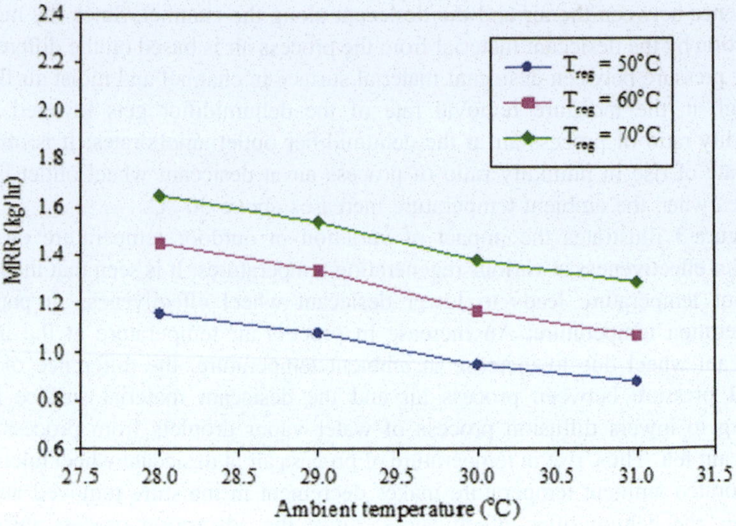

Fig. 4 Effect of variation in ambient temperature on moisture removal rate (MRR) at different regeneration temperatures

Figure 4 shows the impact of variation in ambient dry bulb temperature over moisture removal rate (MRR) at different regeneration temperatures. It is found that due to increase in ambient temperature, MRR decreases at a particular regeneration temperature.

At higher ambient dry bulb temperature, the process air dry bulb temperature at desiccant wheel inlet rises that in turn lowers partial vapor pressure of process air at dehumidifier inlet. Due to this, vapor pressure difference among air and desiccant felt along the channel gets reduced. As the moisture attraction by the desiccant material from the process air depends on the variation in vapor pressure between desiccant felt and moist air flowing across, the moisture removal rate gets reduced. It is also observed from the behavior that MRR is a strong function of regeneration temperature. This is due to the fact that by increasing the regeneration temperature, the dehumidification capabilities of the desiccant wheel rises, and hence the MRR. At constant ambient temperature, increasing regeneration temperature enhances the temperature difference between hot desiccant matrix and humid process air steam that rises adsorption moisture within dehumidifier and leads to higher MRR.

Figure 5 depicts the impact of changes in outdoor temperatures on coefficient of performance (COP) at various regeneration temperatures.

It is seen that COP of the system lowers with the rise in ambient air temperature as well as in regeneration air temperature. It is owing to matter that with an increase in regeneration temperature, sensible heat transferred to the process air stream becomes better, leads to a higher outlet temperature of the process air. This ultimately needs to be cool down by enthalpy wheel and the conventional VCR air-conditioning unit, which lowers the COP. Moreover, at higher ambient

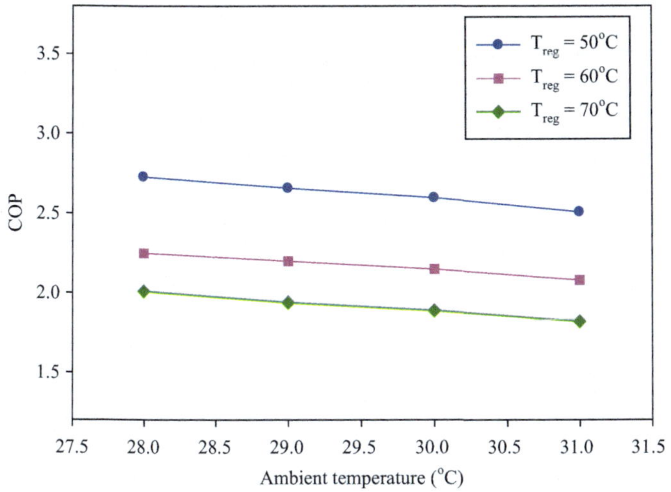

Fig. 5 Effect of variation in ambient temperature on COP at different regeneration temperatures

temperature, the MRR and ultimately the effectiveness of the dehumidifier also get lowered as explained before. That leads to raise the latent cooling load on sensible cooling coil that leads to affect the COP of system. On the other side, while rising the regeneration temperature, the electricity usage of heater rises which ultimately leads to a reduction in system COP.

3 Performance Prediction Using TRNSYS

For the transient system simulation of solid-desiccant vapor-compression based hybrid comfort space cooling system powered by solar energy, a transient energy system simulation software TRNSYS was used [36]. Simulation studio project has been developed to perform simulations of the system. Simulation results were obtained for different ambient conditions of the system at hot and humid season from March to September. The yearly weather data file (TMY) for Roorkee has been taken as input for the TRNSYS model. Other inputs include air flow rates, temperature, relative humidity of air and the regeneration temperature. The outputs of the simulation are temperature and humidity of the process air at each state point of the system. These outputs are further used to calculate the coefficient of performance of the system. Simulation of different projects in TRNSYS for solar assisted hybrid desiccant cooling system gives the performance of individual component and their interaction within the system. The key benefits of simulating the system in TRNSYS are due to its ability to make quick changes in the system configuration and in getting the graphical output simultaneously in the form of time varying quantities. The TRNSYS simulation was mainly used to predict the system

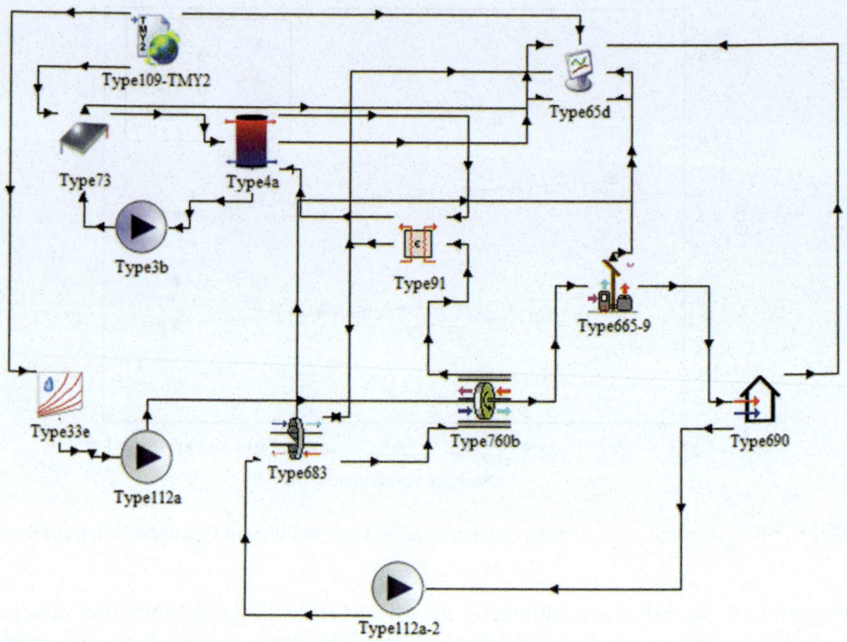

Fig. 6 TRNSYS simulation studio project

performance according to variation in ambient weather conditions to avoid unnecessarily large amount of experimental test results which are costlier and increases engineering effort.

Figure 6 shows the TNRYSYS simulation studio model of solar assisted hybrid solid desiccant—vapor compression comfort space cooling system operating in recirculation configuration by use of TRNSYS software. In TRNSYS simulation studio project, desiccant wheel is modeled as type 683, type 760b is heat recovery wheel, type 665–9 is split conditioner, type 690 is room load, type 33e is psychrometrics for condition calculation, type 73 is solar collector, type 3b is circulation pump, type 4a is tank, type 91 is heat exchanger, type 65d is plotter and type 109–TMY2 is weather data input file, type 112a and type 112a–2 are fans used to circulate process and regeneration air respectively. During TRNSYS modeling, all the simulation models are created in the assembly panel environment of simulation studio which is basically a visual interface for system simulation. Standard library components are brought into the assembly panel with the help of direct access tool bar or pull down menu. Components are connected in such a way that the output of one component goes to the input of next component. This way the components or modules form a flow chart of information going from one component to the other during simulation run.

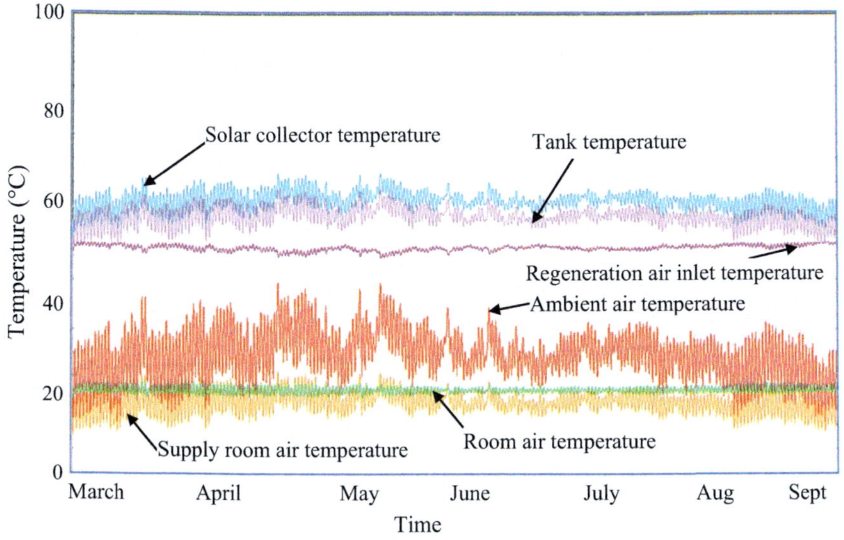

Fig. 7 Variation of temperatures with time at various state points

Figure 7 shows the behaviour of temperatures of solid desiccant - vapor compression hybrid comfort space cooling system for variation with time at various state points. Ambient air temperature, room air temperature, supply room air temperature, solar collector temperature, tank temperature and reactivation air inlet temperature were simulated. Out of them the reactivation temperature is one of the most important temperatures that has a key role in variation in the overall system performance.

Figure 8 shows the behaviour of humidity ratio (g/kg) of solid desiccant vapor compression hybrid comfort space cooling system with time for various state points. Humidity ratio of ambient air, room air, supply room air and regeneration air outlet have been shown. Humidity ratio is the important parameter responsible for removal of moisture from room air in terms of latent heat to obtain required comfort conditions within room.

It can be seen that in some occasions during the month of May, June and July, humidity ratio is very high indicating the hot and humid ambient conditions during those period. Supply air humidity ratio also increases during those period demanding higher regeneration temperature for efficient control over supply humidity ratio. Moreover, that also rises the latent load for cooling over sensible cooling coil that also influence total energy consumption of the hybrid comfort space cooling system. There is almost 50% decrease in humidity ratio of the process air while passing through desiccant wheel is depicted from the simulation results.

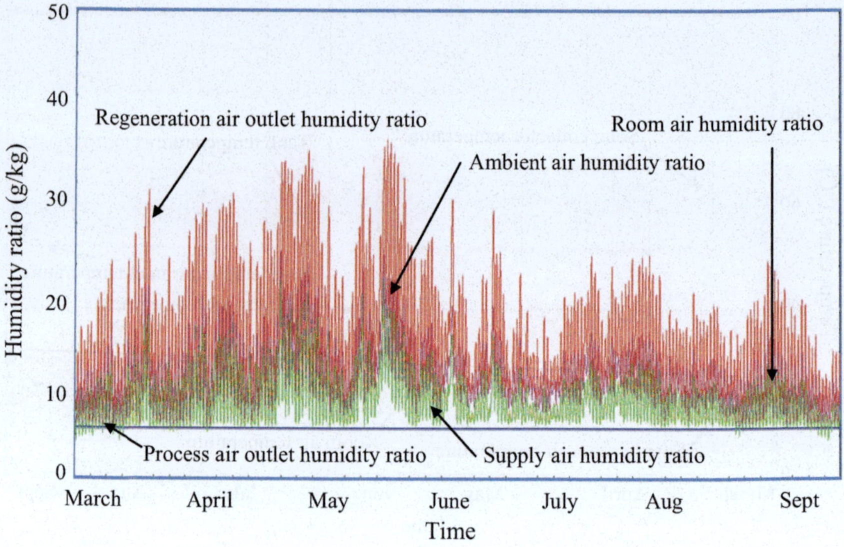

Fig. 8 Variation of sp. humidity ratios with time at various state points

4 Conclusions and Future Prospects

The results of TRNSYS simulation for solar assisted solid desiccant - vapour compression hybrid comfort space cooling system highlights good performance for hot - humid climatic conditions and justifies its use. It is also seen that the coefficient of performance of the system is very sensitive to the variation in reactivation temperature and ambient humidity ratio. Moreover, the changes in the temperatures and humidity ratios of ambient air during the cooling season were used to study its influence on the coefficient of performance of the system. The simulation results mainly give the output parameters or the system performance varying over the selected time period. One needs to fix-up the time for getting the output/performance at that instant.

The use of TRNSYS can best be exploited by the design engineers by considering the extreme range of parameters in their design. Also, the performance can be predicted for such extreme conditions.

References

1. Ahmed MH, Kattab NM, Fouad M (2005) Evaluation and optimization of solar desiccant wheel performance. Renew Energy 30:305–325
2. Alili AA, Hwang Y, Radermacher R (2014) A hybrid solar air conditioner: experimental. Int J Refrig 39:117–124

3. Angrisani G, Minichiello F, Sasso M (2016) Improvements of an unconventional desiccant air conditioning system based on experimental investigations. Energy Convers Manag 112:423–434
4. Angrisani G, Minichiello F, Roselli C, Sasso M (2010) Desiccant HVAC system driven by a micro-CHP: experimental analysis. Energy Build 42:2028–2035
5. Angrisani G, Minichiello F, Roselli C, Sasso M (2011) Experimental investigation to optimize a desiccant HVAC system coupled to a small size co-generator. Appl Therm Eng 31:506–512
6. ASHRAE (2013) Chapter 9, ASHRAE handbook-fundamentals. ASHRAE, Atlanta
7. Baniyounes A, Liu G, Rasul M, Khan M (2012) Analysis of solar desiccant cooling system for an institutional building in subtropical Queensland, Australia. Renew Sustain Energy Rev 16:6423–6431
8. Banks PJ (1972) Coupled equilibrium heat and single adsorbate transfer in fluid flow through porous media—I, characteristic potentials and specific capacity ratios. Chem Eng Sci 27:1143–1155
9. Barlow R (1982) Analysis of the adsorption process and desiccant cooling systems: a pseudo-steady-state model for coupled heat and mass transfer. Technical Report No. SERI/TR-631-1330, Solar Energy Research Institute, Golden, CO
10. Bourdoukan P, Wurtz E, Joubert P (2009) Experimental investigation of a solar desiccant cooling installation. Sol Energy 83:2059–2073
11. Bourdoukan P, Wurtz E, Sperandio M, Joubert P (2007) Global efficiency of direct flow vacuum collectors in autonomous solar desiccant cooling: simulation and experimental results. In: Proceedings: building simulation, pp 342–347
12. Dai Y, Li X, Wang R (2015) Theoretical analysis and case study on solar driven two-stage rotary desiccant cooling system combined with geothermal heat pump. Energy Procedia 70:418–426
13. Dhar PL, Singh SK (2001) Studies on solid desiccant based hybrid air-conditioning. Appl Therm Eng 21:119–134
14. Fong KF, Lee CK, Chow TT, Fong AML (2011) Investigation on solar hybrid desiccant cooling system for commercial premises with high latent cooling load in subtropical Hong Kong. Appl Therm Eng 31:3393–3401
15. Halliday SP, Beggs CB, Sleigh PA (2002) The use of solar desiccant cooling in the UK: a feasibility study. Appl Therm Eng 22:1327–1338
16. Hurdogan E, Buyukalaca O, Yilmaz T, Hepbasli A (2010) Experimental investigation of a novel desiccant cooling system. Energy Build 42:2049–2060
17. Jain S, Dhar PL, Kaushik SC (1995) Evaluation of solid based evaporative cooling cycles for typical hot and humid climates. Int J Refrig 18:287–296
18. Jani DB, Mishra M, Sahoo PK (2013) Simulation of solar assisted solid desiccant cooling systems using TRNSYS. In: Proceedings of the 22th national and 11th international ISHMT-ASME heat and mass transfer conference (ISHMT-ASME-2013), IIT, Kharagpur, Dec 28–31, pp 1–7
19. Jani DB, Mishra M, Sahoo PK (2015) Performance studies of hybrid solid desiccant—vapor compression air-conditioning system for hot and humid climates. Energy Build 102:284–292
20. Jani DB, Mishra M, Sahoo PK (2015). Numerical simulation of rotary desiccant dehumidifier for hybrid solid desiccant—vapor compression air-conditioning system. In: The proceedings of the 24th IIR international congress of refrigeration, Yokohama, Japan, Aug 16–22, pp 1–8
21. Jani DB, Mishra M, Sahoo PK (2015) Experimental investigations on hybrid solid desiccant—vapor compression air-conditioning system for Indian climate. In: The proceedings of the 24th IIR international congress of refrigeration, Yokohama, Japan, Aug 16–22, pp 1–9
22. Jani DB, Mishra M, Sahoo PK (2016) Performance prediction of rotary solid desiccant dehumidifier in hybrid air-conditioning system using artificial neural network. Appl Therm Eng 98:1091–1103
23. Jani DB, Mishra M, Sahoo PK (2016) Solid desiccant air conditioning—A state of the art review. Renew Sustain Energy Rev 60:1451–1469

24. Jani DB, Mishra M, Sahoo PK (2016) Performance analysis of hybrid solid desiccant—vapor compression air-conditioning system in hot and humid weather of India. Build Serv Eng Res Technol 37:523–538
25. Jani DB, Mishra M, Sahoo PK (2016) Performance prediction of solid desiccant—vapor compression hybrid air-conditioning system using artificial neural network. Energy 103:618–629
26. Jani DB, Mishra M, Sahoo PK (2016) Experimental investigation on solid desiccant—vapor compression hybrid air–conditioning system in hot and humid weather. Appl Therm Eng 104:556–564
27. Jia CX, Dai YJ, Wu JY, Wang RZ (2006) Analysis on a hybrid desiccant air-conditioning system. Appl Therm Eng 26:2393–2400
28. Joudi KA, Madhi SM (1987) An experimental investigation into solar assisted desiccant-evaporative air-conditioning system. Sol Energy 39:97–107
29. Kim DS, Ferreira CAI (2008) Solar refrigeration options-a state of the art review. Int J Refrig 31:3–15
30. Koronaki IP, Rogdakis E, Kakatsiou T (2012) Experimental assessment and thermodynamic analysis of a solar desiccant cooling system. Int J Sustain Energy 1–16
31. Kulkarni K, Sahoo PK, Mishra M (2011) Optimization of cooling load for a lecture theater in a composite climate in India. Energy Build 43:1573–1579
32. La D, Dai YJ, Li Y, Wang RZ, Ge TS (2010) Technical development of rotary desiccant dehumidification and air conditioning: a review. Renew Sustain Energy Rev 14:130–147
33. Liu W, Lian Z, Radermacher R, Yao A (2007) Energy consumption analysis on a dedicated outdoor air system with rotary desiccant wheel. Energy 32:1749–1760
34. Maalouf C, Wurtz E, Mora L, Allard F (2005) Optimization and study of an autonomous solar desiccant cooling system. In: Proceedings of the international conference passive and low energy cooling for the built environment, Santorini, Greece, May 1–3, pp. 663–668
35. Subramanyam N, Maiya MP, Murthy SS (2004) Application of desiccant wheel to control humidity in air conditioning system. Appl Therm Eng 24:2777–2788
36. TRNSYS (2006) TRNSYS 16, a transient system simulation program. The solar energy laboratory, University of Wisconsin-Madison, Madison, U.S.A
37. Yadav A, Bajpai VK (2012) Experimental comparison of various solid desiccants for regeneration by evacuated solar air collector and air dehumidification. Drying Technol 30:516–525
38. Yadav YK, Kaushik SC (1991) Psychrometric techno economic assessment and parametric studies of vapour-compression and solid/liquid desiccant hybrid solar space conditioning systems. Heat Recovery Syst CHP 11:563–572
39. Yadav YK (1995) Vapour-compression and liquid-desiccant hybrid solar space-conditioning system for energy conservation. Renew Energy 6:719–723
40. Yong L, Sumathy K, Dai YJ, Zhong JH, Wang RZ (2006) Experimental study on a hybrid desiccant dehumidification and air conditioning system. J Sol Energy Eng 128:77–82

Solar Food Processing and Cooking Methodologies

Abhishek Saxena, Varun Goel and Mehmet Karakilcik

Abstract In this study, a theoretical analysis of food processing (e.g., solar drying), worldwide cooking pattern, and cooking methods by using the solar energy has been reviewed. Solar food processing method is applied as direct absorption, air heater, and a combination of direct and indirect drying by solar radiation. Therefore, this process is one of the most accessible and hence the most widespread processing technologies. Traditional solar drying involves keeping products in the direct sunlight. Solar drying and cooking processes take place at different temperatures and timescales, and it depends on the nature of the food or substance. The amount of solar energy that reaches to the system and design parameters determines the performance of food processing and cooking systems. The time duration of drying and cooking depends on the temperature of heated air and environment. The temperature distributions, mass, and ingredient of food have an important role in the performance of dryers and cooker boxes. For a better understanding of the system parameters, the concept of solar food processing has been discussed thermodynamically. Energy saving by using solar systems has also been discussed.

Keywords Solar food processing · Solar drying · Cooking methodologies Energy analysis

A. Saxena (✉)
Mechanical Engineering Department, Moradabad Institute of Technology, Moradabad 244001, India
e-mail: culturebeat94@yahoo.com

V. Goel
Mechanical Engineering Department, National Institute of Technology, Hamirpur, India

M. Karakilcik
Faculty of Sciences and Letters, Department of Physics, University of Cukurova, Adana 01330, Turkey

H. Tyagi et al. (eds.), *Applications of Solar Energy*, Energy, Environment, and Sustainability, https://doi.org/10.1007/978-981-10-7206-2_13

Nomenclature

Symbols

PBP	Payback period
x	Interest rate
y	Inflation rate
Z	Manufacturing cost of cooker
M	Maintenance cost
E	Energy saving per year
FC	Fuel combusted per year
OF_{fc}	Fuel oxidation factor
HHV_{fc}	Fuel higher heating value
CCC_{fc}	Carbon content coefficient
MW_{CO2}	Molecular weight of CO_2
MW_c	Molecular weight of carbon
E	Energy
zdT	Drop in temperature
MC	Heat capacity
Q	Heat
F	Heat exchanger efficiency factor
L	Latent of heat vaporization
W	Weight of water evaporated
LCV	Lower calorific value of fuel
P_f	Power of fan
Mb	Mass of fuel consumed
Asc	Crop surface area
K_f	Thermal conductance of air
K_m	Mass transfer coefficient of vapor
P_v	Vapor pressure

Subscripts

o	Output
in	Input
avg	Average
w	Water
wf	Final
if	Initial
th	Thermal
amb	Ambient
pm	Plate mean
i	Insulation
p	Plate
c	Glass cover
w	Wind

L Loss
a Air
d Dried substance
sc Solar collector
e Equilibrium

Greek letters

ε Emissivity
τ Drying time
η Efficiency
λ Latent heat evaporation

1 Introduction

Sun is a very large, inexhaustible source of energy. The power from the sun intercepted by earth is approximately 1.8×10^{11} MW which is many thousands of times larger than the present consumption rate on earth of all commercial energy sources. Smoke caused by cooking and heating is the main cause of many respiratory diseases altogether with environmental pollution. The World Bank's Development Report—1993—reports that eliminating indoor smoke could cut childhood pneumonia by half and reduce the burden of other diseases by 5%. It has been estimated that at present more than 15 million hectares of forest are lost per year in developing countries mainly due to consumption of firewood. According to UN Food and Agricultural Organization, some 2400 million people are expected to face acute fuelwood shortage by the end of century with serious nutritional and health consequences. So we need to move to renewable sources of energy in which solar energy is the most promising one. Thus, in principle, solar energy could supply all present and future energy needs of the world on a continuing basis. Also in India, energy subsidies are provided for petroleum products including kerosene, diesel, and LPG as well as for electricity. In 2010–2011, this accounted to more than ₹ 25,000 crores in total LPG subsidies (IISD 2012). So, this huge amount of money can be utilized to develop and improve methods to harness solar energy, i.e., solar cell, solar cooker, solar drier.

1.1 History

Food processing times back to the early age when the crude processing containing different types of cooking, such as over biomass firing, smoking, baking and steaming, fermenting, solar drying, solar cooking and preserving with some salts was in practice. Food preserved in this manner was a mutual part of 'soldiers' and 'sailors' diets. These crude processing methods were continued to be the same until the initiation of the industrial revolution [1]. In 1809, Nicolas Appert had developed a vacuum bottling practice by using heat energy for supplying food to the troops in the French army, which was ultimately led to sacking in tins by Peter Durand in 1810. He developed the theory of food preservation in airtight jars and containers by applying sufficient heat energy, and by this, food would last longer safe from bacterial spoilage. In 1864, Louis Pasteur discovered pasteurization which improved the quality of preserved food. In the nineteenth century, food processing technologies were developed on a large scale to fulfill military needs [2]. Later on, in the twentieth century, the major changes in food habits (eating and cooking both) and the quality awareness of the humans toward the development of food processing have been observed [3]. Apart from this, at present, the trends of cooking and the demand of food have quite changed. Now, ready-to-eat foods are available across the globe with a possibility of cooking/baking or roasting, at almost all locations [2–4]. Some common techniques of food preservation are shown in Fig. 1.

In technical aspect, heating is a common and reliable practice for food treatment. Thermal processing of foods, such as heating, pasteurizing or boiling, drying, baking or roasting, frying and grilling, affects the quality of the food (nutritional values). Thermal processing can be done by both conventional methodology and modern techniques for a quality food [5]. During heating of food (based on Millard reaction), all bacteria, inactive enzymes, and microorganisms are died, while the vitamins, flavors, colorants, and quantity are less affected [6]. In the present modern techniques

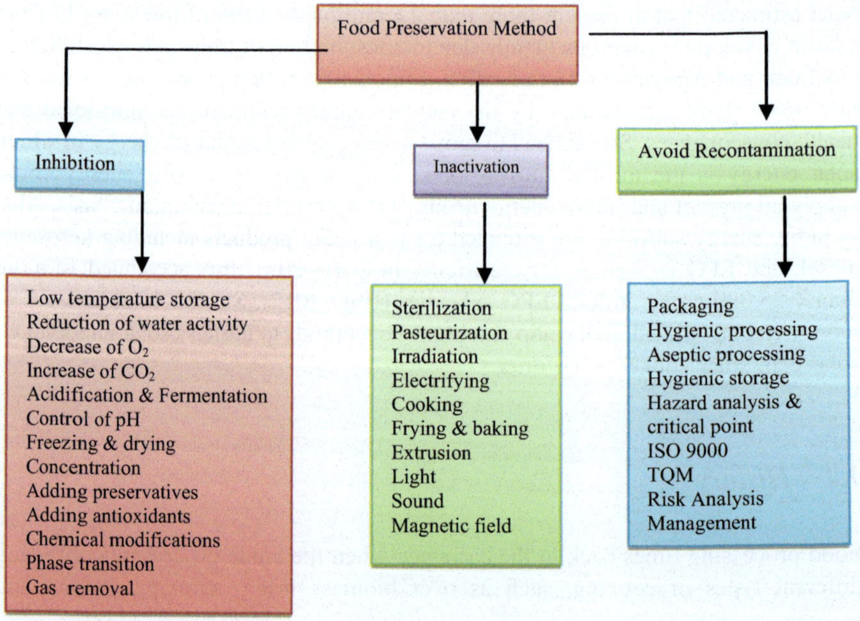

Fig. 1 Major food preservation techniques [5]

or food processing methods, there is no risk of toxicity, bacterial or allergic reactions. Heating also improves palatability, microbiological safety, and shelf life [7].

1.2 Solar Food Processing

Solar food processing is a developing technology which provides quality foods almost at negligible or minimum cost. There are a number of drying techniques for various substance or chemicals [as shown in Table 1]. Solar drying is a common practice for the drying of vegetables, fruits, or chemicals without losing their useful properties [8]. Different designs of solar dryers are used globally according to the substance. Besides this, open sun drying is also used for some foods with low moisture such as chili, pepper, pickle. Forced convection-type solar dryers are comparatively better to open sun drying for dehydration [9]. In general, solar food processing fetches in two emergent concepts together to resolve the two major issues of the twenty-first century: First one is how to produce sufficient energy for global population, and second is how to meet their demands of feeding and living. Obviously, the reason is population blast and unequal wealth distribution in different nations [9, 10]. Food production per capita above poverty line is necessary for a sustainable life. But, the problem still prevails in major portions of the different countries because of lack of education and unemployment [11]. By knowing the easy and a safe way of food preservation and full utilization of reliable energy sources (such as solar energy), people can sort out this problem at their end, especially in rural areas.

If one can talk about the effect of heat on microorganisms, then it is notable that the preserving result of heat dispensation is just because of denaturation of proteins, which terminates enzyme commotion and enzyme-controlled absorption in microorganisms. The level of demolition is a first-order reaction, i.e., when food is heated up to a temperature at which contaminating microorganisms destroy; the same proportion of microorganism decease in a particular period of heating [5]. The demolition of microorganisms depends upon the temperature; cells decease more swiftly at peak temperatures. There are many factors that control the heat resistance of microorganisms, but common reports of the result of a assumed variables for heat resistance are not continuously conceivable [13]. Given factors are acknowledged for significance.

(i) Nature of microorganisms,
(ii) Cultivation conditions during cell evolution or spore development (temperature, age of the culture, and culture mediocre used),
(iii) Improper heat treatment (i.e., pH value of the food, water activity of the food, composition of the food, and the growing media and cultivation conditions).

Knowledge of the temperature resistance of the different enzymes or microorganisms found in particular food can be used to estimate the temperature conditions desired for destruction of them. In common practice, the most heat-resilient enzyme or microorganism in a specified food is generally used as a base of calculating process conditions. It is supposed that supplementary less heat-resistant species will also be destroyed [14].

Table 1 Summary of novel and conventional drying techniques for food [12]

Technique	Suitability/current usage	Advantages	Disadvantages
Microwave drying and dielectric drying	High value-added products	Low temperature, batch or continuous operation, good quality	Slow and expensive
Microwave-augmented freeze drying	High value-added products	Low temperature, rapid, good quality	Expensive
Centrifugal fluidized bed drying	Small particles, vegetable pieces, powder	Rapid, easy to control	Loss of product integrity, noisy
Ball drying	Small particles, vegetables pieces	Relatively low temperature, rapid continuous	Loss of product integrity, difficult to control
Ultrasonic drying	Liquids	Rapid	Requires low fat solutions
Solar open drying	Fruit, meat, fish plant	Simple, low cost	Large space required, slow, labor intensive, difficult to control
Smoking	Meat, fish	Added flavors	Difficult to control, slow
Convection drying	Low value-added products	Continuous	Difficult to control
Drum drying	Liquids, gelatin	Continuous	Modification of liquid
Freeze drying	Value-added products	Continuous, no restriction particle size, low temperature	Slow, expensive
Fluidized bed drying	Small uniform particles, small vegetables	Usually batch operation, uniform drying, rapid	Restriction on particle size
Osmotic drying	Sugar infused fruit	High quality	Two-step process

Besides this, if one takes a look in typical mechanism of irradiance, then ionizing radiation takes the form of gamma rays from isotopes or from X-rays and electrons. It has already been acceptable in 38 countries to preserve foods by devastation of microorganisms or reserving of biochemical variations [13, 14]. The irradiation procedure includes exposing foods either prepackaged or in bulk, to a determined level of ionization heat. The application of solar irradiance [Table 2] to organic materials has a direct/indirect effect, in which the direct effect is a result of energy deposition by the irradiance in the object molecule, while the unintended effects transpire as a significance of sensitive diffusible free extremists fashioned from the radiolysis of water. Solar irradiance has a very wide scope in food disinfection, shelf life allowance, refinement, and substance quality improvement. All the details are thoroughly investigated through a comprehensive review by Wilkinson and Gould [15] with an importance of solar irradiance. The common benefits of solar irradiance are as follows:

(i) There is only marginal heating of the substance and thus negligible change to physical characteristics.

Table 2 Applications of food irradiation [15]

Application	Dose range (kg)	Examples of foods	Countries with commercial processing
Sterilization	7–10	Herbs, spices	Belgium, Canada, Croatia, Denmark, Finland, Israel, Korea, Mexico, South Africa, USA, Vietnam
	Up to 50	Long-term ambient storage of meat	None
Sterilization of packaging materials	10–25	Wine corks	Hungary
Destruction of pathogens	2.5–10	Spices, frozen poultry, meat, shrimps	Belgium, Canada, Croatia, USA, Denmark, Finland, Israel, Korea, Mexico, South Africa, Vietnam
Control of molds	2–5	Extended storage of fresh fruit	China, South Africa, USA
Extension of chill life from 5 days to 1 month	2–5	Soft fruit, fresh fish, and meat at 0–4 °C	China, France, South Africa, USA, the Netherlands
Inactivation of parasites	0.1–6	Pork	–
Disinfection	0.1–2	Fruit, grain, flour, cocoa beans, dry foods	Argentina, Brazil, Chile, China
Inhibition of sprouting	0.1–2	Potatoes, garlic, onions	Algeria, Bangladesh, China, Cuba

(ii) Packed and freezing foods may be treated.
(iii) Fresh foodstuffs may be conserved in a single process and without using chemical preservers.
(iv) Energy demand is very low.
(v) Variations in nutritive value of foodstuffs are comparable with other approaches of food conservation.
(vi) Processing is habitually organized and has low functioning costs.

1.2.1 Concepts of Thermal Processing

Concepts of thermal processing clearly show that the traditional thermal food processing depends on certain old ideas and keys, which are employed well, while current food study has raised some queries in contradiction to those perceptions. The traditional theory of the D value and of the z value is illustrated in Fig. 2a, b. To define a D value of a specific strain of microorganisms, illustrations of these microorganisms are bare to a higher temperature for a figure of time steps. The

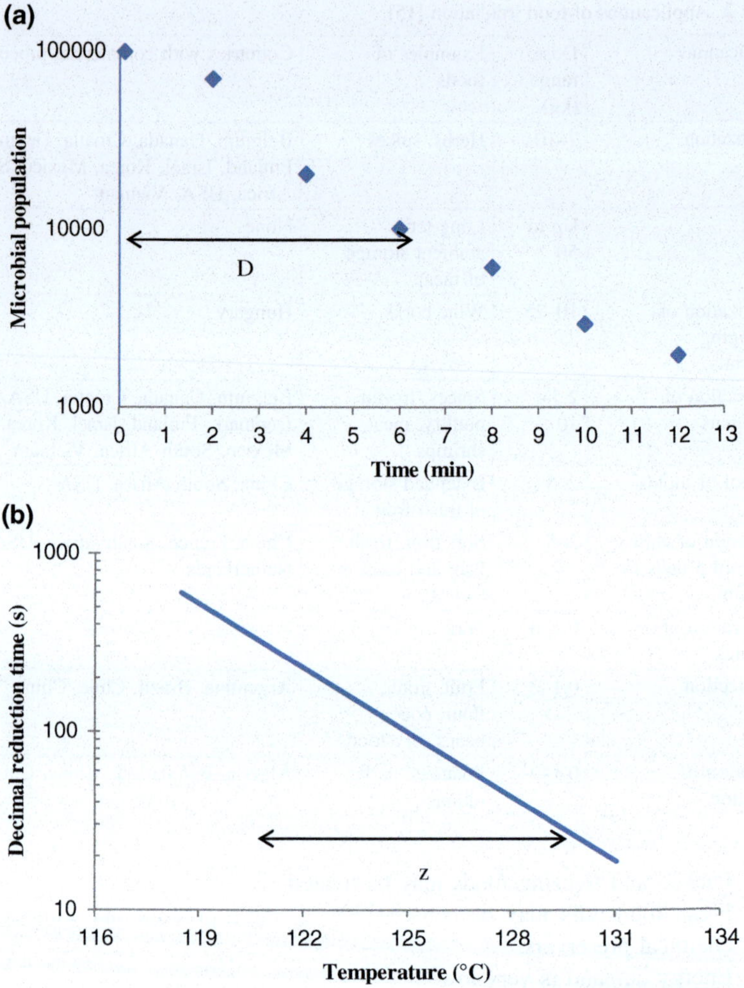

Fig. 2 **a** Microbial survivor curve on semi-logarithmic coordinates (D value) [10], **b** A plot of logarithm of D versus temperature used to determine the thermal resistance constant (z) [10]

residual amount of microorganisms is plotted on a log-scale versus treatment phase, and the data points bring into line linearly. The *D* value can be defined as the time required for reducing the sum of microorganisms at that particular temperature by one log cycle. The *z* value defines the necessity of the *D* value on a fixed temperature [16]. These standards are usually used to define the heat inactivation of bacteria.

Yet, all thermal falling-off reactions, including the obliteration of heat-sensitive food constituents, follow these laws. Figure 2b shows a mutual concept of thermal food processing, i.e., a high-temperature short-time treatment (HTST). Besides this, as per the '*D*' and '*z*' values, various time–temperature configurations can be established, which are really adequate to attain the same consequence of pasteurization.

1.2.2 Effect of Heat on Nutritional and Sensory Characteristics

The obliteration of various vitamins, aroma amalgams, and stains by heat follows a similar first-order reaction to microbial demolition. Commonly, both the values are greater than those of microorganisms and enzymes. As an outcome, nourishing and sensory properties are improved and retained by the usage of high heat (temperature) and short span of time during heat treating [17]. It is thus conceivable to select specific time–temperature permutations from a thermal death time curve (all of which attain the same degree of enzyme or bacterial demolition), to optimize a method for nutrient preservation of desired sensory potentials. This conception forms the basis of specific rapid blanching, high temperature, short-time pasteurization, high heat sterilization, extrusion, etc. [18].

Heat treatment is an important method, generally used in food processing, due to the desired belongings on eating quality (different eatables are consumed in cooking processes such as baking yield flavors which cannot be produced by other method), but due to a preserving effect on diets by the annihilation of enzymes, microorganisms, pests, and organisms. Some other main advantages of heat treatment of foods are as follows [16]:

(i) Relatively a simple mechanism of dispensation conditions,
(ii) Competence to develop shelf-stable foods which do not need freezing,
(iii) Obliteration of anti-nutritional issues (e.g., trypsin inhibitor in some pulses),
(iv) Improvement in the obtainability of certain nutrients (e.g., better digestibility of proteins, gelatinization of starches).

Besides this, the some common nutritious benefits of thermal treatment of food are [10]:

(i) Removal of harmful constituents (bacterial contamination, venom, enzyme inhibitors, or allergens protection of food);
(ii) Modifications in food environment erection and texture (advancement in digestion, improved bioavailability);
(iii) Generation of valuable new combinations (smell, antioxidants, etc.).

1.3 Cooking Scenarios in Different Countries

Cooking is a primary need, or it is a major activity of every household and commercial places (such as hotels, hostels, hospitals, restaurants, small roadside shops) and in some common transportation models such as trains and airplanes. It is also notable that nowadays, ready-to-eat meal (light and heavy stuffed food) is available in the market, which is usually carried by various persons to their working places, and through a heating process, this will be ready within few minutes for eating. This is the actual practice for eating and selling of different kinds of food, across the globe [4, 19]. Also in some countries like, India, Pakistan, Nepal, Bangladesh, Bhutan etc., where some religious trends such as celebrations of festivals (Diwali, Holi, Eid, Lohri, X-Mas, etc.), marriages, other functions/parties or even a funeral gathering, when food is

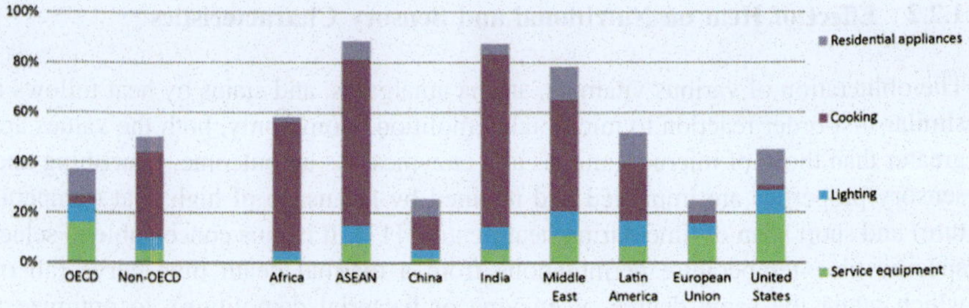

Fig. 3 Energy consumption of different households for some different countries [19]

cooked and served to all gathered people [20]. The food in these activities is generally cooked in open or closed places (like an open field or ground and a banquet hall) for a long time through various cooking fuels such as LPG, firewood, or coal.

In this section, the efforts have been made to highlight the present scenario of cooking foods with their methodologies and obviously the fuel used for cooking in different countries or region.

In emerging countries, at present around 2.6 billion people rely on biomass in rural areas to meet their daily energy needs for cooking or heating [19]. Figure 3 shows the energy consumption for different household activities in which energy consumption for cooking is much higher than other activities, across the globe. Apart from this, it is also notable that these resources account for more than 90% of household energy consumption in many countries. If the problem remain same, i.e., lack of certain new policies, then the number of people depending on biomass will rapidly increase to over 2.7 billion by 2030 due to unstoppable population growth. Statistics shows clearly that one-third population of this world will still be depending on these conventional fuels [21].

There are few corresponding methods which can improve this condition [19–21]:

(i) Encouragement of more effective and sustainable use of conventional biomass (fuel) as well as encouraging people for switching to updated cooking fuels and skills.
(ii) Dynamic and rigorous government plan and action is required to accomplish this target, organized with improved funding from both the public and private sectors/sources.
(iii) Strategies to encourage, more efficient cleaner fuels and technologies for cooking necessity to report barriers to access, adoptability or affordability and supply chain, and to formulate a dominant module of larger development policy.

Lack of new policies for the development of cleaner fuels and new technologies of cooking is the major components in consumption of conventional fuels, while another major issue is world's population blast and continuously increasing daily higher energy demands for cooking and heating across the globe. The government of different countries (or state-wise) should develop new policies for cleaner fuels such as LPG, electricity.

1.3.1 Household Cooking Patterns, Fuel Switching, and Policies

Commonly, there are three types of household in every country that are categorized according to particular household income, viz higher income household (HIH), medium income household (MIH), and lower income household (LIH), and they can also be categorized as rural or urban households in some developing countries. Each household has their own choice of fuel for cooking and heating according to their need and affordability [22]. According to the energy ladder [23], LIHs generally use the wood (including wood chips, straw, shrubs, grasses, and bark), charcoal, agricultural residues, and dung, while MIHs use wood, residues, dung, kerosene, and biogas and lastly HIHs use wood, kerosene, biogas, LPG, natural gas, electricity, and coal for cooking in developing countries [24].

Available literature and reports by World bank, IEA (International Energy Agency), WHO, and energy outlooks show that households normally use a combination of different fuels for cooking and heating (see Table 3 for example) which can be characterized as traditional (such as dung cakes, crop residues, and fuelwood), transitional (such as charcoal, lignite, and kerosene), or modern (such as LPG, biogas), while electricity is used for small appliances for cooking and heating but mainly for lighting. The share of population relying on the traditional biomass for cooking in 2012 was: 67% for Africa, 80% for sub-Saharan Africa, 51% for developing Asia, 15% for Latin America, and 4% for Middle East. Table 4 shows the traditional biomass consumption for cooking in different countries [26, 27].

The excessive dependence on solid fuels for cooking is an indicator of energy deficiency. It is familiar that access to modern energy facilities (clean fuels such as electricity) is important to the success of Millennium Development Goals. The burden on women and children for collecting fuelwood (wood pellets) can be reduced by applying the modern energy services (MES). By this, not only the time will save, but also one can be educated or get some other employment opportunities. There will also be a notable effect on child mortality [28]. It is also well known that traditional biomass fuels generate a bundle of pollution. These harmful pollutants have a poor effect on our health and sometimes become a major cause of death [29]. Jain [30] had discussed the energy security issues for all types of households by focusing on clean energy fuels for cooking in India. It was mentioned that the clean energy fuels (LPG, electricity, and kerosene) must be affordable for LIHs for cooking and heating and the policies should be revised for the same.

Pandey and Chaubal [31] had developed a logit model for rural household (India) fuel choices for cooking by using database of around 403, 207 observations. Authors focused on some major indicators such as females and children's education, regular income of the particular household, and possession of ration card (BPL—below poverty line type, especially). It was concluded that clean fuel such as kerosene should be provided in sufficient quantity for cooking and lighting. Akpalu et al. [32] had discussed that the almost all energy policies in Ghana or other developing countries are focused primarily on industrial energy consumption instead of household energy use. The socioeconomic factors such as household income and fuel switching cost were considered. It was concluded that a good subsidy on LPG and kerosene is a better and effective option to a LIH and MIH for fuel switching option. It was noticed that LPG is primarily chosen for cooking in Ghana's households.

Table 3 Combination of cooking fuels in Nigeria (2015) [25]

Fuels used	Population share (%)
Wood + Kerosene	17.71
Wood + Kerosene + Charcoal	3.21
LPG + Kerosene + Solar	8.21
LPG + Kerosene	1.04
LPG + Electricity	4.27
Kerosene + Electricity + Wood	4.27
LPG + Wood + Charcoal	3.13
LPG + Wood	5.21
Wood + LPG + Kerosene + Charcoal	10.42

Table 4 People using traditional biomass for cooking [28]

Region/Country	2009	2015	2030
	(Millions)		
Africa	657	745	922
Sub-Saharan Africa	653	741	918
Developing Asia	1937	1944	1769
Other developing Asian countries	659	688	709
India	855	863	780
China	423	393	280
Latin America	85	85	79
Total	2679	2774	2770

Wickramasinghe [33] had carried out a research project in Sri Lanka to extend the appreciative of human aspect of energy access. A questionnaire was prepared to survey around 2269 households for gathering a data on socioeconomic situations and matters influencing a transition toward clean cooking services. The results of the present study show that to develop a transition (switching of fuels), two domains must be addressed: risk capability and livelihood (Fig. 4).

The results have also shown that the transition is hampered by the following aspects:

(i) Lack of motivation and financing;
(ii) Burden for switching over to cleaner fuels;
(iii) Lack of updates of energy technology options and other supports;
(iv) The financial risks (the results also reveal that there is a need for a policy agenda linking the stakeholders, funding, and standardized technologies).

Foell et al. [34] had discussed the use of biomass fuel by the people of world and the impacts of biomass burning used for household cooking systems (China). This chapter focuses on the framework and policy development for clean cooking accesses, to develop a strategy for suppliers of clean fuels, and to revise the failures policies to improve the program. Maes and Verbist [35] had reviewed the biomass cooking literature and discussed the sustainability of domestic cooking pattern in developing countries and how this can be improved. As a solution to air pollution, the use of energy ladder and switching of fuels (traditional to cleaner) was discussed with a result-oriented fashion, in which it was shown that the health damages can be reduced, and time and money can be saved as well as the green environment. Duan et al. [36] had focused on same points and discussed that the household fuels are intensely associated with numerous effects comprising the air quality, health, and a district environment change. The study reveals the results of first Chinese environmental exposure-related human activity patterns survey (CEERHAPS), carried out among 91,121 households located in 9745 villages and towns. The objective of the study was to investigate the cooking and heating pattern in China. It was observed that the LPG and biomass fuels are principal energy fuels for cooking, used by 45% (approximately) and 32% of households, respectively. Biomass (47.6%) has been observed as a main fuel for cooking in rural households, while urban population was more likely to cook with gas (65.8%). A model was developed to WHO for indoor air pollution (IAP), health effects, and burden of disease.

There are numerous data available on cooking fuel choice and modern stoves [37–43], fuel switching theory and models [44–51], energy ladder [52–55], IAP [56–60], respired illness, and burden of disease and death [61–72] in developing countries. As reference to the above literature, the daily demand of cooking energy fuels with increasing world's population is very tough to adopt the clean cooking fuels by MIHs and LIHs, while the new improvements in cookstoves are introducing day by day to achieve more efficient performance of cooking systems.

Government of different nations are promoting the clean fuels and providing various offers (subsidies, special discounts, etc.) on them. According to a report of the World Bank, different designs/types of cookstoves are used by different households (here for MIH and LIH) which are commonly associated with particular fuel types such as a three-stone type for biomass fuel, a traditional clay pot/simple ceramic liner-type stove for dung cakes or wood pellets, a round cylindrical stove for charcoal, and gasifier stoves. For a HIH or a rural area population, modern

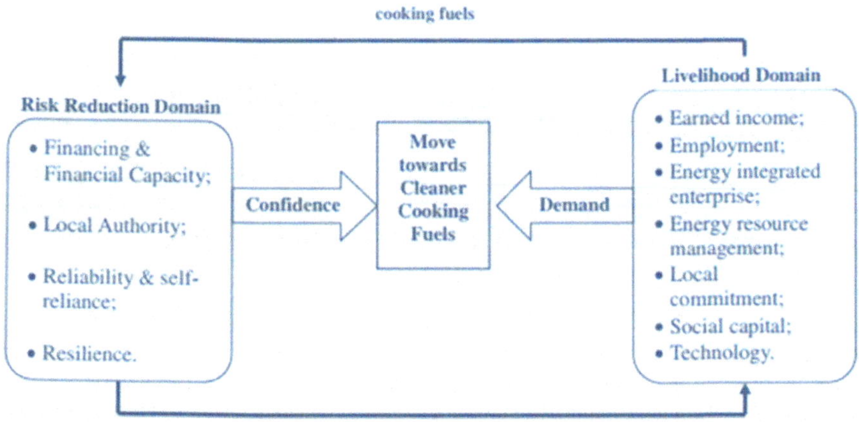

Fig. 4 Risk capacity and income for switching cleaner cooking fuels and stoves [33]

Fig. 5 Different designs of modern cooking stoves for clean cooking: **i** microwave oven, **ii** LPG stove (5 burners), **iii** electricity induction stove, **iv** modern coal stove (Angeethi), **v** modern kerosene stove, **vi** large wood pellets stove (big Chulha), **vii** modern single burner small family LPG stove, **viii** compact wood pellets stove for grilling and roasting, **ix** solar box cooker, **x** solar dish cooker, **xi** modern biomass stove, **xii** solar evacuated type cooker for roasting [139]

cooking stoves (see Fig. 5) are used in developing countries because of using clean cooking fuels such as LPG (1–5 burner type cooking stove), natural gas, electricity (a microwave oven, induction stove, or an electric kettle). The conversion efficiency of these stoves is much better over previous designs of cooking stoves. Apart from this, in modern era, biogas cookstoves and solar cookers are also getting a good attention in rural areas of emerging places. It is notable that solar cookers are not popular in only rural area but some urban households are using them well in both developing and developed countries because of their simplicity, low maintenance, no carbon emissions, and obviously performing on free fuel from the sun. Apart from this, by using of solar cookers one can save the limited fossil fuels and power energy for a fair amount annually.

2 Solar Drying Technologies

Solar drying is simple and an economic way to preserve the food since nineteenth century. This treatment removes water or heavy moisture presents in various ingredients and prevents fermentation. Solar food dryer (SFD) presents a significant progress upon this primeval method of drying foods. SFD has an initial expense, but it provides a better taste and a safe nutritious food and their marketability. They perform a fast and safe drying process more efficiently than other traditional sun drying methodologies. Solar drying can be categorized into two simple categories: open sun drying and cabinet solar drying. Solar cabinet dryer produces a high-quality dried foodstuff more quickly in humid or arid climates. Dried stuffs contain a high value of vitamin C content [73].

In recent years, several attempts have done across the globe to design or develop novel solar dryer for various activities such as for agriculture or industrial sector. The previous research shows that the efforts has been made not only to improve the efficiency of the SFD but also for cost and design optimization as well as to improve the year-round performance by making a hybrid SFD. In the next section, the discussion is made on various types of solar dryers and their performance.

2.1 *Types of Solar Dryers*

Solar dryer simply utilizes solar radiant energy to heat the air which flows over the substance placed inside the SFD. Warm or dry air that flows through the system carries away the moisture contents from the different substances through evaporation. A solar dryer generally consist of a solar air collector or air heater [74], a drying unit and an air handling unit. The main components of a solar drying are shown in Fig. 6. It is notable that the efficiency of the SFD majorly depends on the efficiency of air heater. Classification of the solar dryers is shown in Fig. 7, and some commonly used solar dryers are discussed in Table 5.

Fig. 6 A solar chimney fruit dryer with a solar air heater (consisting of single or double glazing, a solar collector, an air blower, ducts for air supply, and sometime a PV module in case of hybrid solar dryer)

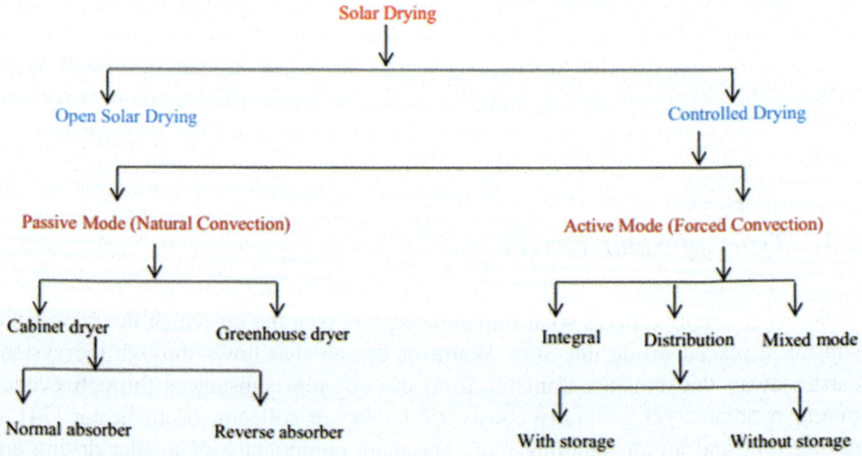

Fig. 7 Classification of solar drying [75]

Table 5 Types of solar drying

Types	Working
Open sun drying	In this, the substance is spread in the form of a thin layer on the ground. Solar radiant energy directly incident on the substance, and the moisture contents removed through the natural convection with low wind velocity. In this phenomenon, the present substance also absorbed radiation which results in a high vitamin C value. This drying is still most common drying in rural areas. The ambient parameters have a major significant effect on open sun drying
Cabinet solar dryer	A cabinet solar dryer is a hot box-type dryer in which fruits, herbs, spices, vegetables, etc., can be dehydrated. It consists of a long rectangular basin, an insulated base, and a single or doubled glazing. Irradiance passes through the glazing and absorbed by blackened surface (absorber) and raises the internal temperature. Small holes are made in the cabinet for providing ventilation to induce fresh air inside the cabinet
Greenhouse dryer	It is a low cost, simple in design, and easy to fabricate. It is used across the globe for crop dehydration. It has a parabolic shape and covered by a polycarbonate sheet. The base of the dryer is generally a concrete floor. The substances are placed in a thin layer on the black concrete floor. Solar irradiance passing through the polycarbonate sheets heats the air and stuffs inside the dryer, as well as the absorber floor. The heated air, when pass over the stuffs, absorbs moisture from the stuffs. A direct exposure to irradiance of the stuffs and the heated air augment the drying rate
Hybrid solar dryer	A hybrid solar dryer generally performs on two or more fuels for drying. This is because the said systems become more efficient by operating on two or more fuels, simultaneously. This system is capable of generating an adequate and continuous flow of hot air for a higher temperature range. Likely hybrid dryers used biomass or electricity along with solar energy

2.2 *Solar Open Drying*

Solar drying is an important application of the sun energy. Earlier farmers were fully dependent on solar energy for drying agriculture crops, herbs, or spices. For this, a ground or a bed of concrete was prepared to achieve a high temperature range through solar energy. The ambient temperature and wind velocity support this as natural convection. Obliviously, the process is much slow; therefore, it generally took 3–4 days to dry the substance. The initial and operation cost of open sun drying is almost negligible. Volume wise and despite a very simple process, the open sun drying is still a most common technique of drying [76]. Table 6 shows the various studies conducted on open sun drying of commonly used crops.

2.3 *Solar Cabinet Drying*

It is a single- or double-glazed rectangular cabinet (Fig. 8). Solar irradiance directly incident on the substance, and interior surface is blackened to enhance the heat

Table 6 Various studies conducted on open sun drying of commonly used crops

Reference	Research work outcome
Anwar and Tiwari [77]	An attempt had been made to estimate the value of 'hc' for six different crops, such as green chillies and green peas under open sun drying conditions. The value of 'hc' was varied from crop to crop because of differences in their properties such as moisture content, porosity, shape, and size of the crop
Mulokozi and Svanberg [78]	In the present work, 8 vegetables commonly consumed in the living areas were traditionally sun-dried and the pro-vitamin 'A' carotenoid was counted. It was remarked that the amount of pro-vitamin A-carotenes in conventionally dried vegetables was much reduced by open sun drying
Kabasa et al. [79]	The present work focused on the availability of vitamins A and C in vegetables and fruits. These vitamins were found out for a major cause of blindness for 54% in Uganda. The government of Uganda promotes solar drying as an economic and reasonable substitute. The effect of 3 drying methods (open sun drying, visqueen-covered solar dryer, and greenhouse solar drying) was investigated for vitamins A and C contents in fruits and vegetables
Jain and Tiwari [80]	In the present work, open sun drying was carried of green pea and chillies, white gram, potatoes, cauliflower, and onions to investigate the thermal behavior of the said vegetables. A mathematical model was developed to estimate the temperature and moisture removal rate from species. The value of 'hc' was estimated for different drying times with respect to different vegetables under open sun drying, and a significant variation was found in this with a change of species. The moisture removal rate for cauliflower and potato slices was significantly high in comparison with other crops
Akpinar and Cetinkya [81]	In the present work, drying response of parsley leaves was carried with forced and natural convection (open sun drying) conditions. It was observed that no constant drying falling rate was occurred during performing on both the modes
Prasad et al. [82]	The efforts have made to investigate the drying characteristics of ginger under open sun drying and a hybrid drying system (biomass based). It was concluded that drying depends upon the product thickness and ambient conditions of the drying place under open sun drying. In comparison with open sun drying, the hybrid solar drying system was found better for drying other stuffs, vegetables, and fruits
Al-Mahasneh et al. [83]	The open sun drying and forced convection drying have been carried of sesame hulls at Jordan University of Science and Technology. In total, six mutual thin-layer drying prototypes were tailored to the experimental data. The experimental results were compared with theoretical models, and it was found that open sun drying took place in nearly 180 min while forced convection drying needed 120–250 min. Both the drying techniques were found adequate for drying sesame hulls
Kooli et al. [84]	The open sun drying and greenhouse drying of red pepper were carried out inside a laboratory by using a 1000 W lamp for radiant energy. The effect of drying was investigated for moisture contents and drying rate by repeating the experiments under open conditions. It was

(continued)

Table 6 (continued)

Reference	Research work outcome
	concluded that experiments carried out in laboratory overestimate the drying process with respect to time
Tripathy and Kumar [85]	In the present article, the experiments have been conducted for drying of potato slices to find out the optimum fast rate drying methodology. The testing was carried out for open sun drying and mix-mode sun drying to observe the drying characteristics such as surface color, texture, and drying rate. It was concluded that the water captivation ability of done product is partial mostly by the variation in rehydration temperature tracked by sample geometry
Gudapaty et al. [86]	An LPG-based drier has been developed for drying of 50 kg of Indian gooseberry, and the comparisons were made with open sun drying of the same. The quality and rehydration characteristics of LPG-based dried stuff were higher and free from impurities. Beside this, the value of vitamin C in the dried product was high rather than dried by open sun technique
Akpinar [87]	Drying characteristics of mint leaves were was studied and investigated in a chimney type solar dryer and under open sun drying. Energy and exergy analysis was performed for both conditions. It was concluded that there was no change in dried mint leaves either by open drying or by solar cabinet drying
Hande et al. [88]	In the present work, the study of drying characteristics for open sun drying of Kokum rind was carried out. The important parameters were the temperature of crop surface, moisture removal rate, and drying time. The value of 'hc' was estimated on the value of constants 'C' and 'n' which were calculated by regression analysis. Some other important parameters were also considered for study such as acidity, pH value, non-reducing sugar contents, carbohydrates, protein ash, calorific value, and color, before and after open sun drying of kokum rind

transfer. For ventilation, small holes are drilled to the bottom surface to induce the fresh air inside the box. Opening ports are positioned on the top segments of the sideways and rearmost panels of the cabinet frame. As the box temperature rises, the hot air passes out of these openings by means of natural convection. In this process, both the heat and mass transfer take place [89]. The heat is transferred to the stuff through radiant energy or energy from any other heating source, while the mass transfer of moisture is from the entire stuff to the surface and then to the surrounding air. The objective of the cabinet drying is to supply the more heat to the crop than open sun drying. A cabinet dryer can also be a forced convection or hybrid SFD, which results in a fast moisture removal rate, less drying time, quality drying characteristics of the matter, and an economic process. Some novel designs of cabinet-type solar dryers are shown in Table 7.

Fig. 8 Different designs of solar dryers [95, 135, 140, 141]

Table 7 Research outcome of some novel solar cabinet dryers

Reference	Research work outcome
Datta et al. [90]	In the present work, a solar cabinet drier has been developed and transient analysis of the SFD was carried out with some applied assumptions. The model was feasible to predict the instant temperatures inside the cabinet, the drying rates, and the moisture contents. Experiments were conducted for no load and for load conditions (10–40 kg of drying substance). It was concluded that up to 20 kg/m^2 of wheat can be dried per day
Sharma et al. [91]	A SCD was fabricated, and thermal performance was carried out with help of energy balance equations at New Delhi. The collector temperature was observed around 85 °C on no load, while around 50 °C was observed for load conditions (20 kg of wheat). The cauliflower, turnip, and green peas were also dried for a comparison at small load conditions. It was concluded that the cabinet dryer was much efficient than a open sun dryer and the product dried in cabinet dryer was superior in quality
Ampratwuma and Dorvlo [92]	In the present work, a reverse FPC has been used for air heating for drying of agriculture stuffs inside a SCD in climatic conditions of Delhi. Energy balance equations were developed for thermal modeling of the system. The design optimization of SCD was done for parametric studies over ambient conditions. It was concluded that new type SCD was much efficient than conventional SCD

(continued)

Table 7 (continued)

Reference	Research work outcome
Goyal and Tiwari [93]	A prototype SCD was fabricated and tested at no load as a solar air-heating system. The system was operated for 28 days of May 1996 in Oman. During the testing, the SCD achieved a temperature around 81.3 °C. The system was observed for temperature variations inside SCD during the duration of 28 days, but no significance difference was observed in inside temperature of SCD during 10:00–15:00 h. The rate of solar radiant energy absorbed by SCD was approximately 0.90 kW sq/m
Adapa et al. [94]	An SCD integrated with a dehumidifier loop was designed and fabricated to study the drying characteristics of a highly moisture content (70%) alfalfa. The stuff was spread on the trays inside the SCD and operated for 10% of moisture content alfalfa. The system was operated on the forced convection for a temperature range of 25–45 °C. The drying time for alfalfa chops was observed from 4 to 5 h. The explicit moisture removal rate was observed from 0.35–1.02 kg/kWh
Srikumar et al. [95]	A new type of SCD has been developed and tested for drying fruits and vegetables. In this design, the stuff was placed beneath the absorber to avoid the problem of discoloration due to irradiance. Two small fans used for forced convection accelerate the drying rate. The system was found adequate for solar drying by removing 90% moisture from the stuff (4 kg) within 6 h. The economic analysis has also been done, drying cost was found around Rs. 17.52 for 1 kg of bitter gourd, and the payback was estimated around 3.26 years
Rawat et al. [96]	The energy analysis of an economic, simply designed, natural convection SCD was studied for India. Present work focused on manufacturing sectors who wish to know energy requirement and to reduce greenhouse emissions. Grown chillies were taken as drying object. Energy payback period was estimated for different conventional fuels. The overall energy input to the system was estimated around 2744.61 MJ, while the energy payback was found in the range of 1–2 years (i.e., depended on the drying characteristics of object)
Signh and Pandey [97]	In the present work, the drying properties of sweet potato (*Ipomoea batatas* L.) were investigated in a SCD. The convective drying has been carried out under the five different temperatures and the air velocities for different sample thicknesses of sweet potatoes. Results shown that the drying of samples took place in the falling rate phase and the diffusivity was observed to be increased with increasing temperature. The effective moisture diffusivity of the taken samples was observed for a range 1.26×10^{-9} to 8.80×10^{-9} m^2/s, while the activation energy was estimated around 11.38 kJ/mol
Demiray and Tulek [98]	In the present work, the effect of the drying behavior of garlic slices has been investigated in a SCD at a constant velocity (2 m/s) and a temperature range of 55–75 °C. The effective moisture diffusivity of the taken samples was observed for a range of 2.22×10^{-10} to 2×10^{-10} m^2/s, while the activation energy was estimated around 30.58 kJ/mol

(continued)

Table 7 (continued)

Reference	Research work outcome
Sallam et al. [99]	In the present work, two similar solar dryers (direct and indirect) were used for drying the mint by operating it on natural and forced convection. In the case of forced convection, velocity of the air was around 4.2 m/s. The effect of flow mode was investigated with the significant effect on drying kinetics of the object. The results shown that the drying of mint was occurred in a falling rate period with different climatic conditions (with no constant rate period of drying). The drying rate was observed to be high in forced convection operation. The effective moisture diffusivity of the taken samples was observed for a range of 1.2×10^{-11} and 1.33×10^{-11} m^2/s
Yahya et al. [100]	In the present work, a solar-assisted pump dryer (SAHPD) and a SCD were investigated for solar drying of cassava chips. The moisture content of cassava was notified to be reduced from 61 to 10.5% with a flow rate of 0.125 kg/s. Results shown that drying rate by using SAHPD was quite higher than SCD. Thermal efficiencies were notified as 25.6% for SCD and 30.9% for SAHPD with an average solar fraction as 66.7% for SCD and 44.6 for SAHPD

3 Solar Cooking

Food is the prime need of humans. Various types of fuels are used to cook the food worldwide. Due to limitations of using the fossil fuels for cooking as well as to reduce the environmental pollution and need of safe cooking, the concept of solar cooking was generated in 1951. But, the efforts toward solar cooking were successfully evaluated in 1767 by H.D. Saussure, a French-Swiss Physicist [101]. After this, efforts were continued by various pioneers of the field toward the effective solar cooking in eighteenth century and later. At present, various designs are available for solar cookers for efficient solar cooking across the globe. These solar cookers are not only simple in design but economic, accident, and pollution-free cooking, easy to maintain and feasible to perform round the year even in low ambient conditions. Indian government programs like Jawaharlal Nehru National Solar mission (JNNSM) ambitious target to generate 100 GW solar energy by 2022. Also Off-grid and Decentralized Solar Cooker Program (phase 2, JNNSM) promotes off-grid cooking application such as cooking/baking/frying using solar device with central financial assistance from Ministry of New and Renewable Energy (MNRE).

3.1 Types of Solar Collector and Solar Cookers

Solar collector—a solar-type collector is a flat wooden or metallic box which generally composed of a transparent glass cover, some glass or metallic tubes carrying coolant or heat-carrying working fluid, and an insulated back plate. The working principle is a simple greenhouse effect; whenever solar beams incident

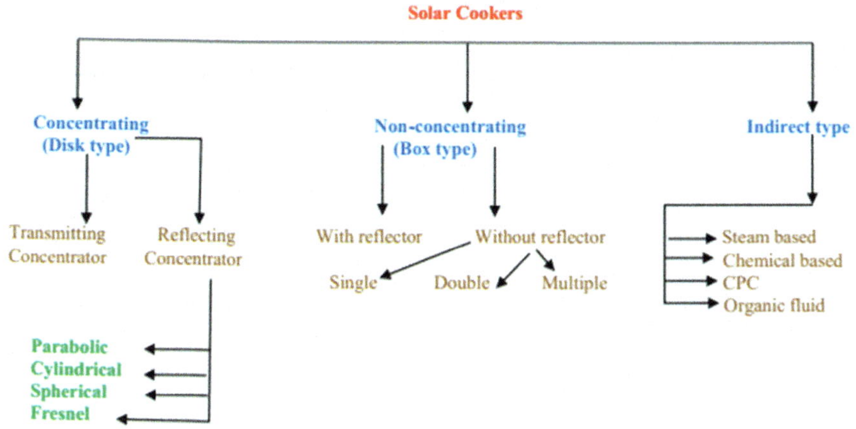

Fig. 9 Types of solar cookers

upon the glazing, the direct radiation strikes on the blackened surface of solar collector through passing the transparent cover and the heat energy is trapped by the collector. There are mainly two types of solar collector, i.e., flat plate collector and concentrating collector. Concentrating collector can further be classified as:

(a) Flat plate collector with plane reflectors;
(b) Compound parabolic collector;
(c) Cylindrical parabolic collector;
(d) Collector with fixed circular concentrator and moving receiver;
(e) Fresnel lens concentrating collector;
(f) Parabolic dish collector.

From the design prospects, solar cookers can generally be categorized into two types: tracking solar cookers and non-tracking solar cookers (Fig. 9). But, in present scenario, the research on solar cooking and standard of solar cookers have broaden the categories of solar cooker as following.

3.1.1 Tracking Solar Cookers

It is a solar cooker which consists of a parabolic dish (combination of small reflectors), a pressure cooker, a continuous tracking mechanism, and a rigid stand for holding both the elements. The sun beams incident on the cooker directly, and reflectors reflect the energy to the cooking vessel placed at focal point. The said cooker requires a continuous movement to follow the sun (each 10 or 15 min) and to retain the focus image on the object. These types of cookers give the highest cooking efficiency (Fig. 5i).

3.1.2 Non-tracking Solar Cookers

It is a simple box-type solar cooker which consists of a rectangular or square well-insulated blackened box, 2–4 cooking utensils, and single or double glazing with a reflector mirror booster. The stuff is placed inside the cooking vessel, and the glazed cover is then easily closed. The sunlight directly falls on the blackened surface (aperture area), on the top surface of vessel, and on the mirror booster. In this manner, conduction and convection take place inside the cooker through the solar radiant energy and the stuff get cooked. These cookers are used on a large scale around the world because of less time taken in quality cooking (Fig. 5j).

3.1.3 Indirect-Type Solar Cookers

These cookers are quite different in design and use. In these types of solar cooker, obviously the heat source is solar radiant energy but it is utilized indirectly for cooking purpose. Generally, they consist with a solar collector integrated with heat pipes, a solar boiler, etc. The food is cooked with the help of steam produced through the solar boiler inside the cooking chamber. The use of PCM has been observed effective in these types of cookers. Sometimes, Fresnel lens or additional mirror boosters can also be used to enhance the thermal performance of the unit (Fig. 10).

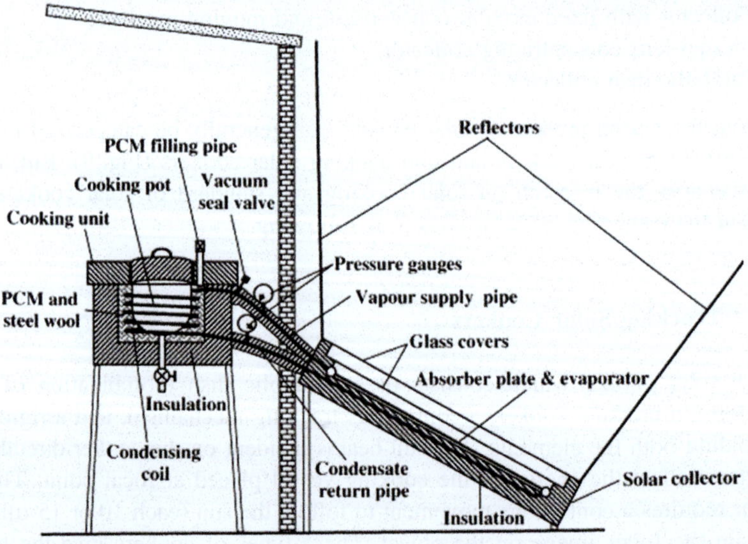

Fig. 10 Schematic diagram of indirect solar cooker [102]

Fig. 11 Experimental setup of a hybrid solar cooker [103]

3.1.4 Hybrid Solar Cookers

These types of solar cookers are more efficient in comparison with other solar cookers because of operating on dual fuel. These cookers are operated on dual inputs, say solar energy and electricity or solar energy and LPG or any other fuel along with solar energy. The system decently performs on solar energy, when the ambient conditions are good, but as they became low or poor, the other fuels initiate and maintain the cooking speed without any interruption. It results in timely and safe cooking (Fig. 11). They often provide a fast thermal response of solar cooking.

3.2 Cooking Methodologies and the Use of Solar Cookers

Cooking is a major end-use activity in which we find robust and frequently high specific energy carrier priorities. The extent and the form of energy carrier used by an individual family depend on the capability of the household to recompense the subsequent fuel costs as well as the lifestyle. The cause of this change is simple; i.e., low-income families cannot invest in high energy efficiency technology because of unawareness toward the costs of energy and poor information barriers. Accessibility and cost of the cooking fuels clearly demonstrate the reason of fuel choice for cooking and heating. Households that change the cooking fuels for a clean environment were much aware, and this also shows their capability toward paying a higher cost for the replacement (especially shifting to LPG from kerosene). Besides this, the cooking methodology is adopted by the various households by taking into account the economical use of food and resource fuel as well as the safety of food. Following are cookery methods for different households of different countries (Fig. 12).

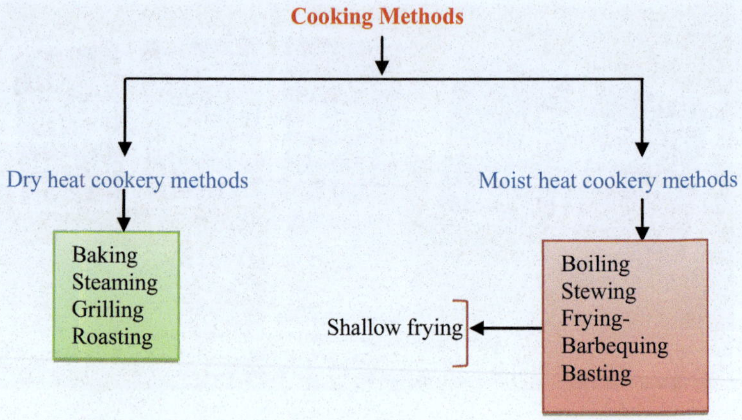

Fig. 12 Cooking methodologies across the globe

No doubt that a large figure of households instantaneously uses a different cooking pattern or variety of fuels for cooking, but another considerable factor is the choice of cooking fuels according to their cost that suits their family budget. If one can talk about the fuel choice for cooking than in rural households, biomass (wood pellets, charcoal, dung cakes, dry grass, etc.) is common fuel for these activities. On the other hand, LPG and kerosene are used for the same purposes in urban households. Besides this, clean energy fuels such as electricity and solar energy are also used for cooking at different locations of the various countries in the present era.

However, it is difficult to measure the exact involvement of solar energy in cooking or heating activities in developed or developing countries, but people from different countries are aware about this source of free energy. Countries such as China, India, Turkey, South Africa are paying attention toward solar energy applications and continue for promoting it by launching new schemes and by providing attractive financial aids. People from these countries are using solar cookers along with LPG and electricity in urban areas while along with kerosene and biogas in rural areas. Still that biomass fuel is used by some rural households for cooking and heating. Both the households (rural and urban) have a different pattern for cooking and use the cooking fuels accordingly. Both the households use solar cookers for cooking pulses and rice, to boil eggs and potatoes, and other light meals or the meals which require less time for cooking. It is also notable that the use of solar cookers is not limited to rural or urban households in different countries but they are also used in hospitals, institutes, hotels, hostels, and many other places where the sunshine is available in plenty [104–108].

Generally, rural households start their routine life early in the morning with preparing breakfast. At this time, solar radiant heat is not available in appropriate amount. During the solar noon (from 11:00 AM to 15:00 PM), one dish for the lunch and one dish for dinner (such as pulses or rice) are prepared by solar cooking.

Table 8 Cooking schedule of different households in Moradabad [109]

Family	Time of cooking	Cooking by solar cooker
A (05 members)	Breakfast (7:30–8:00 h), lunch*2 (12:30–14:00 h), dinner*2 (19:00–20:15 h)	lunch*2 (10:00–13:00 h) and (13:15 h–up to the finishing of one dish for dinner)
B (07 members)	Breakfast (7:30–8:15 h), lunch*2 (12:30–14:15 h), dinner*2 (19:00–20:15 h)	lunch*2 (10:00–12:45 h) and (13:00 h–up to the finishing of one dish for dinner)
C (07 members)	Breakfast (7:30–8:15 h), lunch*2 (12:30–14:15 h), dinner*2 (19:00–20:15 h)	lunch*2 (10:15–12:45 h) and (13:10 h–up to the finishing of one dish for dinner)
D (06 members)	Breakfast (7:30–8:00 h), lunch*2 (12:30–14:00 h), dinner*2 (19:00–20:15 h)	lunch*2 (10:00–12:30 h) and (13:00 h–up to the finishing of one dish for dinner)
E (08 members)	Breakfast (7:30–8:25 h), lunch*2 (12:30–14:15 h), dinner*2 (19:00–20:15 h)	lunch*2 (10:15–13:15 h) and (13:30 h–up to the finishing of one dish for dinner)

*shows the number of dishes to be cooked for both lunch and dinner

For this, a dish cooker, a box cooker, or an especially designed solar cooking unit can be used. A solar dish cooker is more efficient than a box cooker to cook the stuff which requires high temperature (potatoes, eggs, meat). On the other hand, box cooker is used for pasteurization of milk or water, for cooking rice or pulse, etc. Other cooking materials are prepared by biomass or kerosene (such as chapatis, some dry vegetables, or other stuff required hard boiling or deep frying).

Table 8 shows the cooking schedule of five different families in which three from rural area (families A, B, and C) 'Naya-Gaon' and two from the town (families D and E) in Moradabad district. This table demonstrates the time of cooking and use of solar cooker by particular family. It is also clear from Table 8 that the people from rural and urban areas use the solar cooker only for cooking two dishes, i.e., one for the lunch and one for dinner.

3.3 *Energy Savings Through Solar Cooking*

Estimation of energy savings through the use of solar cookers is a simple exercise in which the amount of fuel saved per year by solar cooking is calculated. These savings depend on the efficiency of solar cooker, design parameters of cooker, ambient parameters during solar cooking as well as the nature of substance to be cooked inside the cooker. For this exercise, a survey-based research is necessary by considering some basic assumptions with the effective cooking time. Let us assume that a simple household of 5 family members uses an electric oven (1200 W) for cooking, then what would be the amount of electricity consumed for cooking? This can be simply estimated by knowing the cooking pattern of the same household.

Assumptions are as follows:

1. For a single dish, the amount of electricity is consumed 0.25 kW.
2. For a single dish, the amount of LPG is consumed 0.20 L.
3. A single dish means 600 g of vegetables or cooking substance (not non-veg).
4. Solar cooking is performed during the sunshine hours.

Now, for the estimation of electricity consumption by considering the cooking pattern of the said household:

Electricity consumed for cooking breakfast, the lunch, and dinner by oven = 1 kW,

Electricity consumed for cooking the light snacks by oven = 0.25 kW.

Electricity consumed on ideal conditions by oven = 0.10 kW.

Total electricity consumed for cooking by oven = 1 kW + 0.25 kW + 0.10 kW = 1.35 kW.

If the electricity connection is through the hydel board of electricity, then the cost of total electricity consumed is (by ₹ 4.40/unit) around ₹ 6.00. It means that the monthly bill is around ₹ 180 for cooking on electricity.

Similarly in the case of LPG, the estimation of LPG consumption by considering the cooking pattern of the said household is:

LPG consumed for cooking breakfast, lunch, and dinner by LPG stove = 0.50 L.

LPG consumed for cooking the light snacks by stove = 0.10 L.

Total LPG consumed for cooking by stove = 0.50 ltr + 0.10 ltr = 0.60 L.

If the LPG connection is through the authorized gas agency by government, then the cost of total LPG consumed is (by ₹ 32.50/unit) around ₹ 19.50/-. It means that the monthly bill of LPG is around ₹ 604.50/- for cooking.

From Table 8, it is shown that a solar cooker is feasible to cook the two dishes per day, i.e., one for lunch and one for the dinner. Now, one can accept that the electricity consumption for cooking -two dishes is 0.50 kW (from the assumptions) and the price for this amount is ₹ 2.20/day which can be ₹ 66.00/month or ₹ 742.00/year. Similarly, for the LPG the saving amount will be ₹ 13.00/day and this will around ₹ 390/month or ₹ 4680/year. But these savings are not the actual because it includes the PBP of the cooker. If one has to estimate the net savings of fuels or the amount which is spent to buy the cooking fuels, then the PBP of the cooker has to also be estimated and subtracted from this savings amount.

The PBP of the cooker can be calculated by considering some economic factors as follows [110]:

$$N = \frac{Log\left[\frac{(E-M)}{(x-y)}\right] - Log\left[\frac{(E-M)}{(x-y)} - Z\right]}{Log\left[\frac{(1+x)}{(1+y)}\right]}$$

Alternatively, the PBP of the cooker can also be estimated by a simple formula without considering the various additional cost such as inflation rate, interest, maintenance cost [111]:

$$Payback\ Period = \frac{Manufacturing\ cost\ of\ solar\ cooker}{Cost\ of\ energy\ saved\ per\ year}$$

By this way, the net savings can be easily estimated through the solar cooking at any place. By this way, people who used biomass fuel for cooking and heating can not only save the limited fossil fuels but also some amount of money, some amount of time, and obviously their health also. Besides this, one can also estimate the annual CO_2 emission by using the different conventional fuels with the help of following formula [112], which produced while cooking on conventional fuels:

$$CO_2 emission = FC \cdot OF_{fc} \cdot CCC_{fc} \cdot HHV_{fc} \cdot \left(\frac{MWco_2}{MW_c}\right)$$

4 Hybrid Systems for Solar Drying and Cooking

Solar hybrid systems are those systems which operate on more than one fuel or input. The optimum design has been selected according to the need and ambient conditions to get desired output. In solar energy systems, electricity (AC or DC) is commonly used as second fuel in most applications. The great advantage of the system is that it can also work in night or under poor climatic conditions. These systems are more efficient in comparison with pure solar thermal systems and can be used at any location of developed or developing countries. Although many households use more than one cooking fuel for cooking such as LPG + kerosene, LPG + electricity, electricity and biomass, LPG + electricity + solar energy, LPG + solar energy, as the people are convenient or due to availability of fuels or resources in their particular region.

Another advantage is that the baking, frying, water or room heating, and high-level drying are also possible with these systems. If we talk about solar cookers, then this practice is more convenient to box cookers in comparison with dish-type solar cookers. Dish-type solar cookers require a continuous tracking of the sun; therefore, the position of the cooker is changed repetitively, but box cookers are installed at one place and there is no need to change the position; hence, second fuel is easily supplied than a dish cooker. Heat is supplied through an electric coil (sufficiently of 10 W) beneath the solar collector from the grid supplied electricity or from the photovoltaic modules by means of battery storage. In designing solar thermal hybrid cookers or dryers, the metal of solar collector is generally copper. The heat in this system is highly controllable, and around 91% heat is supplied to the substance in the cooking vessel (in case of cooker) or around 80% supplied to the substance to be dried (in case of dryer). Overall, solar hybrid systems are more convenient to operate as well as for savings of fuels or to reduce CO_2 emissions.

In daily practice of solar cooking and drying, it has been observed that the using solar cooking or drying, systems alone are low in efficiency and fully depend on sunshine conditions. To overcome this problem, a hybrid solar cooker-cum-drying system was developed in Costa Rica by Nandwani [113] for cooking and baking of almost all edibles. As per the convenience of the consumer, the cooker can be shifted to the electrical input during or before the cooking. An electric back plate of 1500 W was used for heating up the cooking vessels to cook the food. Hussain et al. [114] had carried out experiments in cloudy days for cooking by incorporating an electrical coil beneath the solar collector plate inside a solar hybrid cooker. Through this mechanism, the cooking was possible in low ambient conditions in Bangladesh. In comparison with Nandwani's model, the present model consumed only 900 W of electricity. The model also satisfied the figures of merit for box cookers [115]. Later on, Chaudhuri [116] had estimated the required electrical back for a SBC. However, BIS does not specify any standard for hybrid solar cooker, but it was concluded that 160 W of electrical backup is sufficient for cooking in a SBC. Besides this, all of the authors [113–116] did not specify the design specifications of cookers, cooking methodologies, cooker size, or anything about the cooking stuffs that were considered for performance evaluation of hybrid model (Fig. 13).

Later on, Nandwani [117] had designed and tested a multipurpose (four in one) solar hybrid device for cooking, drying, and water heating-cum-distillation unit in Costa Rica. This device uses the electrical energy and solar energy to operate. Experimental studies were carried out for each one operation of the said device such as; cooking, drying, pasteurization and distillation and obtained thermal efficiency of the box as a cooker was 23% as a dryer it was 32.3% and as a still it was come out to around 27%. Ali et al. [118] had developed a hybrid solar cooker as a substitute to conventional energy and to reduce the environmental pollution. The cooker was operated on electricity and solar energy simultaneously. A matrix of parabolic reflectors was used to boost up the solar irradiance and thereby thermal efficiency. The performance of the hybrid cooker was evaluated with the help of

Fig. 13 A Hybrid solar multipurpose device [117]

energy balance equations. The system was feasible to reduce the cooking times in low ambient conditions. Joshi and Jani [119] had developed a small PVT box hybrid cooker with the help of five solar panels of 15 W each. During the experimentation, the value of first figure of merit (as per BIS) was found to close as per standard. Efficiency of the cooker was around 38%, and the estimated cost was around ₹ 7200/-, which can be reduced for consumers who already have solar panels for solar lights, etc.

Solar food drying is another form of solar food processing in which the heated air by solar energy is used to dehydrate food for storage and consumption. In conventional method, some thin slices of the substance are placed outdoor on a well-ventilated surface or on the rooftop of buildings for drying through hot air produced by solar irradiance. The heated air flows over the surface of the food stuff and carries a large amount of heavy moisture contents present in food through natural convection. For an advance option, these thin slices are placed in an enclosure or closed drying cabinets that utilize solar energy with forced convection methods in which mass flow rate is controllable (because the system operated on electricity or on PV modules). This method of drying is more effective in comparison with open sun drying, the system is easy to design and maintain, and food is quite safe from animal and birds attacks. By this, the substance or food got dehydrated with a high rate in less time. The quality of food stuff is not affected anyhow.

About storing of the dried stuff, clean plastic freezer bags (or glass jars with tight lids) are quite safe and promising for a heavy duty and can be simply transported. They can also be labeled or tagged with a code or date. All the packed stuff should be periodically checked special for weevils (a small and clean insect) or webbed cocoons. If the insects found, then pasteurization method can be adopted to kill the weevils. Before cooking the dried and stored food stuff, it should be hold for 15 min in hot water (70–80 °C) for a safe and healthy cooking (Fig. 14).

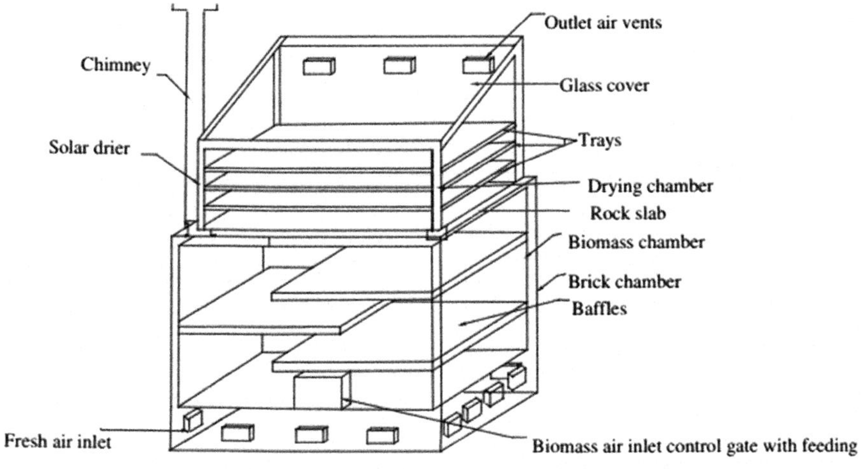

Fig. 14 A solar-cum-biomass drier (hybrid dryer) [121]

Akyurt and Selcuk [120] had developed a prototype solar food dehydrator (tunnel type) in which bell peppers and sultana grapes were dried at various mass flow rates. An additional auxiliary heating system was integrated with drying unit for a continuous operation. Economic analysis also showed that the present system can also be undertaken for an effective drying of fruits and vegetables. Prasad et al. [121] had developed a solar-cum-biomass hybrid dryer which performed on natural convection for drying of turmeric rhizomes. The objective of the experiment was to reduce the consumption of petroleum fuel or electricity, especially in rural areas. The system was found adequate to supply the hot air around 60 °C. The system was found for a good satisfaction comparatively with open sun drying. The overall efficiency of the drying unit was around 29%.

Ferreira et al. [122] had experimentally discussed that hybrid solar dryers are good option to reduce the drying costs of food in comparison with conventional solar dryers. It was also discussed that hybrid dryers (solar energy + electricity) improve the quality of dried product because of the controlled thermal drying state. Obtained results were in the favor of solar hybrid dryer for drying of banana slices in comparison with natural sun drying and artificial drying. Hossain et al. [123] had developed a hybrid solar dryer which consists of a concentrating collector, heat storage, drying unit with auxiliary unit. The system was tested in different operating conditions for drying 20 kg fresh tomato slices for different physical characteristics and flavonoids. The maximum efficiency was achieved around 29%. It was observed that this efficiency can be increased by 5–6% by using a solar reflector. It was recommended that if tomato is dried at 45 °C of the hot air, then no pre-treatment is necessary to prevent infectious growth. Amer et al. [124] had designed and constructed a new hybrid solar dryer with a heat exchanger. Water was used as heat storage media which release the stored heat at night or in poor ambient conditions for drying light substances with the help of electric heaters. The maximum efficiency of the drier was achieved by recycling around 65%. The system was feasible to provide the hot air about 40 °C more than ambient air. The experiments were conducted for drying of 30 kg of banana slices. The physical characteristics were found better in comparison with open sun drying (Fig. 15).

Lopez-Vidana et al. [125] had developed a solar-LPG hybrid drier with an auxiliary heating mode. The significance of design parameters was noticed to be affected through variations in input to the system. Mass flow rate was measured as a major parameter that affects energetic efficiency very much. The thermal efficiency was noticed as 26%, while the drying efficiency was noticed maximum for 71%. It was concluded that the system was found satisfactorily operating for drying tomatoes. Reyes et al. [126] modified and tested a solar hybrid dryer integrated with PV panels and a phase change material. The system was tested for mushroom drying in the form of thin slices (8–12 mm). The thermal efficiency was noticed as between 22 and 62%, while the efficiency of accumulator panel was noticed maximum for 21%. The PCM had a significant effect on the thermal efficiency of the system. Yaseen and Al-Kayiem [127] had utilized the thermal energy of flue gas from a biomass unit to develop a hybrid solar thermal drying unit. The system was tested for a day and night performance through a hybrid mode and thermal mode alone. The system was integrated

Fig. 15 Hybrid dryer test rig [127]

with an additional recovery drier for an optimum design and effective performance in poor ambient conditions. The results showed that the overall drying efficiency was around 13% with recovery drier. The novelty of the hybrid system was its recovery unit which enhances the overall efficiency of the system to be 30%.

5 Thermal Analysis of Solar Cookers

Thermal performance of solar cookers depends on many parameters: climatic parameters such as solar irradiance, wind velocity, ambient temperature, as well as design parameters such as shape, size, material of collector, number of glazing, vessel design, insulation, selective surfaces; in general terms, three parameters are essential for experimentation:

1. Time–temperature curve of cooker at ideal conditions (i.e., no load);
2. Cooking duration for variety of foods;
3. Time required for a fix volume of water up to boiling temperature.

A complete thermal analysis of solar cooker is quite complex because of 3D transient heat transfer involved. Solar energy cookers are the topic of research for the last 160 years but yet now there is no standard thermal rating is available for solar cookers.

5.1 *Thermal Performance of Box Cooker*

With the help of the first law of thermodynamics and the energy balance equations for the box cooker [101], energy input can be estimated as:

$$E_{in} = I_{avg} \cdot A_{sc}$$

while the energy output (on load conditions) for the SBC can be estimated through above equation as:

$$E_o = \frac{m_w c_{p-w}(T_{wf} - T_{if})}{t}$$

Now, having the value of above parameters in above two equations, one can easily estimate the value of thermal energy efficiency of the present system by:

$$\eta_{th} = \frac{m.c_p \cdot \Delta T}{t.I_{avg} \cdot A_{ap}}$$

Along with this, the cooking power has been estimated by [128]:

$$P_{sbc} = m.c_p \frac{(T_{wf} - T_{iw})}{600}$$

Equation for cooking power is divided by 600 to account for the number of seconds in each 10-min interval as per recommendation [128]. The figures of merit (F_1 and F_2) for the box cooker suggested by Mullick et al. [129] had computed by:

$$F_1 = \frac{(T_p - T_{amb})}{I}$$

where I (W/m^2) is the insolation on a horizontal surface (taken at time of stagnation testing)

$$F_2 = \frac{m_w c_{p-w}}{A_{sbc} t} F_1 \ln\left[\frac{1 - \frac{1}{F_1}\left(\frac{T_{wi} - T_{amb}}{I}\right)}{1 - \frac{1}{F_1}\left(\frac{T_{wf} - T_{amb}}{I}\right)}\right]$$

Now, heat transfer coefficient can be obtained by:

$$h = \frac{Q_U}{A_p(T_p - T_f)} = \frac{\tau . I_{avg} . A_p}{A_p(T_p - T_f)}$$

Overall heat loss coefficient has been obtained by summing the top loss coefficient and bottom loss coefficient [130], while the side losses are assumed negligible:

$$U_L = \left[\frac{2.8}{\frac{1}{\varepsilon_P}\left(\frac{1}{N_c^{0.025} + \varepsilon_c} - 1\right)} + 0.825(x_m)^{0.21} + aV_{win}^b - 0.5\left(N_C^{0.95} - 1\right) \right] \left(T_{pm} - T_{amb}\right)^{0.2} + \frac{k_i}{t_i}$$

5.2 *Thermal Performance of Dish Cooker*

During the experimentation on dish-type solar cooker, examining over a minute time interval during the sensible cooling of fluid or water, the total time taken dτ for a drop of dT_w (–ve) in water temperature is [131, 132]:

$$d\tau = -\frac{(MC)'_w dT_w}{Q_L} = -\frac{(MC)'_w dT_w}{A_t F' U_L (T_w - T_a)}$$

Now assuming U_L and T_a are constant during the cooling test, the equation for total time taken can be integrated over the 'τ' during which the T_{wo} falls to $T_{w:}$

$$\tau = -\frac{(MC)'_w}{A_t F' U_L} In \left[\frac{(T_w - T_a)}{(T_{wo} - T_a)} \right]$$

Now rewriting the equation, we get

$$(T_w - T_a) = (T_{wo} - T_a) e^{-\tau/\tau_o}$$

where the time constant for cooling can be obtained by:

$$\tau_o = (MC)'_w / A_t F' U_L$$

Now, during the sensible heating, the time taken (dτ) for a temperature rise (dT_w) of water is:

$$d\tau = \frac{(MC)'_w dT_w}{\dot{Q}_u} = \frac{(MC)'_w dT_w}{F' [A_p I_b \eta_o - A_t U_L (T_w - T_a)}$$

Now, if the insolation (I_b) and T_{amb} are constant over a fixed interval of time (τ), then the above equation can be integrated over (τ) for a water temperature rise from T_{w1} to T_{w2} and given as

$$\tau = -\tau_o ln \left[\frac{F'\eta_o - \frac{F'U_L}{C}\left(\frac{T_{w2}-T_a}{I_b}\right)}{F'\eta_o - \frac{F'U_L}{C}\left(\frac{T_{w1}-T_a}{I_b}\right)} \right]$$

Now, with the help of above equation, the cooker parameter ($F'\eta_o$) can be obtained as:

$$F'\eta_o = \frac{F'U_L}{C} \left[\frac{\left(\frac{T_{w2}-T_a}{I_b}\right) - \left(\frac{T_{w1}-T_a}{I_b}\right)^{e^{-\tau/\tau_o}}}{1-e^{-\tau/\tau_o}} \right]$$

Now, the time required for sensible heating can be obtained by rewriting the equation for the cooker parameter ($F'\eta_o$) as:

$$\tau_{boil} = \tau_o ln \left[\frac{1}{1 - \frac{F'U_L}{F'\eta_o} - \frac{1}{C}\left(\frac{100-T_a}{I_b}\right)} \right]$$

6 Thermal Analysis of Solar Dryers

There are various design and types of solar dryers such as natural circulation dryers, mixed mode solar dryers, and forced convection-type solar dryers. Energy balance equations are much helpful for the thermal modeling of the system or to estimate the necessary parameters such as heat transfer coefficient, heat loss coefficient, food temperature, moisture evaporation rate, heat utilization factor, coefficient of performance [133, 134]. One can easily estimate all those parameters if the design of system is known, say if the system is a solar cabinet dryer or shelf-type dryer.

The efficiency of a solar energy dryer can be obtained through the ratio of energy consumed for moisture evaporation by the substance (to be dried) to net input supplied energy to the system as [135]:

$$\eta_d = \frac{m_d[M_o - M_\tau].\lambda}{S.A_{sc}.\tau}$$

The system efficiency for a natural convection solar dryer can be obtained by [136]:

$$\eta_s = \frac{WL}{IA_{sc}}$$

The system efficiency for a forced convection solar dryer can be obtained by [136]:

$$\eta_s = \frac{WL}{(IA_{sc} + P_f)}$$

The system efficiency for a hybrid convection solar dryer can be obtained by [136]:

$$\eta_s = \frac{WL}{(IA_{sc} + P_f) + (m_b.LCV)}$$

Specific energy consumption is another major factor and can be obtained by the given relationship between useful thermal energy for evaporation and unit mass of moisture [135]:

$$\text{Specific energy consumption} = \frac{Q_u}{[m_d(M_o - M_\tau)]}$$

Now, for the estimation of annual thermal energy that is required for drying the substance, the following equation can be used

$$Q_{u,ann} = 365\left(\frac{f.N_d}{\tau}\right).Q_u$$

Another important characteristic for solar food drying process is the drying rate. When a food product is continuously heated at constant moisture content, then its vapor pressure is increased. The moisture flow rate is proportional to its vapor pressure transformation with environment due to crop resistance to moisture flow. Following are the two main drying rate regimes [137]:

(a) Constant drying rate—In this, drying of the object takes place from the surface (simply the evaporation of moisture contents from free water surface). This drying rate mainly depends on the environmental or surrounding conditions, and only the nature of the substance (hard or light stuff) affects this.

$$\frac{dw}{dt} = \frac{K_m A_{sc}}{R_o T}(P_v - P_{va}) = K_f \frac{A(T_a - T_s)}{L}$$

(b) Thin-layer drying—It can define as the drying rate that is only dependent on the physical characteristics of seed, i.e., type and size of seed, moisture contents, and drying temperature. Now, by rewriting the constant drying rate equation, we get

$$\frac{dM}{dt} = -\frac{(P_v - P_{va})}{(1/K_m A_{sc}) R_o T} = -\frac{K_m A_{sc} (P_v - P_{va})}{R_o T}$$

Drying rate may also be expressed from the equation of thin-layer drying, such as [138]

$$\frac{dM}{dt} = -k(M_t - M_e)$$

Now, the moisture contents present in the food substance or product to be dried at any time 't' can be expressed as:

$$M_t = \left[\frac{W_i - W_d}{W_d}\right]$$

7 Conclusion

The present chapter focuses on the solar food processing and cooking methodologies in some developing and developed countries. Many types of cooking fuels are available, but still biomass covers a major portion of various households which generates a lot of pollution and cause fire accidents. To overcome problem, solar cooking is a good option that not only saves the fossil fuels but also a promising candidate for safe cooking. Also, solar food processing found to be the most accessible option for the people and hence the most widespread processing technology. Solar energy is used for drying the food or other products by keeping them under open sun conditions or some cabinet or greenhouse-type dryer. Under the sunlight, some notable thermodynamically or chemically process takes place and the moisture of the product is evaporated by means of solar irradiance and natural or forced convection.

In the present work, various models of solar food processors, solar dryers, and solar cookers have investigated theoretically and their testing analysis. A solar cabinet dryer has been found superior among the various models for a fast drying of products, while the box cooker was found a more appropriate solar energy application for clean cooking because of low thermal losses in comparison with dish cooker. Solar air heater has been observed as an important key element for large size of solar dryers where the high mass flow rate is essential for the product for drying. The natural convection-type solar dryer is found suitable in remote rural areas because of economic and easily accessible. Hybrid-type solar dryers or solar cookers are found more efficient (multipurpose, high efficiency, minimum losses, and working in low ambient conditions) in comparison with conventional systems because of performing on additional energy source along with solar energy.

The performance analysis of both the systems shows the thermal behavior of both the systems under various testing conditions. This analysis is very helpful for the readers to design an efficient solar dryer, food processor, or solar cooker. Besides this, the energy saving analysis will also support the readers to develop a cost optimized solar system.

References

1. Claflin K, Schollers P (2012) Writing food history—a global perspective. Berg Publications
2. Source. http://en.wikipedia.org/wiki/food-processing
3. Viet ZH (2013) Modern food—moral food: self-control, science and the rise of modern American eating in the early 20th century. Carolina Publications
4. Contini C, Romano C, Scozzafava G, Casini L (2016) Food habits and the increase in ready-to-eat and easy-to-prepare products. In: Food hygiene and toxicology in ready-to-eat foods. Academic Press
5. Eisenbrand G (2007) Thermal processing of food: potential health benefits and risks. Wiley
6. Jaeger H, Janositz A, Knorr A (2010) The Maillard reaction and its control during food processing. Pathol Biol 58(3):207–213
7. Motarjemi Y (2014) Encyclopedia of food safety. Academic Press
8. Aravindh MA, Sreekumar A (2015) Solar drying—a sustainable way of food processing. In: Energy sustainability through green energy. Springer, India, pp 15–46
9. Eswara AR, Ramakrishnarao M (2013) Solar energy in food processing—a critical appraisal. J Food Sci Technol 50(2):209–227
10. Ahmed J, Rahman MS (2013) Handbook of food process design. Wiley-Blackwell Publications
11. Tiwari BK, Norton T, Holden NM (2014) Sustainable food processing. Wiley-Blackwell Publications
12. Cohen JS, Yan TCS (1995) Progress in food dehydration. Trends Food Sci Technol 61:20–25
13. Toledo RT (2007) Fundamentals of food process engineering, 3rd edn. Springer, India
14. Singh RP, Heldman DR (2014) Introduction to food engineering, 5th edn. Elsevier Publications
15. Wilkinson VM, Gould G (1996) Food irradiation, 1st edn. Woodhead Publications
16. Fellows PJ (2000) Food processing technology—principles and practice, IInd edn. Woodhead Publications
17. Hayhurst AN (1997) Chemical engineering for the food processing industry. Blackie Academic and Professional, London
18. Al-Baali, AGA, Farid MM (2006) Sterilization of food in retort pouches. Springer Publications (US)
19. The World Bank (2015) The state of the global clean and improved cooking sector, Technical report 007/15
20. Al-Qattan MM, Al-Zahrani K (2009) A review of burns related to traditions, social habits, religious activities, festivals and traditional medical practices. Burns 35:476–481
21. Renewables 2015, global status report (REN-21) (2016)
22. Ekholm T, Krey V, Pachauri S, Riahi K (2010) Determinants of household energy consumption in India. Energy Pol 38:5696–5707

23. Leach G (1992) The energy transition. Energy Policy 20(2):116–123
24. Pachauri S (207) An energy analysis of household consumption: changing patterns of direct and indirect use in India. Springer Publications, India
25. Bisu DY, Kuhe A, Iortyer HA (2016) Urban household cooking energy choice: an example of Bauchi metropolis, Nigeria. Energy Sustain Soc 6:15–21
26. World energy outlook (2015)
27. International Energy access report (2012)
28. Ekouevi K, Tuntivate V (2012) Household energy access for cooking and heating: lessons learned and the way forward. In: Energy mining sector board discussion paper, no. 23, (The World Bank) June 2012
29. Saxena A, Lath S, Agarwal N (2013) Impacts of biomass burning on living areas. TIDDE 12 (1):1–10
30. Jain G (2010) Energy security issues at household level in India. Energy Pol 38:2835–2845
31. Pandey VL, Chaubal A (2011) Comprehending household cooking energy choice in rural India. Biomass Bioenerg 35:4724–4731
32. Akpalu W, Dasmani I, Aglobitse PB (2011) Demand for cooking fuels in a developing country: to what extent do taste and preferences matter? Energy Policy 39:6525–6531
33. Wickramasinghe A (2011) Energy access and transition to cleaner cooking fuels and technologies in Sri Lanka: issues and policy limitations. Energy Policy 39:7567–7574
34. Foell W, Pachauri S, Spreng D, Zerriffi H (2011) Household cooking fuels and technologies in developing economies. Energy Policy 39:7487–7496
35. Maes WH, Verbist B (2012) Increasing the sustainability of household cooking in developing countries: Policy implications. Renew Sustain Energy Rev 16:4204–4221
36. Duan X, Jiang Y, Wang B, Zhao X, Shen G, Wang L (2014) Household fuel use for cooking and heating in China: results from the first Chinese environmental exposure-related human activity patterns survey (CEERHAPS). App Energy 136:692–703
37. Ramakrishna J, Durgaprasad MB, Smith KR (1989) Cooking in India: the impact of improved stoves on indoor air quality. Environ Int 15:341–352
38. Tripathi AK, Iyer PVR, Kandpal TC (1999) Biomass gasifier based institutional cooking in India: a preliminary financial evaluation. Biomass Bioenerg 17:165–173
39. Dixit CSB, Paul PJ, Mukunda HS (2006) Experimental studies on a pulverized fuel stove. Biomass Bioenerg 30:673–683
40. Mercado IR, Masera O, Zamora H, Smith KR (2011) Adoption and sustained use of improved cook-stoves. Energy Policy 39(12):7557–7566
41. Chandrasekaran S, Ramanathan S, Basak T (2013) Microwave food processing—a review. Food Res Int 52:243–261
42. Joshi SB, Jani AR (2015) Design, development and testing of a small scale hybrid solar cooker. Sol Energy 122:148–155
43. Shinde YH, Gudekar AS, Chavan PV, Pandit AB (2016) Design and development of energy efficient continuous cooking system. J Food Eng 168:231–239
44. Pokharel S, Chandrashakhar M, Robinson JB (1992) Inter-fuel and inter-mode substitution for cooking. Energy 17(10):907–918
45. Sharma SK, Sethi BPS, Chopra S (1990) Thermal performance of the 'Rohini'—an improved wood cook-stove. Energy Convers Manag 30(4):409–419
46. Masera OM, Navia J (1997) Fuel switching or multiple cooking fuels. Biomass Bioenerg 12 (5):347–361
47. Stumpf P, Balzar A, Eisenmann W, Wendt S, Vajen K (2001) Comparative measurements and theoretical modeling of single- and double-stage heat pipe coupled solar cooking systems for high temperatures. Sol Energy 71(1):1–10
48. Almeida ATD, Lopes A, Carvalho A, Mariano J, Nunes C (2004) Evaluation of fuel-switching opportunities in the residential sector. Energy Build 36:195–203
49. Heltberg R (2004) Fuel switching: evidence from eight developing countries. Energy Ecol 26:869–887

50. Park H, Kwon H (2011) Effects of consumer subsidy on household fuel switching from coal to cleaner fuels. Energy Policy 39:1687–1693
51. Yangka D, Diesendorf M (2016) Modeling the benefits of electric cooking in Bhutan: a long term perspective. Renew Sustain. Energy Rev 59:494–503
52. Hosier RH, Dowd J (2014) Household fuel choice in Zimbabwe: an empirical test of the energy ladder hypothesis. Res Energy 9:347–361
53. Masera OR, Saatkamp BD (2000) From linear fuel switching to multiple cooking strategies: a critique and alternative to the energy ladder model. World Dev 28(12):2083–2103
54. Hosier RH (2004) Energy ladder in developing countries. Encycl Energy 2:423–435
55. The World Energy book (2007) Let everyone climb the energy ladder
56. Raiyani CV, Shah SH, Desai NM, Venkaiah K, Patel JS, Parikh DJ, Kashyap SK (1993) Characterization and problems of Indoor pollution due to cooking stove smoke. Atmos Environ Part A 27:1643–1655
57. Hall DO, Scrase JI (1998) Will biomass be the environmentally friendly fuel of the future. Biomass Bioenerg 15(4/5):357–367
58. Parikh J, Balakrishnana K, Laxmi V, Biswas H (2001) Exposure from cooking with biofuels: pollution monitoring and analysis for rural Tamil Nadu, India. Energy 26:949–962
59. Yu F (2011) Indoor air pollution and children's health: net benefits from stove and behavioral interventions in rural China. Environ Resource Econ 50:495–514
60. Van-Vliet EDS, Asante K, Jack DW, Kinney PL, Whyatt RM, Chillrud SN, Abokyi L, Zandoh C, Owusu-Agyei S (2013) Personal exposures to fine particulate matter and black carbon in households cooking with biomass fuels in rural Ghana. Environ Res 127:40–48
61. Ekwo EE, Weinberger MM, Lachenbruch PA, Huntley WH (1983) Relationship of parental smoking and gas cooking to respiratory disease in children. Chest 84(6):662–668
62. Jedrychowski W, Adamczyk BT, Flak E, Mroz E, Gomola K (1990) Effect of indoor air pollution caused by domestic cooking on respiratory problems of elderly women. Environ Int 16(1):57–60
63. Siddiqui AR, Lee K, Gold EB, Bhutta ZA (2005) Eye and respiratory symptoms among women exposed to wood smoke emitted from indoor cooking: a study from southern Pakistan. Energy Sustain Dev 9(3):58–66
64. Rangel LG, Bouscoult LT (2011) Pollution/biomass fuel exposure and respiratory illness in children. Pediatric Res Rev 12(1):S40–S42
65. Jayes L, Haslam PL, Gratziou CG, Bee JL (2016) Smoke hazards: systematic reviews and meta-analyses of the effects of smoking on respiratory health. Chest 150:164–179
66. Mabrouk A, Badawy AE, Sherif M (2000) Kerosene stove as a cause of burns admitted to the Ain Shams burn unit. Burns 26:474–477
67. Smith KR, Mehta S (2003) The burden of disease from indoor air pollution in developing countries: comparison of estimates. Int J Hygiene Environ Health 206(4–5):279–289
68. Kim KH, Jahan SA, Kabir E (2011) A review of diseases associated with household air pollution due to the use of biomass fuels. J Hazard Math 192(2):425–431
69. Mestl HES, Edwards R (2011) Global burden of disease as a result of indoor air pollution in Shaanxi, Hubei and Zhejiang, China. Sci Total Environ 409(8):1391–1398
70. Ahuja RB, Dash JK, Shrivastava P (2011) A comparative analysis of liquefied petroleum gas (LPG) and kerosene related burns. Burns 37:1403–1410
71. Epstein MB, Bates MN, Arora NK, Balakrishnan K, Jack DW, Smith KR (2013) Household fuels, low birth weight, and neonatal death in India: the separate impacts of biomass, kerosene, and coal. Int J Hygiene Environ Health 216(5):523–532
72. Milovantseva N, Ogunseitan OA (2015) Composite measures of the environmental burden of disease at the global level (Reference module in earth systems and environmental sciences). Encycl Environ Health 813–821
73. Gregcire RG (1984) Understanding solar food dryers. ASHRAE report
74. Sabarez H (2016) Reference module in food science
75. Sharma A, Chen CR, Lan NV (2009) Solar-energy drying systems: a review. Renew Sustain Energy Rev 13(6):1185–1210

76. Kalogirou S (2009) Solar energy engineering: processes and systems, 1st edn. Elsevier
77. Anwar SI, Tiwari GN (2001) Evaluation of convective heat transfer coefficient in crop drying under open sun drying conditions. Energy Convers Manag 42:627–637
78. Mulokozi G, Svanberg U (2003) Effect of traditional open sun-drying and solar cabinet drying on carotene content and vitamin—A activity of green leafy vegetables plant foods for human nutrition. 58:1–15
79. Kabasa JD, Ndawula J, Byaruhanga YB, Meulen U (2003) Implications of open-sun drying and visqueen-covered and polyethylene-covered solar-drying technology on fruit and vegetable vitamins A and C content. In: Proceeding of international conferences on technology and institute innovations sustainable rural development. Deutscher Tropentag, Gottingen, 8–10 October 2003
80. Jain D, Tiwari GN (2003) Thermal aspects of open sun drying of various crops. Energy 28:37–54
81. Akpinar EK, Cetinkaya YBF (2006) Modelling of thin layer drying of parsley leaves in a convective dryer and under open sun. J Food Eng 75:308–315
82. Prasad J, Prasad A, Vijay VK (2006) Studies on the drying characteristics of Zingiber officinale under open sun and solar biomass (hybrid) drying. Green Energy 3(1):79–89
83. Al-Mahasneh MT, Rababah M, Al-Shbool MA, Yang W (2007) Thin layer drying kinetics of sesame hulls under forced convection and open sun drying. J Food Process Eng 30:324–337
84. Kooli S, Fadhel A, Farhat A, Belghith A (2007) Drying of red pepper in open sun and greenhouse conditions- mathematical modeling and experimental validation. J Food Eng 79 (3):1094–1103
85. Tripathy PP, Kumar S (2009) Influence of sample geometry and rehydration temperature on quality attributes of potato dried under open sun and mixed-mode solar drying. Green Energy 6(2):143–156
86. Gudapaty P, Indavarapu S, Korwar GR, Shankar AK (2010) Effect of open air drying, LPG based drier and pretreatments on the quality of Indian gooseberry (aonla). J Food Sci Technol 47(5):541–548
87. Akpinar EK (2010) Drying of mint leaves in a solar dryer and under open sun: Modelling, performance analyses. Energy Conver Manag 51:2407–2418
88. Hande AR, Swami SB, Thakor NJ (2016) Open-air sun drying of Kokum (Garcinia indica) rind and its quality evaluation. Agri Res 1–11
89. Lawand TA (1966) A solar cabinet dryer. Sol Energy 10(4):158–164
90. Datta G, Garg HP, Ray RA, Prakash J (1988) Performance prediction of a cabinet-type solar drier. Solar Wind Technol 5:289–292
91. Sharma S, Sharma VK, Jha R, Ray RA (1990) Evaluation of the performance of a cabinet type solar dryer. Energy Convers Manag 30(2):75–80
92. Ampratwuma DB, Dorvlo ASS (1998) Evaluation of a solar cabinet dryer as an air-heating system. Appl Energy 59(1):63–71
93. Goyal RK, Tiwari GN (1997) Parametric study of a reverse flat plate absorber cabinet dryer. Sol Energy 60(1):41–48
94. Adapa PK, Sokhansanj S, Schoenau GJ (2002) Performance study of a re-circulating cabinet dryer using a household dehumidifier. Drying Technol 20(8):23–28
95. Sreekumar A, Manikantan PE, Vijayakumar KP (2008) Performance of indirect solar cabinet dryer. Energy Convers Manag 49:1388–1395
96. Rawat BS, Pant PC, Joshi GC (2009) Energetics study of a natural convection solar crop dryer. Ambit Energy 30(4):193–198
97. Singh NJ, Pandey RK (2012) Convective air drying characteristics of sweet potato cube 90:317–322
98. Demiray E, Tulek Y (2014) Drying characteristics of garlic (Allium sativum L) slices in a convective hot dryer. Heat Mass Trans 50:779–786
99. Sallam YI, Aly MH, Nassar AF, Mohamed EA (2015) Solar drying of whole mint plant under natural and forced convection. J Advert Res 6:171–178

100. Yahya M, Fudholi A, Hafizh H, Sopian K (2016) Comparison of solar dryer and solar assisted heat pump dryer for cassava. Sol Energy 136:606–613
101. Saxena A, Varun Pandey SP, Srivastva G (2011) A thermodynamic review of solar box type cookers. Renew Sustain Energy Rev 15(6):3301–3318
102. Hussein HMS, El-Ghetany HH, Nada SA (2008) Experimental investigation of novel indirect solar cooker with indoor PCM thermal storage and cooking unit. Energy Convers Manag 49:2237–2246
103. Joshi SB, Jani AR (2015) Design, development and testing of a small scale hybrid solar cooker. Sol Energy 122:148–155
104. Duan X, Jiang Y, Wang B, Zhao X, Shen G, Huang N (2014) Household fuel use for cooking and heating in China. Appl Energy 136:692–703
105. Clean, Affordable and sustainable cooking energy for India, CEEW Report, New Delhi (2015)
106. Puzzolo E, Pope D (2017) Reference module in earth systems and environmental sciences. Elsevier
107. Musango JK (2014) Household electricity access and consumption behaviour in an urban environment: The case of Gauteng in South Africa. Energy Sustain Dev 23:305–316
108. Desmarchelier P (2014) Food safety management. Elsevier
109. Saxena A, Deval N (2016) A high rated solar water distillation unit for solar homes. J Eng 1–8:2016
110. Nahar NM (2001) Design, development and testing of a double reflector hot box solar cooker with a transparent insulation material. Renew Energy 23:167–179
111. Saxena A, Srivastava G (2011) Potential and economics of solar water heating. MIT Int J Mech Eng 2:97–104
112. Panwar NL, Kothari S, Kaushik SC (2013) Techno-economic evaluation of masonry type animal feed solar cooker in rural areas of an Indian state Rajasthan. Energy Policy 52:583–586
113. Nandwani SS (1989) Design construction and experimental study of electric cum solar oven —II. Solar Wind Technol 6(2):149–158
114. Hussain M, Das KC, Huda A (1997) The performance of a box-type solar cooker with auxiliary heating. Renew Energy 12:151–155
115. Saxena A, Varun, Srivastava G (2012) Performance testing of a solar box cooker provided with sensible storage material on the surface of absorbing plate. Int J Renew Energy Technol 3:165–173
116. Chaudhuri TK (1999) Estimation of electrical backup for solar box cooker. Renew Energy 17:569–572
117. Nandwani SS (2007) Design construction and study of a hybrid solar food processor in the climate of Costa Rica. Renew Energy 32:427–441
118. Chaouki A, Kamel R, Slimen A, Ibrahim J (2010) Theoretical study and design of a hybrid solar cooker. IJRRAS 3(3)
119. Joshi SB, Jani AR (2015) Design, development and testing of a small scale hybrid solar cooker. Sol Energy 122:148–155
120. Akyurt M, Selcuk MK (1973) A solar drier supplemented with auxiliary heating systems for continuous operation. Sol Energy 14:313–320
121. Prasad J, Vijay VK, Tiwari GN, Sorayan VPS (2006) Study on performance evaluation of hybrid drier for turmeric (Curcuma longa L) drying at village scale. J Food Eng 75:497–502
122. Ferreira AG, Charbel ALT, Pires RL, Silva JG, Maia CB (2007) Experimental analysis of a hybrid dryer. Engenharia Térmica (Thermal Eng) 6(02):3–7
123. Hossain MA, Amer BMA, Gottschalk K (2008) Hybrid solar dryer for quality dried tomato. Drying Technol 26:1591–1601
124. Amer BMA, Hossain MA, Gottschalk K (2010) Design and performance evaluation of a new hybrid solar dryer for banana. Energy Convers Manag 54:813–820
125. Lopez-Vidana EC, Mendez-Lagunas LL, Rodrıguez-Ramırez J (2013) Efficiency of a hybrid solar–gas dryer. Sol Energy 93:23–31

126. Reyes A, Mahn A, Vásquez F (2014) Mushrooms dehydration in a hybrid-solar dryer, using a phase change material. Energy Convers Manag 83:241–248
127. Yassen TA, Al-Kayiem HH (2016) Experimental investigation and evaluation of hybrid solar/thermal dryer combined with supplementary recovery dryer. Sol Energy 134:284–293
128. ASAE S580 (2003) Testing and reporting solar cooker performance
129. Mullick SC, Kandpal TC, Saxena A (1987) Thermal test procedure for box type solar cooker. Sol Energy 39(4):353–360
130. Channiwala SA, Doshi NI (1989) Heat loss coefficients for box type solar cookers. Sol Energy 42:495–501
131. Mullick SC, Kandpal TC, Kumar S (1991) Thermal test procedure for a paraboloid concentrator solar cooker. Sol Energy 46:139–144
132. Mishra D, Tekchandani CK, Nahar NM (1987) Performance and testing of a parabolic spheroidal concentrator type solar cooker. Reg J Energy Heat Mass Trans 9(2)
133. Tiwari GN (2008) Solar energy—fundamentals, design, modeling and applications. Narosa Publications
134. Garg HP, Prakash J (2009) Solar energy—fundamentals and applications. Tata McGraw Hill
135. Singh S, Kumar S (2013) Solar drying for different test conditions: Proposed framework for estimation of specific energy consumption and CO_2 emissions mitigation. Energy 51:27–36
136. Leon MA, Kumar S, Bhattacharya SC (2002) A comprehensive procedure for performance evaluation of solar food dryers. Renew Sustain Energy Rev 6:367–393
137. Ekechukwu OV (1999) Review of solar-energy drying systems I: an overview of drying principles and theory. Energy Convers Manag 40:593–613
138. Singh SP, Jairaj KS, Srikant K (2012) Universal drying rate constant of seedless grapes: a review. Renew Sustain Energy Rev 16:6295–6630
139. Source. https://www.google.co.in/search?q=clean+cooking+stoves&rlz
140. Rathore NS, Panwar NL (2010) Experimental studies on hemi cylindrical walk-in type solar tunnel dryer for grape drying. Appl Energy 87(8):2764–2767
141. Kumar A, Tiwari GN (2006) Effect of shape and size on convective mass transfer coefficient during greenhouse drying (GHD) of jaggery. J Food Eng 73(2):121–134

Visual Comfort Based Algorithmic Control for Roller Shade and Assessment of Potential Energy Savings

Lakshya Sharma and Dibakar Rakshit

Abstract Offices which are mostly operated during day-lit hours are fascinated toward incorporation of solar daylighting systems so as to get benefitted of energy savings along with intangible benefits like good health, well-being, and productivity of their occupants. But discomfort glare is usually ignored in front of rigorous energy load optimization practices and researches, or even if considered, is not properly quantified, which leads to a false evaluation of the performance of daylighting in the context of lighting energy savings. Blinds when operated with optimum visual comfort prevention measures, then only can create a functional day-lit environment. The present study includes modeling and analysis of an east facing office building located in New Delhi, India. It utilizes drafting tool Rhinoceros 3D 5.0 and simulation tool DIVA 4.0 to mathematically analyze the utilization of natural lighting for the office. The analysis includes properly arranged viewpoints similar to that experienced by any occupant, being placed in simulation environment so as to visualize and evaluate realistic glare scenario. The study further deals with the observation and analysis of the illuminance patterns inside the test space and glare values for different blind positions modeled. An algorithm has been developed to analyze the required visual comfort level of the space and suggests an appropriate blind position for every occupancy hour that eventually leads to estimate the potential energy savings through the utilization of daylight. It was observed that even after providing visual comfort to the occupants, the reduction in savings was marginal, measured to be 1% as compared to the case when visual discomfort is overlooked in daylighting utilization.

Keywords Daylight · DIVA · Roller blinds · Glare · Energy savings

Nomenclature

L_S	Brightness of glare source
L_B	Background luminance

L. Sharma · D. Rakshit (✉)
CES, IIT Delhi, New Delhi, India
e-mail: dibakar@iitd.ac.in

H. Tyagi et al. (eds.), *Applications of Solar Energy*, Energy, Environment, and Sustainability, https://doi.org/10.1007/978-981-10-7206-2_14

ω	Solid angle size of glare source
P	Position index
G	Discomfort glare
DGI	Daylight glare index
VCP	Visual comfort probability
UGR	Unified glare rating
DGP	Daylight glare probability
LPD	Lighting power density
NBC	National building code

1 Introduction

1.1 Energy Efficiency in Buildings

The building sector is experiencing rapid growth in developing countries like India, and energy demand is the most essential for this. Therefore, energy efficiency has turned into a major area of concern among builders, occupants as well as the governmental bodies in this sector. According to a report by Bureau of Energy Efficiency (BEE), almost 34% of total electricity consumption in India is contributed by buildings. Out of the total electricity consumption, approximately 55% of the electricity is consumed in maintaining thermal comfort, i.e., reducing heating/cooling load, as desired in an area, after which comes the electricity consumed in lighting [1]. There have been many studies which focus on assessing the heating/cooling load and analyzing the impact of various parameters on the thermal load, so that an appropriate solution can be provided to obtain the desired results. One such study has been done by Kulkarni et al. where they studied the effects of various types of glazing, exterior, and interior insulation of wall, along with effect of the replacement of the fluorescent tube lamp with the CFL on the cooling load requirements in a building [2]. In another such study by Jani et al., the effectiveness of replacing vapor compression system with the solid desiccant cooling system in a building electricity consumption due to HVAC system has been analyzed [3, 4]. Since lighting energy consumption comes only second in major electricity consumption pattern, as provided in a report by BEE, so studies should focus on reducing the lighting energy consumption as well. Thus, this study focuses on lighting energy savings in an office building, as lighting corresponds to approximately 25% of total electricity consumption in a commercial building, as depicted in Fig. 1 [1].

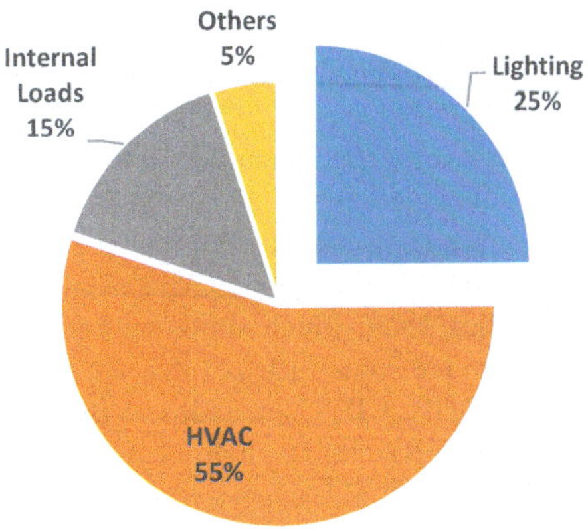

Fig. 1 Energy consumption pattern of commercial buildings in India (2005) Source: Bureau of Energy Efficiency [17]

1.2 Daylighting: Pros and Cons

Nowadays, optimized building design, construction, and operational practices are being implemented so as to increase energy efficiency, along with providing better environmental quality to the occupants. Daylighting incorporation in buildings not only improves the quality of the indoor visual environment but also reduces the lighting energy consumption. In India, a major portion of the total electricity is generated through thermal power plants, thus artificial lighting contributes in a huge amount of CO_2 emissions in the environment [5]. Daylighting utilization is trending these days among building designers and occupants due to its huge energy-saving potential. Other than this, there are other intangible benefits like good health, well-being, and better mental and visual stimulations of the occupants [6].

1.3 Glare: Definition and Measurement

The energy utilization pattern of any space is highly influenced by the thermal and visual conditions required by the occupants. But rigorous energy efficiency studies and critical design practices still overlook visual discomfort caused due to daylighting incorporation [7]. As per the study on "Windows and Offices," glare has been found as one of the responsible factors reducing the performance of the office workers by almost 18–21% in a test carried out on the basis of the mental functioning of humans [8]. Therefore, studies based on lighting energy efficiency in buildings should also incorporate visual discomfort caused due to excessive daylight.

Glare is generally categorized as disability glare and discomfort glare. Disability glare harms vision with the reduction in visibility of any object, caused by scattering of light from any light source somewhere in space that affects contrast that is required to see the object properly, while discomfort glare causes sensation and irritation to the occupant due to highly bright or illuminated objects [9].

$$G = \sum_{i=1}^{n} \frac{L_{s,i} \times \omega_{s,i}}{L_b \times P_i} \quad (1)$$

The expression mentioned above is a basic discomfort glare equation which shows that glare is mostly related to the brightness of glare source (L_S), background luminance (L_B), solid angle size of glare source (ω), and position index (P) of the occupant. Other researches have also witnessed dependency of glare on observer's adaptation level, which varies with person to person and increases with age [10].

There are some matrices that are used for glare calculation like Daylight Glare Index (DGI), CIE Glare Index, Visual Comfort Probability (VCP), Unified Glare Rating (UGR), and Daylight Glare Probability (DGP). An experimental study done by Jakubiec et al. suggests that Daylight Glare Probability (DGP) metrics is the most robust metrics when quantitative glare prediction is to be done. The results are less sensitive to varying solar conditions as Daylight Glare Probability (DGP) includes the comparison of illuminated areas with vertical eye illuminance for a viewing hemisphere. Glare prediction using DGP classifies glare level as following (Table 1) [10].

1.4 Shading Systems

Most common objectives that are needed to be considered with the utilization of daylighting system are:

- Illuminating the maximum area through natural light to the required level
- Improving visual comfort including glare control
- Maintaining thermal gain

Shading systems are primarily designed to address these objectives along with protecting the privacy of the occupants. Conventional solar shading systems include pull-down shades, e.g., roller blinds, which also provide glare control while maintaining the appropriate interior illumination level. There are many types of window blinds usually made up of either horizontal/vertical slats or piece of fabric by materials like wood, plastic, metals. [11]. These can be classified into various categories considering different aspects:

Table 1 Glare classification by DGP

Discomfort classification	DGP range
Intolerable	>0.45
Disturbing	0.4–0.45
Perceptible	0.35–0.4
Imperceptible	<0.35

(a) **Inside and Outside blind**

Outside blinds are usually made up of hard materials as they have to cope with the outside conditions, e.g., rain, wind, etc., other than blocking solar gains. Inside blinds are not very helpful in providing protection from solar gains as these are applied inside the room so the heat absorbed is radiated inside only. But the inside blinds provide flexible control to allow a certain amount of light and prevent glare.

(b) **Based on structure**

Venetian blinds are horizontal blinds which offer most flexible controls to maintain daylighting and thermal characteristics such as the fraction of the window covered and slope of blades. Roller shades are made of a single piece of material, usually semi-transparent fabric. It reflects or transmits part of energy received based on the properties of the fabric. There are situations when blind operation contradicts daylighting and thermal requirements. This is balanced on the basis of application requirement and energy consumption constraints of the space under consideration [12, 13].

Many researches have been done so far for quantifying energy-saving potential of daylighting incorporation with artificial lighting for the different type of buildings, such as residential buildings, commercial buildings, and auditoriums. But most of the studies, while evaluating daylighting performance, have overlooked essential operational environment requirements of the occupants. These requirements include visual and thermal comfort mainly. Any daylight performance study cannot be considered practical unless it includes discomfort prevention due to glare caused by excessive daylight. Some of the studies have considered glare, but the measure used for quantifying the glare scenario is based on methods that are not sufficient to evaluate glare properly. The present research, therefore, takes into account evaluating performance of an office building by utilizing daylight through glazing, but preferably considering discomfort experienced by an occupant due to glare and its prevention. Glare quantification is done using Daylight Glare Probability (DGP), and roller shades have been considered for the prevention of glare.

2 Methodology

A simple office with one façade available for integration of daylighting measures is modeled to analyze the potential energy savings, with our area of consideration being the first floor of the office building. Modeled room, shown in Fig. 2, having 4 workplaces, is of the dimensions 5 m × 3 m × 2.75 m, and its east facing wall is provided with a single pane glazing. The total area of glazing is 15.9 m^2, with the

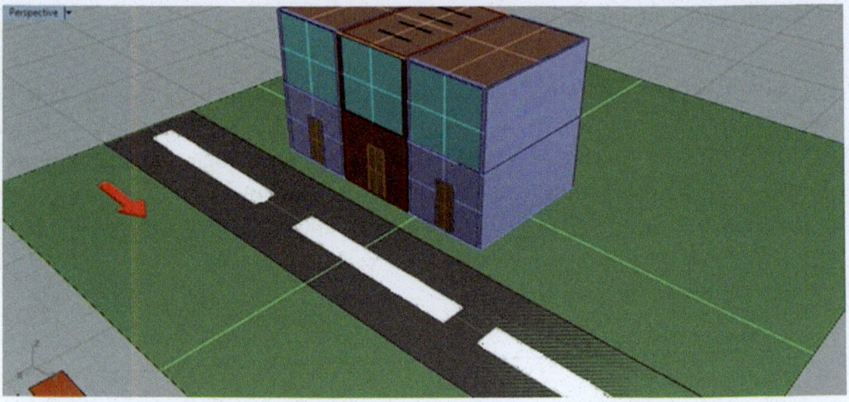

Fig. 2 Isometric view of modeled building

interior area of office being 15 m^2. The walls are having a width of 110 mm. The present study considers typical office operational hours, i.e., 9 AM to 5 PM, six days a week.

New Delhi climate region, which falls under the northern hemisphere, is considered for the study, with the latitude and longitude as 28.38° N and 77.12° E, respectively. Designing of the test building has been done in the drafting tool Rhinoceros 3D 5.0, and DIVA 4.0 has been used for running the simulation.

Optical properties of various components of the building are listed below in Table 2.

There are a variety of shading systems in use for maintaining appropriate thermal and visual comfort conditions inside the space. Here, interior roller blind is modeled with five different operational positions as fully opened, 25% closed, 50% closed, 75% closed, and fully closed, properties of which are stated in Table 2.

For the purpose of validation, the outside illuminance was considered for the comparison. Error percentages observed between the simulated readings and measured data from a lux meter are found to be acceptable as shown in Table 3.

Table 2 Optical properties of various components of modeled building

Sr. No.	Components of geometry	Optical property
	Walls	Reflectivity: 50%
	Ceiling	Reflectance: 70%
	Glazing (single pane)	Visual transmittance: 88%
		Visual transmissivity: 96%
	Outer ground	Reflectance: 10%
	Furniture	Reflectivity: 50%
	Floor	Reflectivity: 20%
	Roller blinds	Total transmission: 0.04
		Direct transmission: 0.01
		Total reflection: 0.41

Table 3 Validation results

Date	Illuminance values (lux) (simulated)	Illuminance values (lux) (measured)	Error %
08-Sep	67010	76150	12.00
09-Sep	62200	67500	7.85

3 Result and Discussions

3.1 Visualization of Test Space

Visualization is run for checking proper modeling and working of designed geometry in DIVA. The visualized image shown in Fig. 3 is simulated for May 01, 11 AM. As the office modeled is east facing, so huge amount of direct sunlight falls on the glazed façade of the building during morning and afternoon hours. Labels with lux levels are clearly depicting the illuminance around the building and on the façade.

3.2 Placing Work Plane Observation Grid

The visualization done above is capable of providing illuminance levels anywhere in the space. However, a summary of illuminance values on a surface or work plane is always preferred for the analysis procedure. So, some node points were required

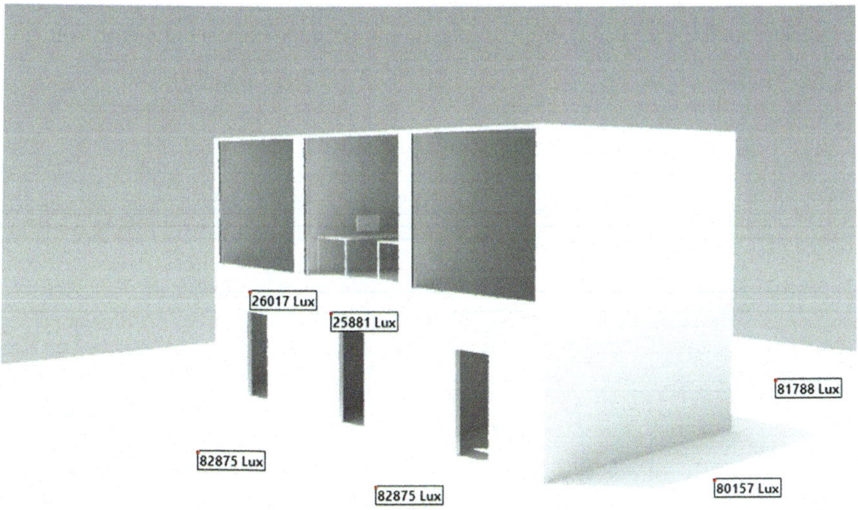

Fig. 3 Visualized image of building under consideration

to be placed to measure illuminance levels in the office space for different operating conditions. So far, almost all of the studies have considered the height of the observation nodes at the standard working plane height, which is typically 2.5 feet or 0.76 m above the floor. In the present study, this height is taken as 0.8 m [14].

$$p = 0.2 \times 5^{\log_{10} d} \tag{2}$$

The equation shown above is used to calculate the optimized distance between the observation nodes for the analysis grid. Here, p gives maximum grid cell size or maximum distance between nodes and d is the longer dimension of the area, such as a room. In this study, the room has dimensions 5 m × 3 m, so the value to be used for d is 5 m, which gives maximum grid cell size as 0.6 m. So, the analysis surface will have grid points at the height of 0.8 m and placed 0.6 m apart. As the total floor area of the room is 16.8 m^2, thus 60 nodes are placed throughout the room, as shown in Fig. 4 [15].

3.3 *Observation of Interior Illumination*

In this process, point-in-time illuminance is measured on the analysis grid placed on the working plane height from the floor. A screenshot of simulation result taken for May 14, 11 AM, is shown in Fig. 5 below. The false color image in the figure depicts the light penetration and distribution inside the room area. A significant decrease in the illuminance value is observed as the node is moved away from the window.

A keen observation of the illuminance pattern of the figures shown above explains that how nodes at farther end from window have a large probability of being underexposed during the evening hours, or after blind application for glare

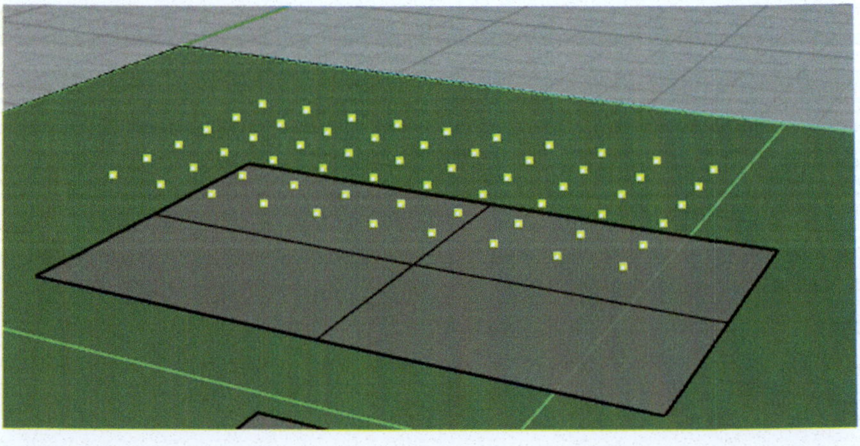

Fig. 4 Observation grid nodes

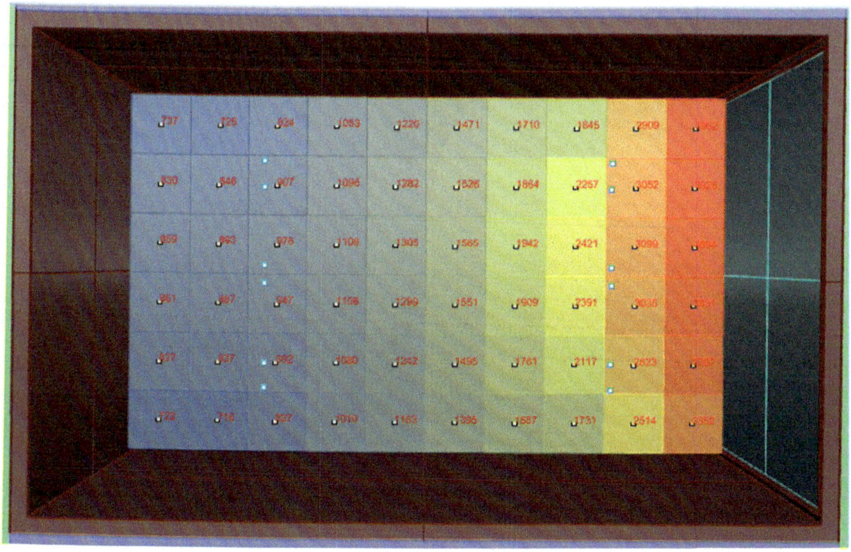

Fig. 5 Interior illumination pattern of office room at 11 AM (number in the blocks are lux values)

prevention. These nodes will require to be lit by artificial lighting, which will become the basis of calculating the energy consumption with various combinations of daylighting levels and glare values. Since the considered test space is an office building, a threshold of 300 lx is mandated by National Building Code (NBC) 2005 [16]. This threshold is used as the basic lighting requirement of an interior space of the considered office building.

3.4 *Observation of Glare*

It has been worked to visualize realistic glare scenario as experienced by employees of the office. For this, a set of 180° fish-eye camera viewpoints have been placed in the simulation tool DIVA for Rhino at the position similar to that of an occupant, shown in Fig. 6. Glare is evaluated for the set perspective view. DIVA utilizes Evaglare to quantify glare. Daylight Glare Probability (DGP) is used to quantify glare in the present study as it is mostly recommended.

3.5 *Formation of Illuminance Matrices*

As shown above, in Fig. 5, point-in-illuminance measurement is helpful to visualize lighting condition of any space for a particular time. But the objective of this study

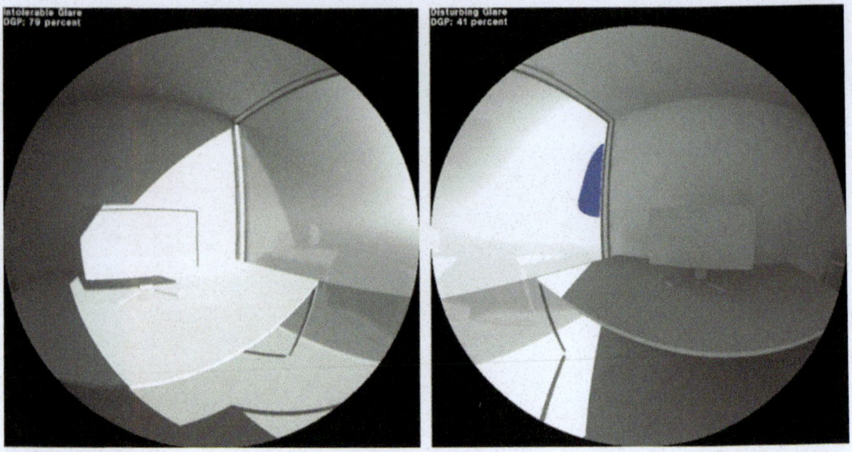

Fig. 6 Fish-eye view of glare scene at system 2 and system 1 (DGP value in percentage)

requires illuminance matrices for every hour throughout the year as per the occupancy behavior considered. Tables 3, 4, 5, 6, 7, 8 are screen captures of excel data sheets for the simulated yearly illuminance values observed for each of the five blind positions. As illuminance values are observed for 60 nodes in the office room under consideration every hour, so five matrices of dimension 2502 × 60 are created corresponding to each of the five blind positions.

3.6 Identifying Glare Scenario

Glare is found to be a function of view of the occupant other than luminance values of the source and background. So, a proper position and view of occupants are needed to be modeled to get the most accurate and actual results. The four workplaces inside the office room have been modeled with 180° fish-eye viewpoints in DIVA, which uses Evaglare to evaluate the glare conditions. Out of the four systems modeled, as shown in Fig. 7, two systems (system 1 and system 2) which are near to the glazing, are prone to a high amount of solar exposure, making the probability of glare higher on these two systems, as compared to the two which are away from glazing. Therefore, these two systems were analyzed for glare for different weather conditions of New Delhi, which are a clear sky, a cloudy sky, and a partly cloudy sky. May, August, and November months have been selected to represent these weather conditions. The observation is done for the 15th day of each of the months considered. Results shown in Fig. 8 explain that system 2 was found to be experiencing the maximum glare in all conditions, which can thus be considered as the worst glare position of the room. Identifying the worst glare scenario

Table 4 Simulated illuminance values (lux) for blind fully open

Month	Day	Hour	Nodes									
			1	2	3	4	5	6	7	8	9	10
1	1	0.5	0	0	0	0	0	0	0	0	0	0
1	1	1.5	0	0	0	0	0	0	0	0	0	0
1	1	2.5	0	0	0	0	0	0	0	0	0	0
1	1	3.5	0	0	0	0	0	0	0	0	0	0
1	1	4.5	0	0	0	0	0	0	0	0	0	0
1	1	5.5	0	0	0	0	0	0	0	0	0	0
1	1	6.5	0	0	0	0	0	0	0	0	0	0
1	1	7.5	80	98	108	122	136	130	78	100	113	126
1	1	8.5	490	608	653	719	770	683	485	624	696	754
1	1	9.5	323	403	416	427	432	362	330	421	451	455
1	1	10.5	337	423	436	447	452	378	346	442	473	475
1	1	11.5	321	403	414	424	428	358	330	421	449	451
1	1	12.5	207	259	266	272	274	229	212	270	288	288
1	1	13.5	225	283	290	297	300	251	232	295	315	315
1	1	14.5	201	252	259	265	268	225	207	263	281	282
1	1	15.5	185	231	238	243	246	206	190	241	258	258
1	1	16.5	146	183	186	189	191	161	149	187	199	197
1	1	17.5	10	12	12	12	12	11	10	12	13	13
1	1	18.5	0	0	0	0	0	0	0	0	0	0
1	1	19.5	0	0	0	0	0	0	0	0	0	0
1	1	20.5	0	0	0	0	0	0	0	0	0	0
1	1	21.5	0	0	0	0	0	0	0	0	0	0
1	1	22.5	0	0	0	0	0	0	0	0	0	0
1	1	23.5	0	0	0	0	0	0	0	0	0	0

Table 5 Simulated illuminance values (lux) for blind 25% closed

Month	Day	Hour	Nodes									
			1	2	3	4	5	6	7	8	9	10
1	1	0.5	0	0	0	0	0	0	0	0	0	0
1	1	1.5	0	0	0	0	0	0	0	0	0	0
1	1	2.5	0	0	0	0	0	0	0	0	0	0
1	1	3.5	0	0	0	0	0	0	0	0	0	0
1	1	4.5	0	0	0	0	0	0	0	0	0	0
1	1	5.5	0	0	0	0	0	0	0	0	0	0
1	1	6.5	0	0	0	0	0	0	0	0	0	0
1	1	7.5	47	59	64	68	77	72	47	55	67	69
1	1	8.5	283	359	382	393	414	346	291	332	404	408
1	1	9.5	186	235	241	240	239	196	198	223	263	255
1	1	10.5	194	245	252	251	250	204	207	233	275	267
1	1	11.5	185	233	239	238	237	194	197	221	262	253
1	1	12.5	119	150	154	153	152	124	127	142	168	162
1	1	13.5	130	164	168	167	166	136	138	156	184	178
1	1	14.5	116	147	150	149	149	122	124	139	164	159
1	1	15.5	107	135	138	137	137	112	113	128	151	146
1	1	16.5	84	106	107	107	107	87	88	99	115	111
1	1	17.5	6	7	7	7	7	6	6	6	7	7
1	1	18.5	0	0	0	0	0	0	0	0	0	0
1	1	19.5	0	0	0	0	0	0	0	0	0	0
1	1	20.5	0	0	0	0	0	0	0	0	0	0
1	1	21.5	0	0	0	0	0	0	0	0	0	0
1	1	22.5	0	0	0	0	0	0	0	0	0	0
1	1	23.5	0	0	0	0	0	0	0	0	0	0

Table 6 Simulated illuminance values (lux) for blind 50% closed

Month	Day	Hour	Nodes									
			1	2	3	4	5	6	7	8	9	10
1	1	0.5	0	0	0	0	0	0	0	0	0	0
1	1	1.5	0	0	0	0	0	0	0	0	0	0
1	1	2.5	0	0	0	0	0	0	0	0	0	0
1	1	3.5	0	0	0	0	0	0	0	0	0	0
1	1	4.5	0	0	0	0	0	0	0	0	0	0
1	1	5.5	0	0	0	0	0	0	0	0	0	0
1	1	6.5	0	0	0	0	0	0	0	0	0	0
1	1	7.5	20	21	23	26	23	26	19	21	24	23
1	1	8.5	120	131	140	150	130	126	118	130	144	138
1	1	9.5	84	91	98	101	90	84	83	93	103	99
1	1	10.5	87	95	101	104	93	87	86	96	106	102
1	1	11.5	82	90	96	99	89	83	82	91	101	97
1	1	12.5	53	58	62	64	57	53	52	58	65	62
1	1	13.5	58	63	67	70	62	58	57	64	71	68
1	1	14.5	52	56	60	62	56	52	51	57	63	61
1	1	15.5	48	52	55	57	51	48	47	53	58	56
1	1	16.5	35	38	40	42	37	35	34	37	41	39
1	1	17.5	2	2	3	3	2	2	2	2	3	2
1	1	18.5	0	0	0	0	0	0	0	0	0	0
1	1	19.5	0	0	0	0	0	0	0	0	0	0
1	1	20.5	0	0	0	0	0	0	0	0	0	0
1	1	21.5	0	0	0	0	0	0	0	0	0	0
1	1	22.5	0	0	0	0	0	0	0	0	0	0
1	1	23.5	0	0	0	0	0	0	0	0	0	0

Table 7 Simulated illuminance values (lux) for blind 75% closed

Month	Day	Hour	Nodes									
			1	2	3	4	5	6	7	8	9	10
1	1	0.5	0	0	0	0	0	0	0	0	0	0
1	1	1.5	0	0	0	0	0	0	0	0	0	0
1	1	2.5	0	0	0	0	0	0	0	0	0	0
1	1	3.5	0	0	0	0	0	0	0	0	0	0
1	1	4.5	0	0	0	0	0	0	0	0	0	0
1	1	5.5	0	0	0	0	0	0	0	0	0	0
1	1	6.5	0	0	0	0	0	0	0	0	0	0
1	1	7.5	8	9	9	9	9	8	8	9	9	9
1	1	8.5	48	57	60	61	57	50	49	58	62	62
1	1	9.5	35	42	45	46	43	38	37	44	47	48
1	1	10.5	36	43	47	47	44	39	38	45	49	49
1	1	11.5	34	41	44	45	42	37	36	43	46	46
1	1	12.5	22	26	28	29	27	23	23	27	29	29
1	1	13.5	24	29	31	31	29	26	25	30	32	32
1	1	14.5	21	26	28	28	26	23	22	27	29	29
1	1	15.5	20	24	25	26	24	21	20	25	26	27
1	1	16.5	13	16	17	17	16	14	14	16	17	17
1	1	17.5	1	1	1	1	1	1	1	1	1	1
1	1	18.5	0	0	0	0	0	0	0	0	0	0
1	1	19.5	0	0	0	0	0	0	0	0	0	0
1	1	20.5	0	0	0	0	0	0	0	0	0	0
1	1	21.5	0	0	0	0	0	0	0	0	0	0
1	1	22.5	0	0	0	0	0	0	0	0	0	0
1	1	23.5	0	0	0	0	0	0	0	0	0	0

inside the office and setting up blind control strategies with respect to that would eventually take care of the other potential glare-prone spaces of the office.

3.7 *Estimation of Energy Savings Through Lighting*

3.7.1 Collecting DGP Values

After evaluating systems 1 and 2 for glare for different weather conditions, system 2 is selected as the worst glare position of the room. Therefore, the annual glare is observed for system 2 for each of the five blind positions, as shown in Table 9. Daylight Glare probability (DGP) is used for the glare analysis. Figure 9 graphically shows the glare (DGP) behavior with time throughout the day as well as the reduction in glare by setting the blind in different positions. The purpose of

Table 8 Simulated illuminance values (lux) for blind fully closed

Month	Day	Hour	Nodes									
			1	2	3	4	5	6	7	8	9	10
1	1	0.5	0	0	0	0	0	0	0	0	0	0
1	1	1.5	0	0	0	0	0	0	0	0	0	0
1	1	2.5	0	0	0	0	0	0	0	0	0	0
1	1	3.5	0	0	0	0	0	0	0	0	0	0
1	1	4.5	0	0	0	0	0	0	0	0	0	0
1	1	5.5	0	0	0	0	0	0	0	0	0	0
1	1	6.5	0	0	0	0	0	0	0	0	0	0
1	1	7.5	4	4	5	5	4	4	4	5	5	5
1	1	8.5	2	20	32	32	30	26	25	31	33	34
1	1	9.5	18	22	23	23	22	19	19	23	24	25
1	1	10.5	18	22	24	24	23	19	19	24	25	25
1	1	11.5	17	21	23	22	21	18	18	22	23	24
1	1	12.5	11	13	14	14	14	11	11	14	15	15
1	1	13.5	12	15	16	16	15	13	12	16	16	17
1	1	14.5	11	13	14	14	13	11	11	14	15	15
1	1	15.5	10	12	13	13	12	10	10	13	13	14
1	1	16.5	6	8	8	8	8	7	7	8	9	9
1	1	17.5	0	0	0	0	0	0	0	0	0	1
1	1	18.5	0	0	0	0	0	0	0	0	0	0
1	1	19.5	0	0	0	0	0	0	0	0	0	0
1	1	20.5	0	0	0	0	0	0	0	0	0	0
1	1	21.5	0	0	0	0	0	0	0	0	0	0
1	1	22.5	0	0	0	0	0	0	0	0	0	0
1	1	23.5	0	0	0	0	0	0	0	0	0	0

above-mentioned figure was just to show the effectiveness of blind, and the unusual variation in the case of 75% closed position of blind is likely due to an outlier effect, but it does not have any significant effect on the desired result. As the office considered is east facing, a large amount of glare is observed during the first half of the day as direct sunlight appears on the façade.

3.7.2 Decision Flow Chart

Glare matrix consists of the hourly DGP values for each of the five blind positions for all occupancy hours for the whole year, while illuminance matrices will provide the hourly illuminance values for the nodes in office. The flowchart in Fig. 10 provides an appropriate blind position for each occupancy hour of the year for the prevention of visual discomfort.

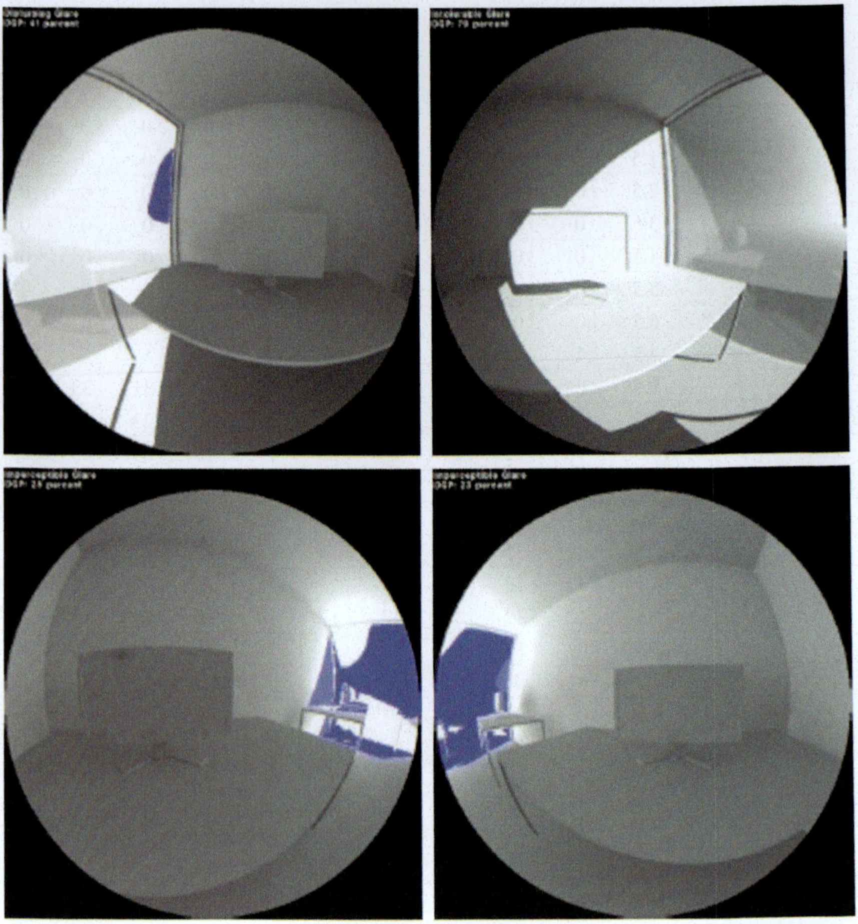

Fig. 7 Glare observed at all four systems on Jan 01 at 10 AM (DGP value in percentage)

3.7.3 Selected Blind Positions

Figure 11 shows the annual glare pattern if a fixed blind position is maintained throughout the year. But fixing the blind to a certain position for the whole year is not recommended as it would result in visual discomfort in the cases where blind is kept opened, and it may also cause unnecessary lighting energy consumption when blind is kept closed.

As per the proposed blind control algorithm, for each occupancy hour, blind was set to the position which is capable of preventing visual discomfort, i.e., maintaining DGP less than 0.40, as shown in Fig. 12. It is the final result after blind control strategy applied, and DGP value for each hour is kept under 0.40. The algorithm selects appropriate blind position for each hour so as to get DGP value

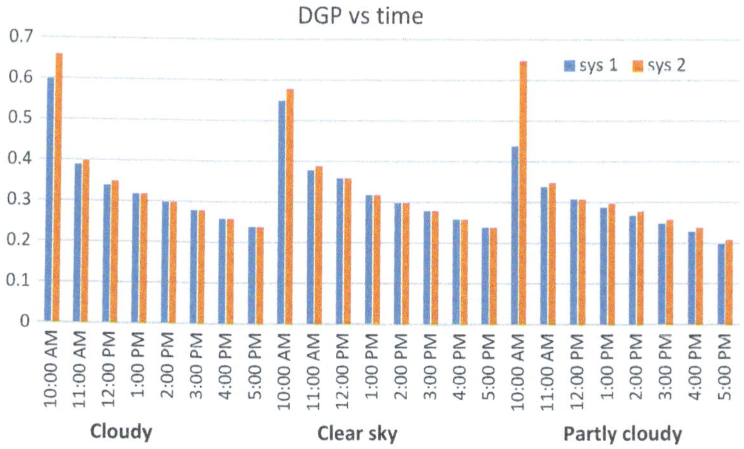

Fig. 8 DGP values for different sky conditions for system 1 and 2

Table 9 Yearly DGP values for system 2 for different blind positions

Month	Day	Time	DGP				
			Blind 0	Blind 25	Blind 50	Blind 75	Blind 100
1	1	9:00 AM	0.3454	0.3173	0.2485	0.1752	0.0603
1	1	10:00 AM	0.3506	0.3221	0.251	0.1779	0.0623
1	1	11:00 AM	0.3422	0.3152	0.2476	0.1697	0.0544
1	1	12:00 PM	0.2852	0.2681	0.2249	0.0799	0.0214
1	1	1:00 PM	0.2946	0.2759	0.2287	0.0988	0.0254
1	1	2:00 PM	0.2825	0.2659	0.2239	0.0776	0.021
1	1	3:00 PM	0.2736	0.2586	0.2207	0.0634	0.018
1	1	4:00 PM	0.245	0.2371	0.2121	0.0234	0.01
1	1	5:00 PM	0.0126	0.011	0.0067	0.0036	0.0036

under visual comfort. Blind positions with their respective operational hours are presented in Fig. 13, from which it is observed that for most of the occupancy hours, blind was in fully opened position, i.e., DGP was well within the acceptable limits, while blind was appropriately set to the required position to get DGP under acceptable limits for the remaining hours.

3.7.4 Calculation of Lighting Energy Consumption

The algorithm mentioned above provides a set of blind position for every occupant hour. Thus, for each hour, blind is set to such a position that results in the desired visual comfort condition. As blind position varies from fully open to the

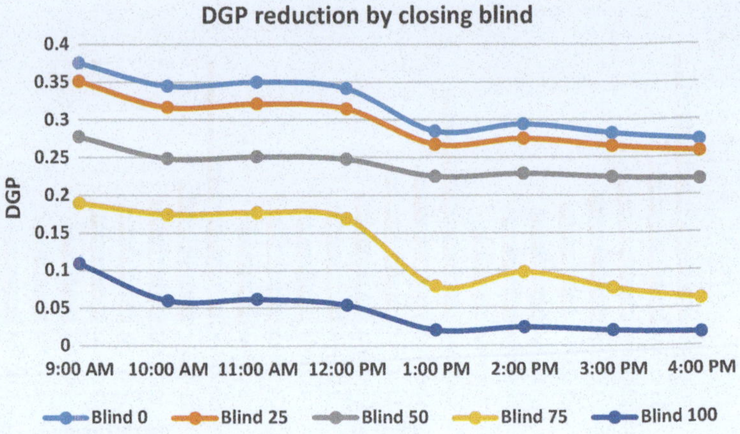

Fig. 9 DGP observed on Jan 01

recommended level as per algorithm, the illuminance values of the interior space of the room vary.

Figure 5 depicts that the nodes that are away from the window might have a deficit of light. The room considered is of area 15 m^2, having a total number of 60 nodes, so every node corresponds to an area of 0.25 m^2. The area of the room which remains underexposed as per NBC 2005 guidelines, i.e., 300 lx [16], requires to be lit through artificial lighting. The room with blinds in fully closed position is considered as the base case in this study, and electricity savings is calculated in comparison to this case. The lighting power density defined by NBC 2005 for an enclosed office room is 11.8 W/m^2. In this study, the lighting energy consumption is calculated on the basis of LPD of the room, i.e., underexposed area is lit by artificial lighting with a wattage density of 11.8 W/m^2. This provides an estimate of the worst-case energy consumption scenario, as generally, buildings have the artificial lighting placed with an LPD less than 11.8 W/m^2.

Table 10 gives the lighting energy consumption for different cases, which show that the maximum energy consumption is estimated in the case when blind is always kept in fully closed position. The blind in fully closed position would result in most of the room area being illuminated below 300 lx, thus the electricity consumption estimated is high, while the case when blind is kept in fully open position results in the least energy consumption, as large glazing area provided would cause the room to be illuminated by daylight for the most of the time. But this case results in a huge amount of glare, which can be seen in the annual glare behavior, as shown in Fig. 12.

The results suggested that the office room with an algorithm-controlled blind positioning maintains the maximum acceptable glare scenario (DGP) throughout the year as well as reduces the lighting energy consumption as compared to the case

Fig. 10 Decision flow chart

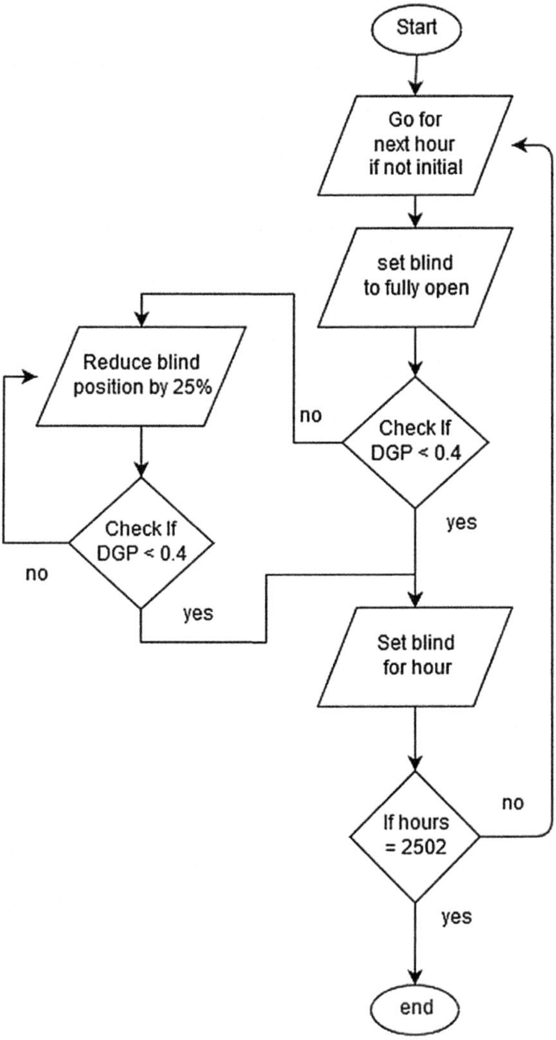

with blind kept in always closed position, with the lighting energy savings of almost 91%. It is also observed that even after implementing the proposed blind control, energy savings only reduced by approximately 1%. Electricity consumption and savings estimated are highly sensitive to the assumed conditions, such as the office area, type, and arrangement of the artificial lighting, etc.

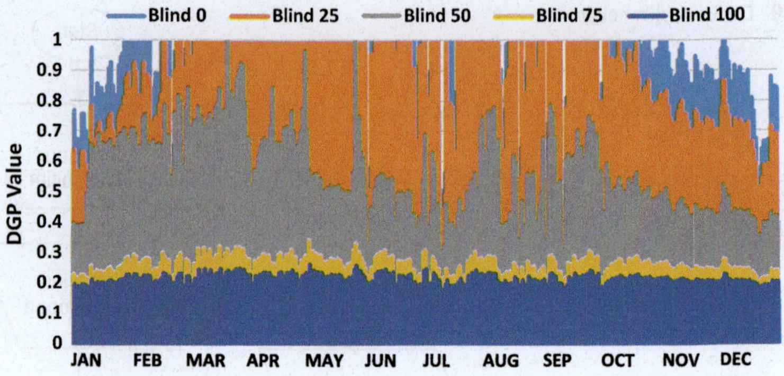

Fig. 11 Glare pattern for different blind positions

Table 10 Lighting energy consumption and savings for cases considered

Considered cases	Lighting energy consumption (kWh)	Energy savings (as compared to base case) (%)
Blind in fully closed position (base case)	4,22,684	–
Blind in fully open position	34,146	91.92
Blind controlled by algorithm	37,497	91.13

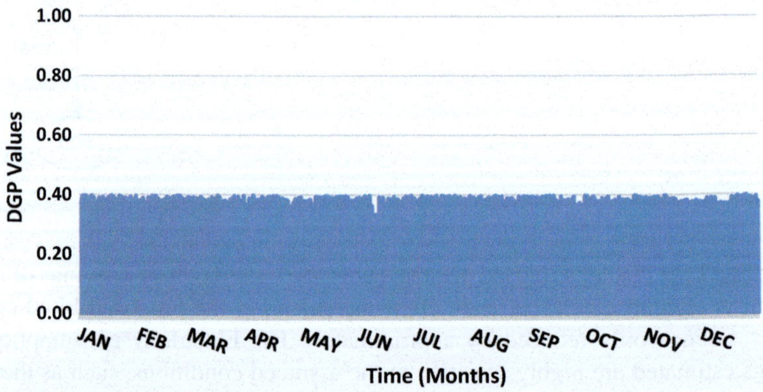

Fig. 12 Glare pattern for proposed blind control

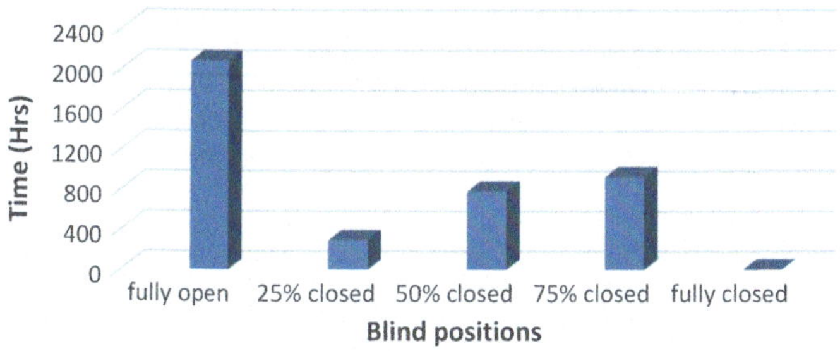

Fig. 13 Blind position and respective operational hours

4 Conclusions

A functional day-lit environment can be created by adapting appropriate blind control for maintaining visual comfort for the occupants. Optimum blind positions were identified based on the algorithm for maximum acceptable glare conditions for each hour throughout the year. The blind positions selected prevent the office space from crossing the non-acceptable glare threshold, and the corresponding potential annual energy savings due to lighting were estimated. It was observed that even after providing visual comfort to the occupants, reduction in savings was marginal, as compared to the case when visual discomfort is overlooked in daylighting utilization.

References

1. Interim Report of the Expert Group on Low Carbon Strategies for Inclusive Growth (2014). Planning Commission, Government of India
2. Kulkarni K, Sahoo PK, Mishra M (2011) Optimization of cooling load for a lecture theatre in a composite climate in India. Energy Build 43(7):1573–1579
3. Jani DB, Mishra M, Sahoo PK (2015) Performance studies of hybrid solid desiccant–vapor compression air-conditioning system for hot and humid climates. Energy Build 102:284–292
4. Batukray JD, Mishra M, Sahoo PK (2013) Simulation of solar assisted solid desiccant cooling systems using TRNSYS. In: Proceedings of the 22nd national and 11th international ISHMT-ASME heat and mass transfer conference, IIT Kharagpur, India, Kharagpur
5. Wymelenberg KVD (2016) The benefits of natural light, 19 March 2014. http://www.archlighting.com/technology/the-benefits-of-natural-light_o. Accessed 14 Sept 2016
6. Al-Ashwal NT, Budaiwi IM (2011) Energy savings due to daylight and artificial lighting integration in office buildings in hot climate. Int J Energy Environ 2(6):999–1012
7. Japee S, Schiler M (1995) A method of post occupancy glare analysis for building energy perfonmance analysis. American Solar Energy Society, Boulder, CO (United States)
8. Heschong L (2003) Windows and offices: a study of office worker performance and the indoor environment. California Energy Commission
9. Dhawan S (2016) Glare, 03 March 2005. http://sdhawan.com/ophthalmology/glare.html. Accessed 11 Sept 2016

10. Jakubiec JA, Reinhart CF (2012) The ‘adaptive zone’–A concept for assessing discomfort glare throughout daylit spaces. Light Res Technol 44(2):149–170
11. Window blinds. https://en.wikipedia.org/wiki/Window_blind. Accessed Sept 2016
12. Meek C, Brennan M (2011) Automated and manual solar shading and glare control: a design framework for meeting occupant comfort and realized energy performance. In: 40th, national solar conference, SOLAR 2011, Raleigh, NC
13. Shading devices CLEAR (Comfort and low energy architecture) (2016). http://www.new-learn.info/packages/clear/interactive/matrix/d/shading/shading_devices.html. Accessed Sept 2016
14. Add calculation points. http://www.elumtools.com/. Accessed Sept 2016
15. Thorn P (2010) Light and lighting—lighting of workplaces, Part 1: Indoor work places, EN12464–1::2011
16. National Building Code of India 2005 (NBC 2005), Bureau of Indian Standards, New Delhi, 2005
17. Bureau of Energy Efficiency, Ministry of Power, Government of India (2005). https://www.beeindia.gov.in/. Accessed Sept 2016

Part VI
Power Generation Using Solar Energy

Solar Updraft Tower—A Potential for Future Renewable Power Generation: A Computational Analysis

Ankit Agarwal, Pradeep Kumar and Balkrishna Mehta

Abstract The full-scale three-dimensional analysis of solar updraft tower power plant in Manzanares, Spain has been performed using commercially available CFD tool ANSYS Fluent. The two-equation k-ε turbulent model with standard wall function has been utilized for the fluid flow. The soil has been modeled with the consideration of the fact that temperature at 20 m depth remains constant throughout the year. The surface-to-surface radiation model is also included in the heat transfer model. The simulation has been performed for the steady state without/with radiation, the transient state with/without thermal storage on 8th of June. In the present simulation, water has been taken as thermal storage. It has been found that results improve considerably by including radiation effect, closely match with the results published from the plant. There is a reduction in the maximum velocity with the thermal storage; however, sufficient energy is available in thermal storage to overcome intermittency of insolation. The strong dependency of the plant on insolation can be reduced with the thermal storage.

Nomenculture

$C_{i\varepsilon}$	Constant (i = 1, 2 and 3)
E_m	Emissive power of surface m
$\hat{e}_t$	Average component of energy
F_{ml}	View factor between surface m and l
$\bar{f}_i$	Vector representing average external forces
G_k	Turbulent kinetic energy due to the mean velocity gradients
G_b	Turbulent kinetic energy due to buoyancy

A. Agarwal · P. Kumar (✉)
Numerical Experiment Laboratory (Radiation & Fluid Flow Physics),
Indian Institute of Technology Mandi, Suran 175005, Himachal Pradesh, India
e-mail: pradeepkumar@iitmandi.ac.in

B. Mehta
Department of Mechanical Engineering, Indian Institute of Technology Guwahati,
Guwahati 781039, Assam, India

H. Tyagi et al. (eds.), *Applications of Solar Energy*, Energy, Environment, and Sustainability, https://doi.org/10.1007/978-981-10-7206-2_15

J_m	Radiosity of surface m
k	Turbulence kinetic energy
N	n^{th} surface
n	Index of refraction
$\bar{p}$	Average pressure
q_j	Time averaged heat flux
$\bar{S}_{ij}$	Mean rate of strain tensor
ΔT	Increase in air temperature in collector
T_h	Temperature of hot wall
T_c	Temperature of cold wall
t	Time
$\hat{u}_i$	Velocity vector in i^{th} direction
$\bar{u}_i$	Average velocity in i^{th} direction
$\bar{u}_j$	Average velocity in j^{th} direction
$\bar{u}_i'$	Fluctuating component of average velocity in i^{th} direction
$\bar{u}_j'$	Fluctuating component of average velocity in j^{th} direction
x_i	Cartesian coordinates

Greek Letters

$\bar{\rho}$	Average density
ρ	Density
ρ_{dif}	Diffuse reflectivity of surface
τ_{dif}	Diffuse transmissivity of surface
μ	Coefficient of viscosity
$\bar{\tau}_{ij}$	Time average shear stress tensor
$\bar{\tau}_{ij}^T$	Transpose of shear stress tensor
μ_t	Turbulent viscosity
σ_k	Turbulent Prandtl number for k
ε	Rate of dissipation of turbulent kinetic energy
σ_ε	Turbulent Prandtl number for ε
ρ_m	Reflectivity of surface m

Subscripts

dif	Diffuse surfaces
m,l	Surface index
i,j	Coordinate index
b	Buoyancy

h	Hot wall
c	Cold wall
t	Turbulent

Superscripts

'	Fluctuation component
T	Transpose

1 Introduction

There are several types of low-grade energies available which are also very low in work potential. Attempts should be made to convert these low-work-potential energies into useful form to meet high-grade energy requirements. Solar updraft tower power plant is one of such efforts to generate electricity from low-grade energy. The plant uses three well-known technologies to produce power (1) greenhouse effect, (2) chimney effect, and (3) wind turbine. The greenhouse effect is generated through the property of glass/plastic which is transparent to the short-wavelength radiation from the sun and is opaque for the long-wavelength radiation (transmissivity around 0.01% [1]) from the moderate temperature objects. This phenomenon creates sufficient temperature difference to generate convection currents inside the greenhouse enclosure. The chimney effect creates a pressure difference which directs these convection currents to the chimney. If convection current is generated over a large area and the chimney size (both diameter and height) is also very big, the tremendous amount of kinetic energy will be available. Now, this kinetic energy can be converted into electricity by deploying a wind turbine. The solar updraft power plant in Manzanares used these principles to generate electricity. These three phenomena are well known to researchers since centuries, but the first successful plant was built using these three phenomena in Manzanares, Spain, in the year 1981. This plant has mainly three components (a) collector which is made of glass or plastic sheet to generate greenhouse effect (b) chimney for chimney effect, and (c) wind turbine. Due to its simplicity in design, broad fluid mechanics and heat transfer phenomena inside the plant and structural difficulty, it is able to attract researchers from many fields. The attempts have also been made to find the optimal size of the plant components; however, it is found that plant output increases with increase in the size of collector and chimney [2]. Table 1 shows the size of components of solar updraft plant for power outputs [3]. With the funding from German Ministry of Research and Technology (BMFT) and Manzanares Spanish utility Union Electrica Fenosa, a solar updraft tower power plant was built for an average capacity of 50 kW for the demonstration of the concept. This plant successfully operated between 1981 and 1989. This prototype also provided the opportunity for the performance evaluation

Table 1 Variation in plant component size with output

Power output (MW)	Collector diameter (m)	Chimney diameter (m)	Chimney height (m)
0.05	244	10.16	194.6
5	1250	45	550
30	2900	70	750
100	4300	110	1000
200	7000	120	1000

Table 2 Main dimensions of plant component and technical data of prototype

Tower height	194.6 m
Tower radius	5.08 m
Mean collector radius	122 m
Mean roof height	1.85 m
Number of turbine blades	4
Turbine blade profile	FX W-151-A
Blade tip speed-to-air transport velocity ratio	1:10
Operating modes	stand-alone or grid-connected mode
Typical collector air temperature increase	$\Delta T = 20$ K
Nominal output	50 kW
Collector covered with plastic membrane	40,000 m^2
Collector covered with glass	6,000 m^2
Upwind velocity on release	15 m/s

of the power plant. The components of the plant and major technical details have been given in Table 2 [3].

HAAF et al. [3] presented the working principle and construction of the Manzanares pilot plant. HAAF [2] published the detailed test results. Many researchers have used different tools like thermodynamic analysis [4], small prototype [5], and computational fluid dynamics [6, 7] to analyze the potential of solar updraft tower. The detailed theory, practical experience, and economy of the solar updraft tower Manzanares, Spain are presented by J. Schlaich et al. [8]. The plastic or glass has been used as collector material to trap the longwave radiation from the ground. This phenomenon creates a small temperature difference, but sufficient to drive fluid flow. The power generated is dependent on the solar irradiance falling on the ground and the overall efficiency which is the product of individual efficiency of the collector, turbine, and chimney. The kinetic energy of air, which is extracted using a wind turbine, is a function of the pressure difference between inlet and exit of the chimney. The thermal storage can be used to create required temperature difference to generate sufficient flow during night. The minimum velocity 2.5 m/s is required in the chimney for the operation of Manzanares plant. On an average, the plant is operational for 8–9 h per day. It was reported that power output from the plant is proportional to the insolation, collector area, and tower height. Thus, the optimal dimensions of the plant can be only based on the cost involved. Hermann F. Fasel et al. [5] studied the power plant using commercial CFD tool, ANSYS Fluent, and also, with in-house developed CFD code. The two-dimensional axisymmetric model of the Manzanares plant and three-dimensional analysis for the geometrical scale of 1:30 of the plant have been taken into consideration. Since there are acceleration and mixing of fluid in all 360° direction involved, the fluid mechanics and heat transfer analysis can provide very good insight into fluid phenomena. They reported Rayleigh–Benard–Poiseuille instability in the fluid flow. The similar studies have been performed by taking the two-dimensional axisymmetric model of the Manzanares solar updraft power plant [9, 10]. They have achieved reasonably good quantitative agreements with the experimental data of the plant. Tingzhen et al. [11] have applied the numerical technique to study the effect of the three turbine blades on the solar chimney power plant. The study is further extended to 10 MW solar updraft power plant with five turbine blades. It has been concluded that there is a decrease in average velocity and mass flow rate in the chimney with the increase in the turbine rotational speed.

The heat transfer by radiation is equally important in natural convection even for small temperature difference from moderate reference temperature like room temperature [12]. The effect of thermal radiation on forced and natural convection has been studied by Kumar and Eswaran [13] and [14], respectively. Kumar [15] has performed the detailed analysis of fluid and radiation interaction in laminar region. The radiation heat transfer changes the fluid flow and temperature profile considerably. At very high optical thickness, the temperature profile of radiation cases reaches to pure convection temperature profile in forced convection and conduction temperature profile in natural convection, but, with a significant difference in fluid flow. Guo et al. [7] have used the radiation model to study the performance of the solar updraft tower using ANSYS Fluent. They have used discrete ordinate model for radiation. Air has been considered as participating medium. The discrete ordinate method of ANSYS Fluent considers the collector as a semitransparent boundary and calculates the reflectivity and transmissivity for the diffused radiation by the following formula [16]

$$\tau_{dif} = 1 - \rho_{dif} \tag{1}$$

where

$$\rho_{dif}(n) = \frac{1}{2} + \frac{(3n+1)(n-1)}{6(n+1)^2} + \frac{n^2(n^2-1)^2}{(n^2+1)^3}\ln\frac{n-1}{n+1} - \frac{(2n^3(n^2+2n-1))}{(n^2+1)(n^4-1)} + \frac{(8n^4(n^4+1))}{(n^2+1)(n^4-1)}\ln(n) \tag{2}$$

where n is the ratio of refractive index of two media separated by the semitransparent boundary.

The transmissivity calculated by above equation will be a nonzero value, whereas transmissivity of the glass is 0.01 [1] for long wavelength. So the glass can be considered as opaque for radiation heat transfer; also the optical thickness of air is quite low (gray absorption coefficient 0.01). Therefore, there would not be any significant difference in results if air is considered as non-participating medium. Considering the facts above, the surface-to-surface radiation model which considers the collector as an opaque boundary, also being computationally less intensive than DO model, is the best-suited radiation model for the present case study.

The above-reviewed articles reveal that the Manzanares solar updraft power plant involves many interesting fluid flow and heat transfer phenomena. It can be also noticed that the studies have been executed either by using geometrical reduction techniques or mathematical modeling of the some of the components of the plant. Also, the heat transfer by radiation is not taken into account in most of the previous studies except Guo et al. [7], where authors have used DO radiation model which is computationally very expensive. The present authors believe that the study of a full-scale model of Manzanares solar updraft would give more insight into the fluid flow and heat transfer phenomena which will further help in optimizing the plant output and design of the large-scale solar updraft power plant. The computationally less expensive surface-to-surface radiation model can fulfill the objective in the current case study. In the present work, the detailed analysis of the solar updraft power plant's characteristic like velocity, temperature has been performed for the steady-state case, with and without radiation model; also the effect of thermal storage on plant characteristics has also been presented.

2 Mathematical Model

The major phenomenon in the plant is buoyancy which is happening in the collector. With the reported value of temperature difference of 20 K and the characteristic length as collector height, the calculated Rayleigh number $Ra = \frac{g\beta\Delta TL^3}{\nu\alpha}$ is approximately $\approx 10^{11}$ which indicates that the fluid flow is turbulent. The turbulent flow is modeled by $k-\varepsilon$ two-equation turbulent flow model. The Boussinesq approximation is used for the buoyancy, and the flow is considered as incompressible. The following equations are solved for the fluid flow and heat transfer in the plant.

Continuity Equation

$$\frac{\partial(\bar{\rho}\bar{u}_i)}{\partial x_i} = 0 \tag{3}$$

Navier–Stokes Equation

$$\frac{\partial(\bar{\rho}\bar{u}_i)}{\partial t} + \frac{\partial(\bar{\rho}\bar{u}_j\bar{u}_i)}{\partial x_j} = \rho\bar{f}_i + \frac{\partial}{\partial x_j}[-\bar{p}\delta_{ij} + 2\mu\bar{S}_{ij} - \rho\overline{u_i^{'}u_j^{'}}] \tag{4}$$

Energy Equation

$$\frac{\partial(\bar{\rho}\hat{e}_t)}{\partial t} + \frac{\partial(\bar{\rho}\hat{u}_i\hat{e}_t + \hat{u}_i\bar{p})}{\partial x_j} - \frac{\partial(\hat{u}_i\bar{\tau}_{ij} + \hat{u}_i\bar{\tau}_{ij}^T)}{\partial x_j} + \frac{\partial(\bar{q}_j + q_j^T)}{\partial x_j} = 0 \tag{5}$$

The second last term of Eq. 4 is turbulent stress which is evaluated by two equations $k-\varepsilon$ turbulence model. The $k-\varepsilon$ turbulence model equations are given by

$$\frac{\partial(\rho k)}{\partial t}+\frac{\partial(\rho k u_i)}{\partial x_i}=\frac{\partial}{\partial x_j}[(\mu+\frac{\mu_t}{\sigma_k})\frac{\partial k}{\partial x_j}]+G_k+G_b-\rho\varepsilon \tag{6}$$

$$\frac{\partial(\rho\varepsilon)}{\partial t}+\frac{\partial(\rho\varepsilon u_i)}{\partial x_i}=\frac{\partial}{\partial x_j}[(\mu+\frac{\mu_t}{\sigma_\varepsilon})\frac{\partial\varepsilon}{\partial x_j}]+C_{1\varepsilon}\frac{\varepsilon}{k}(G_k+C_{3\varepsilon}G_b)-C_{2\varepsilon}\rho\frac{\varepsilon^2}{k} \tag{7}$$

$C_{1\varepsilon}=1.44$, $C_{2\varepsilon}=1.92$, $C_{3\varepsilon}=-0.33$ $\sigma_k=1$ and $\sigma_\varepsilon=1.3$

The default value of closure constants of turbulent equations has been used as in ANSYS Fluent.

The surface-to-surface model requires radiative flux at the surface. The radiosity of radiation model is calculated as

$$J_m=E_m+\rho_m\sum_{i=1}^{N}F_{mj}J_l \tag{8}$$

In the present study, the surface-to-surface radiation model has been used assuming air as transparent and collector as opaque for longwave radiation.

3 Geometry and Grid Generation

The full-scale geometry of the Manzanares plant has been created with dimensions given in Table 2. To control the quality of the mesh, the plant geometry is divided into three zones: collector, fillet, and chimney as shown in Fig. 1. These zones are connected using the mesh interface feature of the ANSYS Fluent. The mixing of the fluid is happening in the fillet zone; thus, the finer mesh is created in the fillet region. The flow would develop after some distance in the chimney; thus, finer mesh is created near the chimney inlet and sufficiently coarse mesh in rest of geometry. Also, the fine mesh is provided in collector near the fillet zone. The present study uses two-equation $k-\varepsilon$ turbulence model which does not resolve viscous sublayer near the wall. To avoid this constraint, the first cell centroid near the wall should lie in the buffer layer, i.e., $50 \leq y^+ \leq 500$ and wall function should be used to apply no-slip boundary condition on the wall. Figure 2 shows the grid points which have been generated by satisfying the constraints of $k-\varepsilon$ turbulence model in the collector, fillet, and chimney zones. The standard wall function has been used to apply the wall boundary condition. The isometric view of the plant geometry is shown in Fig. 3.

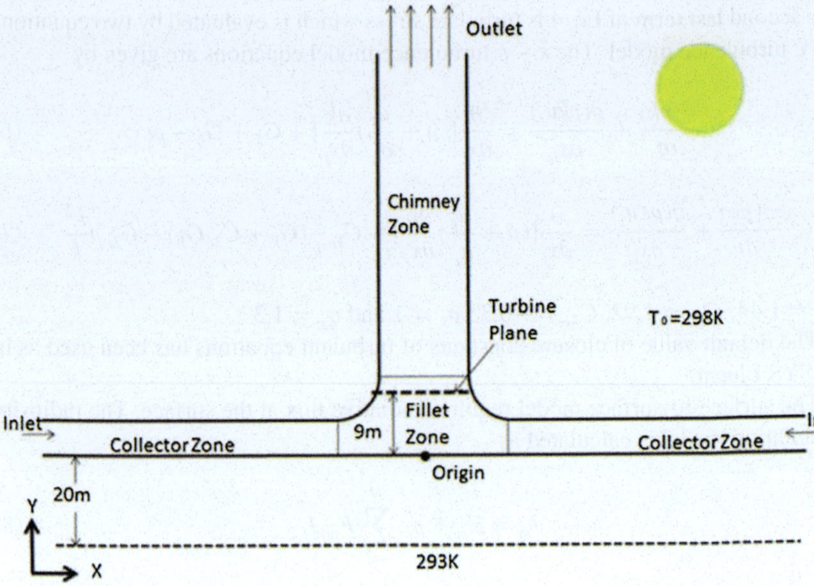

Fig. 1 Front view of solar updraft tower

Fig. 2 Computational grid in collector, fillet, and chimney zones

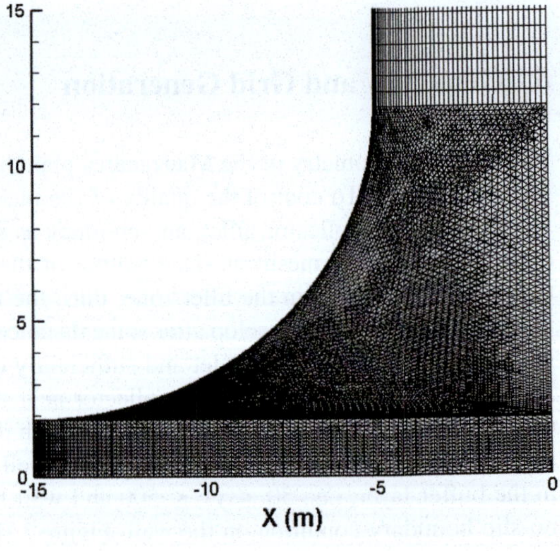

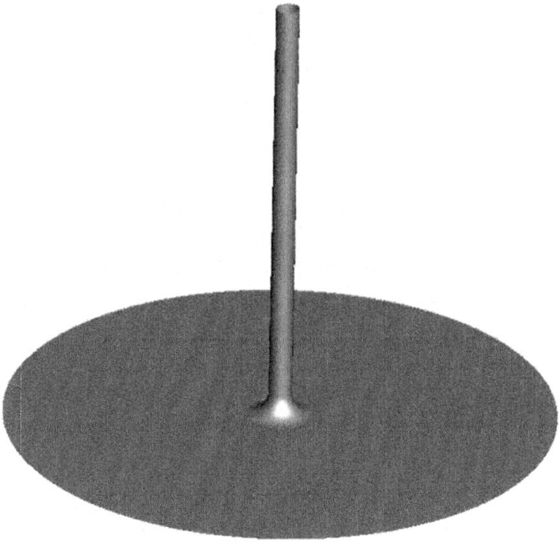

Fig. 3 Isometric view of solar updraft tower geometry

4 Boundary Conditions and Soil Modeling

The boundary conditions for the collector inlet and chimney outlet are pressure inlet and pressure outlet, respectively, with gauge pressure zero. The atmospheric temperature assumed to be at 298 K, and this is also inlet air temperature in the collector. The collector is made of glass with a thickness of 0.006 m. The heat transfer in the glass is modeled by shell conduction feature of ANSYS Fluent. The chimney wall is considered as adiabatic. The solar irradiation is applied using solar load model of ANSYS Fluent. For steady-state calculation, the insolation value is calculated with the solar calculator with input for the location of Manzanares plant, time at 12 noon of 8th of June. The transient calculation starts for 6 am for the same site and same date since these data are available for comparison purpose [8].

The collector glass is considered semitransparent for solar radiation model with the transmissivity of 0.8 for the short wavelength radiation, whereas it is considered

Table 3 Summary of boundary conditions

Components	Flow boundary condition	Temperature boundary condition
Collector inlet	Pressure inlet	298 K
Chimney outlet	Pressure outlet	Not required
Collector surface	No slip	Shell conduction
Ground surface	No slip	Thin wall
Chimney surface	No slip	Adiabatic

as opaque for long-wavelength radiation. The summary of boundary conditions is given in Table 3.

The soil is modeled with the consideration that the temperature remains constant at a depth of 20 m at 20 °C throughout the year [17], i.e., independent of external layer condition of the soil. Also, there is no significant variation of temperature in the lateral directions, i.e., there is only one-dimensional conduction. The geometry is not created for the soil; instead, wall thickness feature of ANSYS Fluent is used which is a one-dimensional conduction model. The absorptivity of the soil is considered to be 0.8. The SIMPLEC algorithm with second-order upwind interpolation scheme for flow and temperature has been used to obtain the convergence of solution.

5 Validation

The validation of turbulent natural convection is performed for the enclosed tall cavity [18, 19]. This case has experimental results which have been used to validate the CFD parameter settings and results. Although there is a difference of the order of magnitude in the size of solar updraft tower and validation case, this validation case provides the information of correct CFD setup for natural convection results. Figure 4 shows the geometry of slender structure with height 2.18 m and thickness 0.076 m. The temperatures at the hot wall and cold wall are 55 °C and 15 °C,

Fig. 4 Geometry of slender cavity

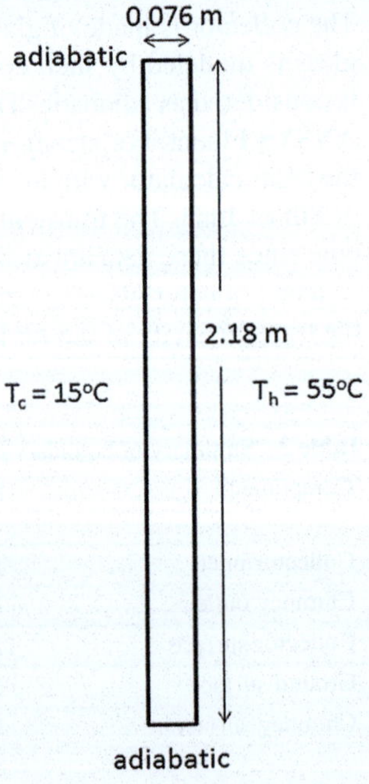

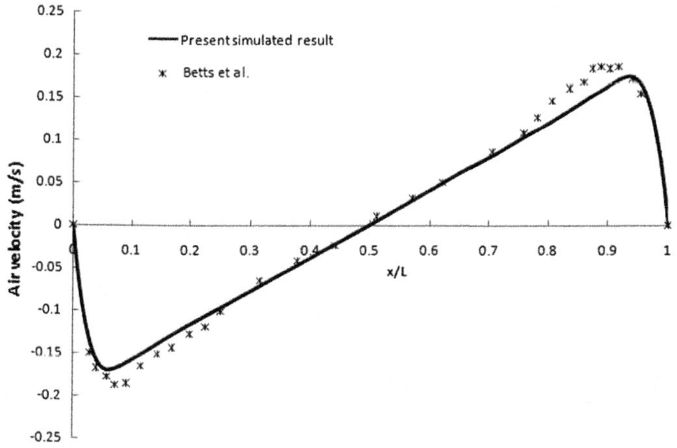

Fig. 5 Variation of velocity profile at mid-cavity height

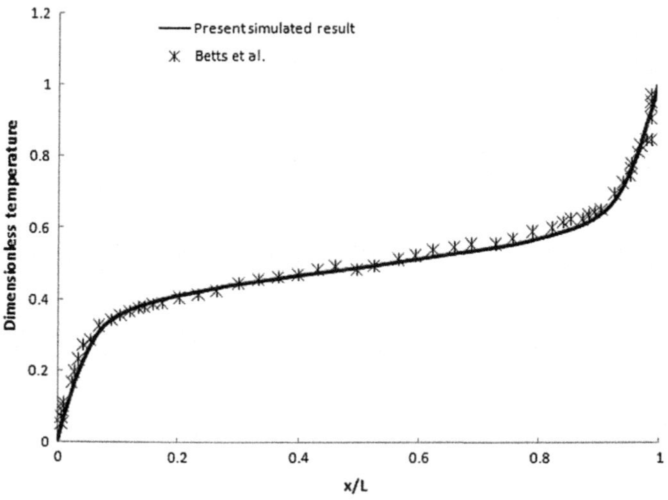

Fig. 6 Variation of dimensionless temperature at mid-height of cavity

respectively. The temperature is normalized by $(T_h - T_c)$. The velocity and dimensionless temperature are plotted at mid-height of the slender cavity. Figures 5 and 6 show the comparison of present simulation results with experimentally measured value of vertical velocity and dimensionless temperature at mid-height of cavity, respectively. There are good agreements in both the results.

Table 4 Velocity and ΔT variation with no. of cells

No. of cells (in millions)	Velocity m/s	ΔT K
1.3	10.99	15.01
1.9	11.18	14.83
2.3	11.11	14.77

6 Grid Independent Test

The grid independent test is performed for the optimal grid size for simulation. As natural convection and mixing of the fluid are the major phenomena taking place in the collector and fillet zone, respectively, the number of grids is varied in these two zones only. The chimney's grid is kept constant for grid independent test. Table 4 shows the velocity at turbine plane and temperature rise in collector with the number of cells in the geometry.

It is found that there is no significant variation after 1.9 million cells. So, this grid is selected for further analysis of solar updraft power plant.

7 Results and Discussions

7.1 *Steady-State Simulation Without Radiation Model and Thermal Storage*

The solar load model of ANSYS Fluent provides the solar flux distribution on the wall. On 8th of June at 12 noon, the value of direct insolation and the value of diffuse insolation are kept at 1200 W/m^2 and 200 W/m^2, respectively. The transmissivity of glass is 0.8. Figure 7a shows the solar flux distribution on the ground. The maximum value of solar flux is 900 W/m^2 falling on most part of the ground. The chimney is considered to be opaque. The shadow of chimney is visible in solar flux contours.

The velocity magnitude is important at the location of wind turbine installation. The wind turbine is installed at a height of 9 m from the ground. The axis of the wind turbine is aligned with the axis of the chimney. The velocity contours at the turbine plane are shown in Fig. 7b. Since the flow in the collector is getting mixed at chimney inlet location and directed to the chimney, the flow is still not fully developed at the turbine location. The velocities are lower in the middle and higher near the chimney wall. This is a good location for the wind turbine since the blades of the turbine are experiencing the maximum velocity. The value of maximum velocity is 15 m/s in the most part of wind turbine location which is also slightly higher than the reported value for the plant. It is reported that velocity on release is 15 m/s [2] which we assume the velocity at the exit of chimney.

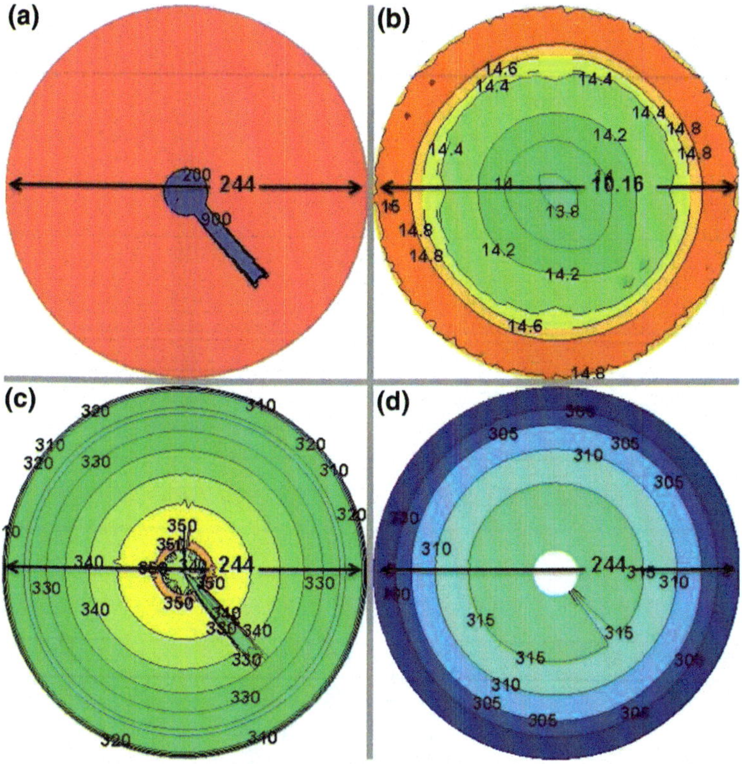

Fig. 7 **a** Solar heat flux (W/m^2) on ground, **b** velocity magnitude (m/s) contours at turbine plane, temperature (K) contour at **c** ground and **d** collector

The temperature contours at the ground and collector are shown in Fig. 7c and 7d, respectively. The ground temperature is in the range of 300–355 K, whereas collector temperature is ranging from 300 to 320 K. The peripheral area of the collector remains at atmospheric temperature which is the expected behavior. This behavior does not appear in the ground temperature contours' plot. As soon as air enters into the collector, it gets heated near the ground due to an irradiation from sun, whereas it remains approximately to the atmospheric temperature, i.e., 298 K up to some distance close to the collector inlet. The maximum temperatures are obtained at chimney's peripheral location at both collector and ground. The shadow zone of collector and ground remains at lower temperature compared to remaining part of collector and ground.

The solar heat flux falling on the ground creates a greenhouse effect in the collector and finally a temperature difference. This temperature difference generates the buoyancy in the collector and a considerable fluid flow in the chimney. The temperature contours and velocity distribution at the vertically middle plane of the plant are shown in Fig. 8. The maximum temperature and velocity are obtained as 355 K and

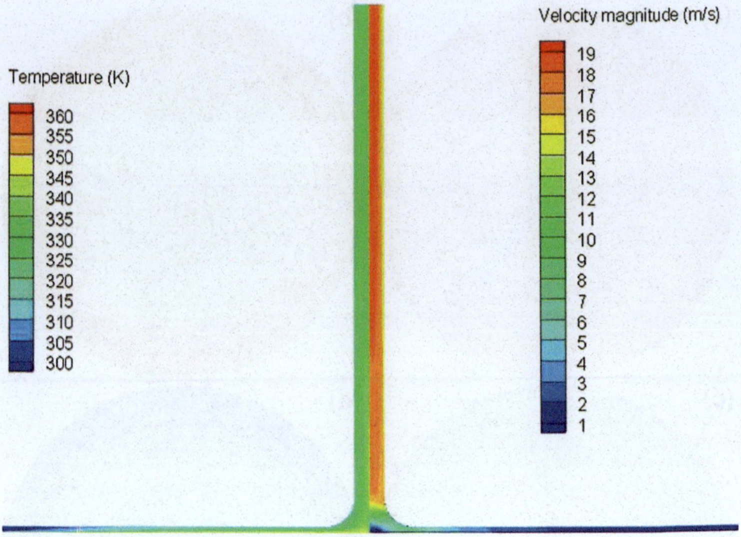

Fig. 8 Temperature and velocity magnitude contours at the vertically middle plane of plant

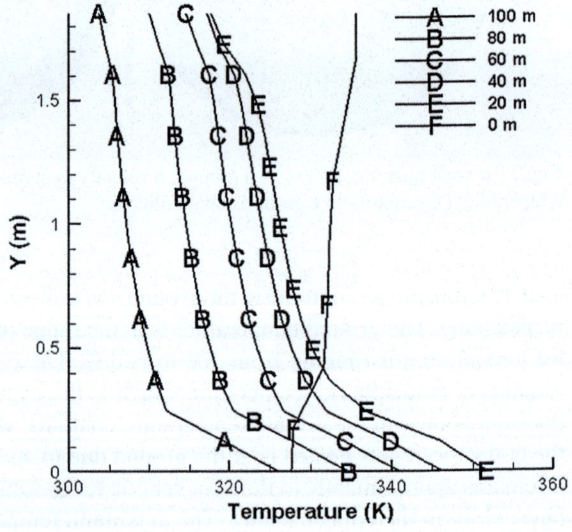

Fig. 9 Temperature variation along height of collector

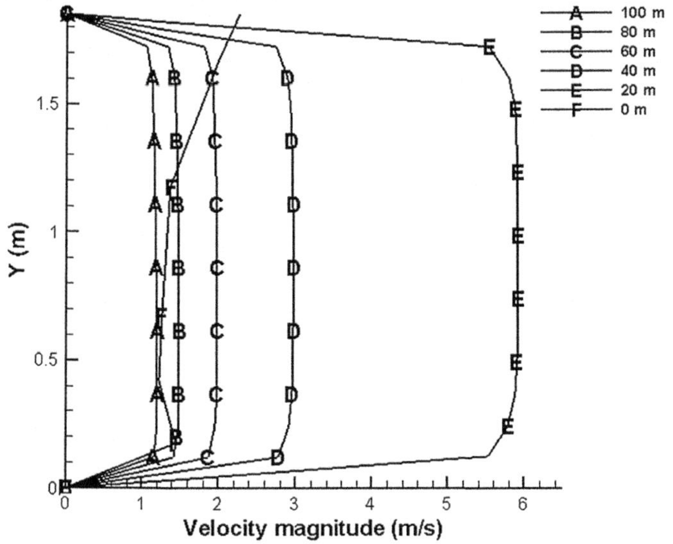

Fig. 10 Velocity profile along collector radius

19.3 m/s, respectively. These maximum values are in the chimney, whereas the average value of temperature and velocity in the collector are 325 K and 3 m/s, respectively. This mean temperature difference in the collector zone is 25 K, which is slightly higher as compared to the values reported from the plant [2]. The velocity at the exit of the chimney by simulation is reported as 18 m/s which is also slightly higher. As the fluid moving from the collector inlet to chimney, it is getting heated from the solar irradiation. Therefore, the average temperature of the air should increase along the radius of the collector. The temperature variations along the height of the collector at various radial locations (distance from chimney axis) are shown in Fig. 9. There is relatively no temperature variation along the height of the collector except near ground.

A sharp temperature gradient is seen near the ground; this gradient is decreasing toward the chimney inlet. The temperature of the ground is higher than the temperature of the collector. The reverse trend for temperature variation along the height of collector is seen at chimney axis. The air temperature is greater at the chimney inlet than that on the ground. The ground temperature is also considerably small at chimney axis.

The velocity variations along the height of the collector at various radial locations (distance from chimney axis) are shown in Fig. 10. The fluid is getting accelerated as it moves toward the chimney. The maximum velocity also increases near the chimney. The flow from all radial direction is getting mixed near the chimney inlet, so there is a deceleration of flow near chimney location. The velocity magnitude is very low at chimney axis.

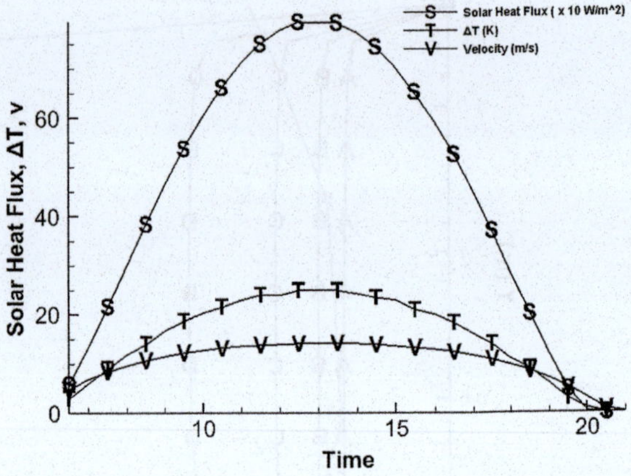

Fig. 11 Variation of insolation, average velocity, and increase in temperature with hours of day (without thermal storage)

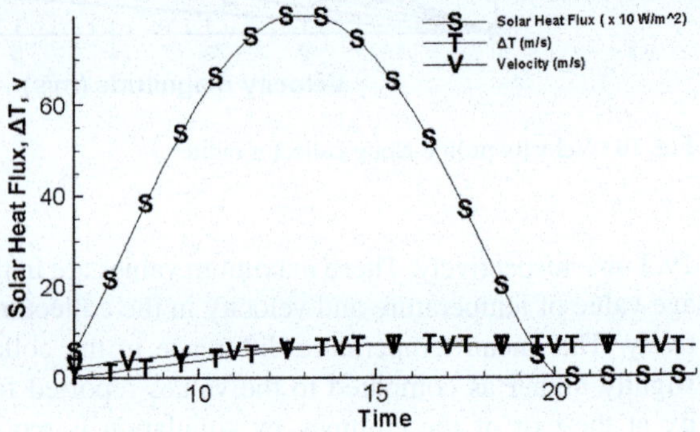

Fig. 12 Variation of insolation, average velocity, and increase in temperature with hours of day (with thermal storage)

7.2 *Transient Simulation Without Radiation Model and Thermal Storage*

The variation of insolation, the average velocity at turbine plane, and average air temperature rise (difference of average temperature at collector inlet and chimney inlet) with hours of the day are shown in Fig. 11. The value of insolation is reported from solar calculator of ANSYS Fluent. The insolation values are divided by 10 for better visualization. The peak values of insolation, temperature rise, and velocity are achieved at the same hour of the day. A strong dependency of velocity and temperature in the plant on insolation is seen. This reveals that the plant can operate only in day hours.

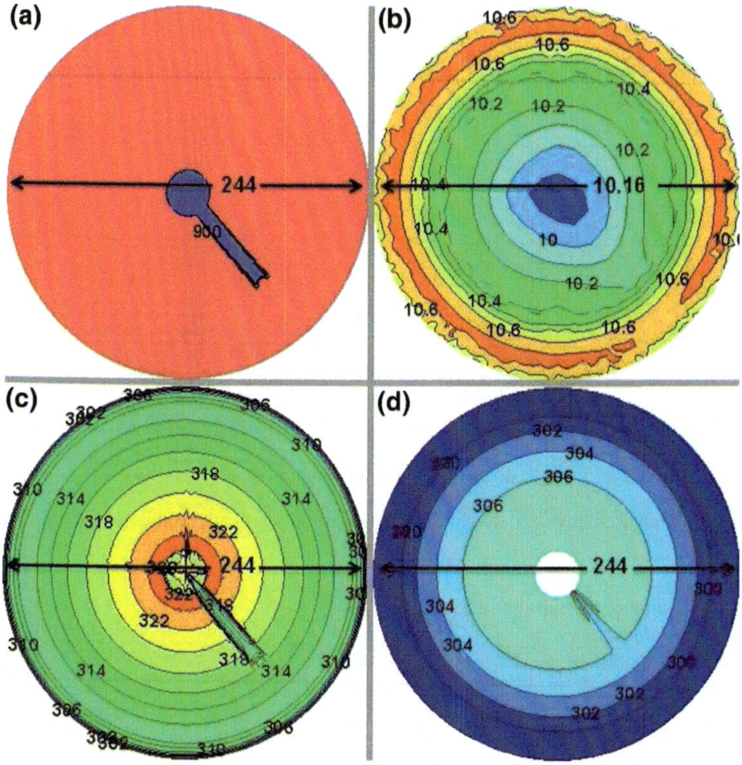

Fig. 13 **a** Solar heat flux (W/m^2) on ground, **b** velocity magnitude (m/s) contours at turbine plane, temperature (K) contour at **c** ground and **d** collector (with radiation model)

7.3 *Transient Simulation Without Radiation Model and with Thermal Storage*

The effect of thermal storage on the plant performance has also been studied. The thermal storage of water over whole collector area with the depth of 10 cm has been considered. A zone for thermal storage has been added to the plant geometry, and material properties of water have assigned to this zone. The convection, evaporation, and any other fluid-related phenomenon are not considered in water zone. The simulation has been started after 9 am so that sufficient insolation is available for the plant and also for thermal storage. Figure 12 shows the variation of insolation, the velocity at turbine plane, and the average increase in air temperature in collector with hours of the day. Although the value of maximum velocity and temperature difference are less compared to without storage, this maximum velocity and increase in temperature remain same and available overnight. The hours of the day, at abscissa of the graph, exceeds 24, which can be considered as hours of the next day. The velocity and temperature rise reach the peak after the solar insolation's peak. The peaks of

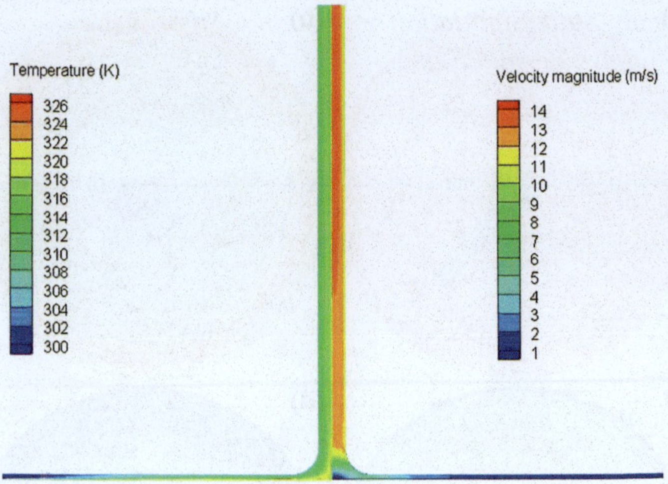

Fig. 14 Temperature contour and velocity contour at the vertically middle plane (with radiation model)

Fig. 15 Temperature profile along height of collector (with radiation model)

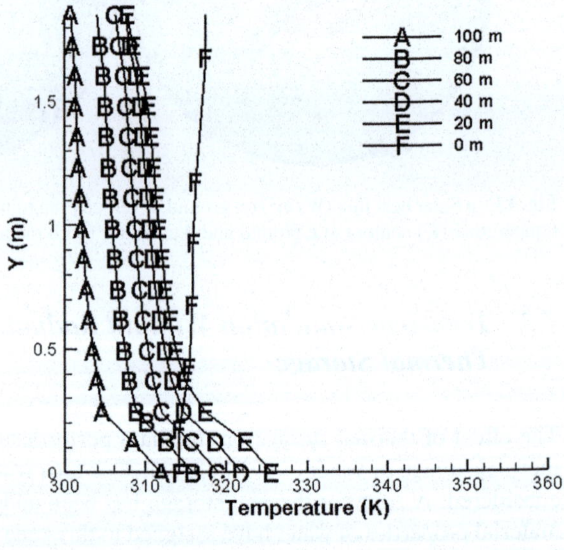

velocity and temperature rise remain constant even after solar insolation is not available. It can be concluded that the plant can be operational at night as well. Therefore, the plant's performance dependency on the insolation can be reduced with thermal storage. It also appears that capacity of thermal storage is higher than needed. The optimized thermal storage capacity can be further studied.

7.4 *Steady-State Simulation with Radiation Model and Without Thermal Storage*

The solar updraft plant has also been studied with thermal radiation model since it is an important mode of heat transfer in natural convection problems. The thermal radiation model is chosen based on the following assumptions

i. The air does not participate in thermal radiation.
ii. The glass/plastic behaves opaque for long-wavelength radiation emission from moderate temperature object.

Also, considering the computation resource requirements of the radiation model, the surface-to-surface radiation model has been selected. The graphs presented in Sect. 7.1 of Results and Discussion section have also been reproduced here with radiation model. Comparing results in Fig. 13 with those in Fig. 7, it is found that the values of velocity at turbine plane, temperature at ground, and temperature at collector are reduced by 3–4 m/s, 15–25 K, and 8–10 K, respectively. These values are close to reported values of the plant which are as follows: 10–12 m/s velocity at turbine plane and 306–312 K at collector zone. Similarly, comparing Figs. 14, 15 and 16 with corresponding Figs. 8, 9 and 10, it is observed that the values of average temperature and maximum velocity are also reduced while incorporating the radiation model. The temperature gradient is seen only near to the ground, whereas temperature remains constant in rest height of collector. The temperature gradient near the ground is also reduced with radiation model compared to non-radiation case. Table 5 provides comparison detail of the results without and with radiation model and also with some published results.

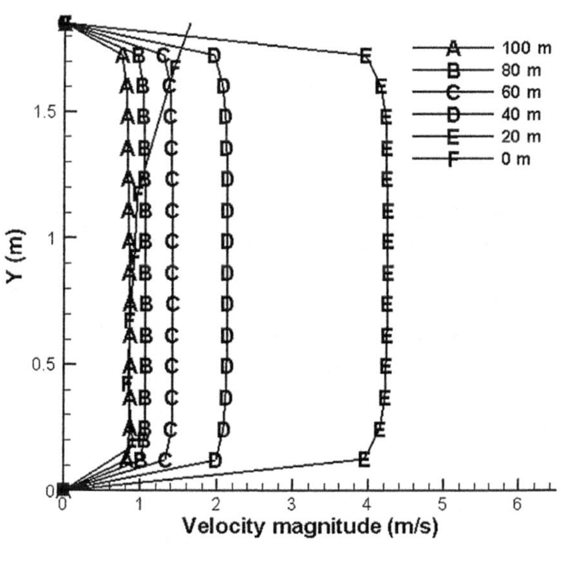

Fig. 16 Velocity profile along collector radius (with radiation model)

Table 5 Comparison detail of the results

Variable	Without radiation model	With radiation model	Haaf et al. [3]	Fasel et al. [5]
ΔT (K)	25	14.8	12	18
Max velocity at turbine plane (m/s)	15	11	9–11	10–11
Max velocity at release (m/s)	19	14	15	-

8 Conclusions

The full-scale three-dimensional computational analysis of solar updraft tower in Manzanares, Spain has been performed using ANSYS Fluent. The simulation has been carried out in the steady state on 8th of June at 12 noon, while the transient case for a day starts from 6 AM without radiation model, whereas thermal storage case simulation starts at 9 AM on 8th of June. The detailed variations of temperature and velocity of the plant have been presented without and with radiation model. This study provides the knowledge about computational models for the solar updraft plant, and the detailed outcome would help in optimizing the natural draft plants and further help in designing the large-scale power plants. The following conclusions have been drawn:

1. The main parameters of the plant, i.e., increase in temperature and velocity, are observed higher than the reported values from plant if radiation model is not considered in the simulation.
2. The surface to surface (non-participating medium model) thermal radiation model uses minimal computational resources than the thermal radiation models for the participating medium. The main parameters of the solar updraft plant like velocity and temperature match very well in the simulation result with the non-participating medium model of thermal radiation.
3. There is decrease in the values of velocity, temperature at ground, and temperature at collector by 3–4 m/s, 15–25 K, and 8–10 K, respectively.
4. There is strong dependency of increase in temperature and average velocity in the plant on insolation without thermal storage.
5. The thermal storage reduces the strong dependency of temperature and velocity on insolation, and plant can also be operated overnight; however, there is reduction in the velocity and temperature rise.
6. The thermal storage considered in the present study appears to be oversized as solar heat flux is not sufficient to generate the required temperature difference. Further, the study needs to be carried out for the optimal size of the thermal storage.

References

1. Holman JP (1997) Heat transfer. McGraw-Hill Inc., New York
2. Haaf W (1984) Solar chimneys Part II: preliminary test results from the Manzanares pilot plant. Int J Sol Energy 2:141–161
3. Haaf W, Friedrich K, Mayr G, Schlaigh J (1983) Solar chimneys Part I: principle and construction of the pilot plant in Manzanares. Int J Sol Energy 2:3–20
4. Gannon AJ, Backström TW (2000) Solar chimney cycle analysis with system loss and solar collector performance. Int J Sol Energy 122:133–137
5. Fasel HF, Meng F, Shams E, Gross A (2013) CFD analysis for solar chimney power plants. Sol Energy 98:12–22
6. Hafizh H, Shirato H (2015) Aerothermal simulation and power potential of a solar updraft power plant. J Struct Eng 61:388–399
7. Guo P, Li J, Wang Y (2014) Numerical simulations of solar chimney power plant with radiation model. Renew Energy 62:24–30
8. Schlaich J, Bergermann R, Schiel W, Weinrebe G (2005) Design of commercial solar updraft tower systems utilization of solar induced convective flows for power generation. J Sol Energy Eng 127:117–124
9. Pastohr H, Kornadt O, Gürlebeck K (2004) Numerical and analytical calculations of the temperature and flow field in the upwind power plant. Int J Energy Res 28:495–510
10. Sangi R, Amidpour M, Hosseinizadeh B (2011) Modeling and numerical simulation of solar chimney power plants. Sol Energy 85:829–838
11. Tingzhen M, Wei L, Guoling X, Yanbin X, Xuhu G, Yuan P (2008) Numerical simulation of the solar chimney power plant systems coupled with turbine. Renew Energy 33:897–905
12. Audunson T, Gebhart B (1972) An experimental and analytical study of natural convection with appreciable thermal radiation effects. J Fluid Mech 52:57–95
13. Kumar P, Eswaran V (2008) The effect of radiation on flow in a conical diffuser. Num Heat Transf Part A: Appl 54:962–982
14. Kumar P, Eswaran V (2010) Numerical simulation of combined radiation and natural convection in a differential heated cubic cavity. Trans ASME J. Heat Tranf 132
15. Kumar P (2009) Radiative heat transfer in a participating gray medium and its interaction with fluid flow. Indian Institute of Technology Kanpur, Kanpur
16. Siegel R, Spuckler CM (1994) Effects of refractive index and diffusive or specular boundaries on a radiating isothermal layer. Trans ASME J Heat Tran 116:787–790
17. Florides G, Kalogirou S (2005) Annual ground temperature measurements at various depths. Pro, CLIMA, Lausanne, Switzerland
18. Betts PL, Bokhari IH (2000) Experiments on turbulent natural convection in an enclosed tall cavity. Int J Heat Fluid Flow 21:675–683
19. Gan G (1998) A parametric study of Trombe walls for passive cooling of buildings. Energy Build 27:37–43

Manufacturing Techniques of Perovskite Solar Cells

Priyanka Kajal, Kunal Ghosh and Satvasheel Powar

Abstract Perovskite solar cells (PSCs) are in focus of the solar cell development research for the last few years due to their high efficiency, cost-effective fabrication, and band gap tunability. Perovskite solar cell efficiency sharply increased from its initial reported efficiency of 3.8% in 2009 to 22.1% in 2016. This makes PSCs as the technology with the fastest growth rate in terms of the efficiency. Different device architectures have also been developed in an attempt to improve the PSC efficiency. At laboratory scale, a spin-coating process is employed to deposit different layers of PSCs. Though spin-coating process helps to achieve high efficiency, for large-scale production viability, researchers are developing different deposition techniques. A broad range of manufacturing techniques for perovskite-based solar cells have been tested and reported comprising drop casting, spray coating, ultrasonic spray coating, slot die coating, electrodeposition, CVD, thermal vapor deposition, vacuum deposition, screen printing, ink-jet printing, etc., with different device architectures. This chapter summarizes different PSC structures along with the corresponding manufacturing techniques.

Keywords Perovskite solar cells · Scale-up · Manufacturing techniques

1 Introduction

Energy demand is rapidly increasing across the globe, and it is estimated that by 2030, the global energy demand will increase to 16 terawatts (TWs) [1]. The growing demand cannot be fulfilled by fossil fuels-based energy sources because of

P. Kajal · S. Powar (✉)
School of Engineering, Indian Institute of Technology Mandi, Mandi 175005, Himachal Pradesh, India
e-mail: satvasheel@iitmandi.ac.in

K. Ghosh
School of Computing and Electrical Engineering, Indian Institute of Technology Mandi, Mandi 175005, Himachal Pradesh, India

H. Tyagi et al. (eds.), *Applications of Solar Energy*, Energy, Environment, and Sustainability, https://doi.org/10.1007/978-981-10-7206-2_16

their harmful impact on the environment. So there is a strong need to harness the energy using renewable sources like solar [2], the wind [3], biogas [4, 5]. Among all other sources, solar is the most abandon source of energy which is available throughout the year. Hence, solar is the best alternative to fossil-based non-renewable energy sources [6]. Solar energy can be produced through thermal and photovoltaic (PV) converters. Out of these, PV is a preferred technique for electricity generation because of direct conversion of sunlight into electricity [7].

Photovoltaic effect was first discovered by Alexandre–Edmond Becquerel in 1839. However, it is in 1954 that Bell Labs first designed the modern solar cell, which is based on separating junction. In a solar cell, when a photon of energy greater than the band gap strikes the material, electron–hole pairs are created. Under the influence of electric field, they are separated and transported to their respective contacts from which they are extracted as current. The fundamental factors which limit the commercialization of solar cells are durability, efficiency, and price, while material availability and its impact on the environment are also considered. Layer thickness, material type, defects, and device architecture directly affect the fill factor (FF), power conversion efficiency (PCE), losses, and absorption capacity of the solar cell. The limiting parameters such as layer thickness, its properties, and defects can be controlled depending upon the type of the manufacturing technique used. Hence, depending upon the designing approaches, solar cell technology may be classified into two categories, i.e., crystalline Si-based technology and thin film-based technology, which are as shown in Fig. 1, and their corresponding merits and challenges are discussed further.

Crystalline silicon (c-Si) solar cell dominates the commercial PV landscape. At laboratory scale, these solar cells exhibit the highest efficiency of 26.3% and achieved efficiency on a commercial scale of about 25% [8]. The basic silicon

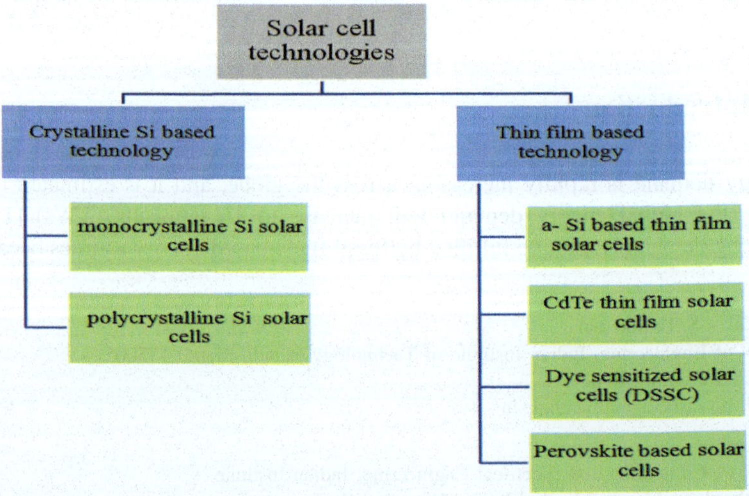

Fig. 1 Classification of solar cells by technologies

(Si)-based *p-n* junction solar cell consists of p- and n-doped Si region and anti-reflective coating with metal contacts in front and back of the solar cell. Depending on the crystal structure, c-Si solar cells are further classified as monocrystalline and polycrystalline. The main advantage of Si-based solar cells is high durability. However, because of the complicated process development, researchers are exploring alternatives to Si-based technology [8, 9].

Thin film-based solar cells are an alternative to crystalline Si solar cell technology [10]. Thin film-based technology is further classified into various categories depending on the material type used as depicted in Fig. 1. In the thin film-based technology, the thickness is from few nanometers to few micrometers, which results in faster charge collection. Inorganic thin film-based solar cells require high vacuum and higher processing temperature which limits their large-scale production for solar cell application [11]. Multiple techniques including a-Si [12], QDSSC [13, 14], DSSC [15], and PSCs [16] exist within thin film-based technology out of which PSCs have shown phenomenal increase in efficiency from 3.8 to 22.1% [17–19] within few years. Keeping all these facts into consideration, this monograph chapter focuses on perovskite solar cells, different perovskite solar cell architectures, and their manufacturing processes.

Perovskite solar cells (PSCs)

Perovskite solar cells have most significant improvement in terms of the efficiency in recent years. These solar cells have shown remarkable efficiency increase from 3.8% reported by Miyasaka and co-workers in 2009 [20] to 22.1% up to 2016 [17–19, 21]; the reported increase in efficiency within these few years makes it a very attractive light absorbing the material. Moreover, perovskites exhibit excellent optical and electrical properties. Thus, Perovskite solar cells exhibit improvement in optical absorption, synthetic feasibility, ease of fabrication, easy charge separation, and slow charge recombination rate [4, 22].

Perovskite is an organic–inorganic hybrid compound having an ABX_3 crystal structure. The basic crystal structure of perovskite following $CaTiO_3$-type structure with lattice parameters $a = 5.404$ Å, $b = 5.422$ Å, and $c = 7.651$ Å drawn using VESTA is as shown in Fig. 2.

In this structure, X represents halogens such as I^-, Br^-, Cl^-, or oxygen, whereas A and B are cations of different sizes having 12 and 6 coordinates with X anions, respectively [23]. Site A is occupied by the elements from the group I or II of the periodic table, whereas the site B is filled by the cations such as transition metals. The valence band maxima of perovskite are made by antibonding hybridization of "B = Pb, Sn" s-state and "X = Br, Cl, I" p-state, whereas conduction band minima are formed by π-antibonding of "B = Pb, Sn" p-state and "X = Br, Cl, I" p-state. Cation A, i.e., methyl ammonium, formamidinium does not directly contribute to valence band maxima and conduction band minima, but it does affect the lattice constants, and it was found that band gap increases with increasing lattice parameters. The ideal structure of the perovskite is similar to bcc with additional anions on the faces of the unit cell.

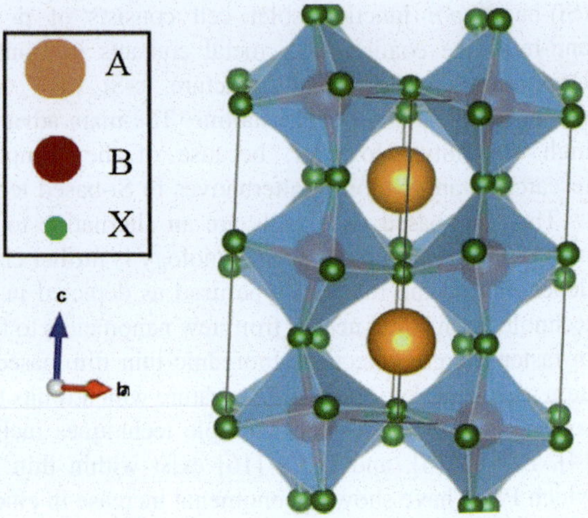

Fig. 2 Crystal structure of perovskite using VESTA

Perovskites are ambipolar and have attracted so much attention within a few years due to their small band gap, high carrier mobility, high absorption, and easy fabrication at low temperature [24–26]. Defects in perovskite solar cells lead to hysteresis effect which acts like traps for both electrons and holes. Less hysteresis indicates more uniform perovskite layer with lesser defects. The stability of perovskites is calculated using *tolerance factor* (t) [23] which is given by:

$$t = \frac{r_A + r_X}{\sqrt{2}(r_B + r_X)} \tag{1}$$

where r_X, r_A, r_B are ionic radii of ions for X, A, B sites, respectively. Tolerance factor affects the size of ionic radius. For $t = 1$, perovskite exhibits perfectly packed cubic structure; for $0.8 < t < 1$, perovskites exhibit highly stabilized structure, and with the increase in the value of t, symmetry value increases, resulting in band gap reduction [27, 28]. Depending on structural configurations, there are five types of PSC architectures, which again are classified into two categories, namely planar and mesoporous. Further, subclassification into various subcategories is as shown (Fig. 3).

Various factors such as device architecture, compositional engineering, and technique of film formation greatly affect the device performance. Film formation technique directly affects the efficiency of the solar cell. Perovskite films can be deposited by a variety of deposition techniques such as spin coating, drop casting, spray coating, ultrasonic spray coating, slot die coating, electrodeposition, CVD, thermal vapor deposition, vacuum deposition, screen printing, ink-jet printing [31–35].

The most common structures of Perovskite solar cells mainly consist of ITO, an electron-transporting layer, perovskite layer, hole-transporting layer, and the metal electrode [36]. Their efficiency depends on the materials used in various layers and

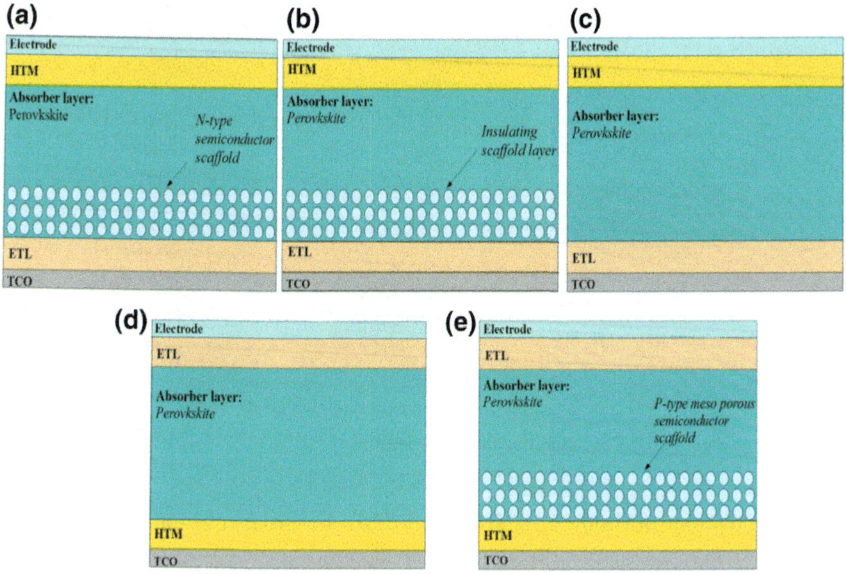

Fig. 3 Different structures of PSC: **a** regular, **b** mesostructured, **c** planar heterojunction, **d** inverted planar heterojunction structure, **e** inverted p-type mesoporous structure [29, 30]

on the type of deposition technique used. This article attempts to report different techniques of depositing perovskite materials and their industrial scalability.

2 Manufacturing Techniques

The primary objective of the PV devices is to generate power efficiently and its large-scale applicability. However, while scaling the dimensions of the PV device, various challenges are encountered which are not found in small-area solar cells [37]. In the case of PSCs, one such difficulty is to control the morphology of perovskites over a large area [38, 39]. Depending on manufacturing technique, layer roughness also gets affected. Hence, manufacturing technique needs to be selected carefully [40, 41]. Thin film-based solar cells are rectangular so as to reduce the length that charge carriers move [42]. The reported scalable techniques for PSCs are solution-based deposition and vapor-assisted deposition techniques, which are further classified into subcategories as shown in Fig. 4.

PSC fabrication may either be carried out using *one-step* or *two-step* deposition technique. Electron lifetime directly depends on the type of the film deposited. One-step deposition of perovskite is the thin film deposition using a mixture of both CH_3NH_3I and PbI_2, while the two-step deposition is carried out using the separate

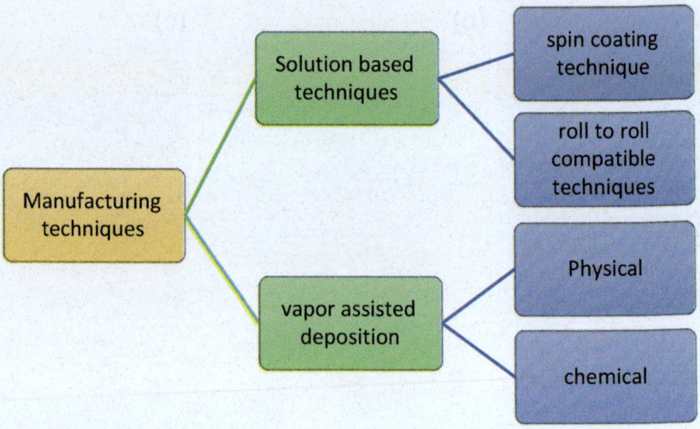

Fig. 4 Classification of PSC manufacturing techniques

layer deposition of both CH_3NH_3I and PbI_2 to produce highly uniform and defect-free layers. Layers obtained by using one-step deposition contain voids which result in faster recombination. Hence, the two-step deposited perovskite layer exhibits electron lifetime of about one order of magnitude greater than one-step deposited films. Also, two-step deposition offers better morphology control over one-step deposition [43]. The solution- and vapor-assisted manufacturing techniques and their subcategories are discussed as follows.

2.1 *Solution Processing-Based Techniques*

Solution processing-based technique is the most widely used technique for depositing perovskite thin films. Solution-processed deposition is carried out using various techniques such as spin coating, dip coating, doctor blade, spray coating, ink-jet printing, screen printing, drop casting, slot die coating. These techniques are illustrated as follows.

Spin-coating technique: Spin coating is the simplest and cost-effective solution-processed technique used for uniform deposition of the perovskite layers of the PSCs. This technique is mainly used for producing small-area solar cells. Generally, after spin coating, the film is baked to produce well-crystallized layers of the perovskite, since the baking results in strong adhesion and bonding between metal cations and halogen anions [44]. By setting up spin speed, acceleration, and time of spin coating, the film thickness and quality can be optimized. Highest recorded efficiency at laboratory scale using this technique is about 22.1% [17–19]. This technique can be used for the fabrication of inverted as well as regular PSC structures [32, 45, 46]. Achieved efficiency using spin-coating technique is very high, but this technique has the limitation of not producing a uniform film on larger

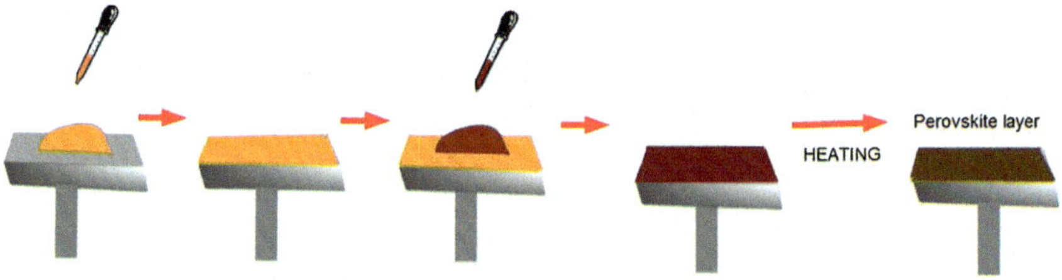

Fig. 5 Sequential steps of spin-coating process [44]

areas. Also, the film thickness varies from center to the end of substrate mostly in case of bigger substrates [47, 48]. The efficiency achieved using this technique is very high with a good level of reproducibility and morphology control in case of smaller sizes of modules. However, because of slow processing technique and also due to more material wastage, this technique is not a good solution for PSC manufacturing at large scale [42]. The two-step deposition of the perovskite using spin-coating technique is as shown in Fig. 5.

Drop-casting technique: Drop casting is a basic and low-cost method for the production of PSCs. This technique is similar to spin coating, but the major difference is that no substrate spinning is required. Also, the film thickness and properties depend on the volume of the dispersion and concentration. Other variables which affect the film structure are substrate wetting, the rate of evaporation, and drying process.

Volatile solvents are generally preferred for this technique which can wet the substrate. One of the advantages over spin coating is less material wastage. On the other hand, there are various shortcomings of this method including difficulty in controlling film thickness and non-uniform film formation on large size wafers [49].

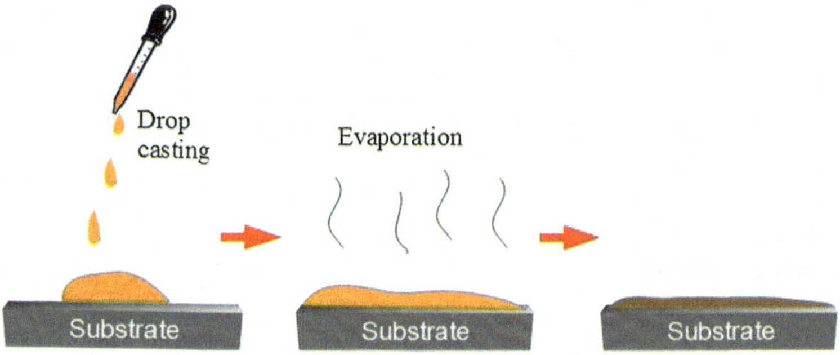

Fig. 6 Perovskite deposition using drop-casting technique [99]

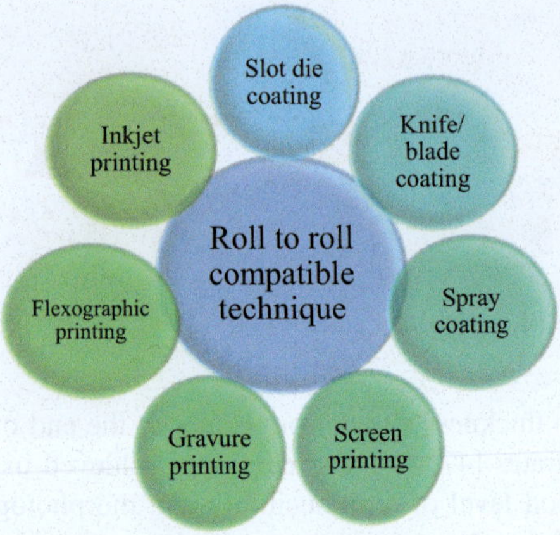

Fig. 7 Types of roll-to-roll fabrication techniques [51]

2.2 *Roll-to-Roll Printing*

Roll-to-roll printing is one of the best solution processing techniques used for the manufacturing of PSCs. This is a highly reproducible technique with an enormous potential of producing flexible PSCs at large scale with high throughput [50]. It includes different coating methods which are used to form a thin film on moving substrate by using rotating rolls. Roll-to-roll compatible printing techniques are shown in the Fig. 7.

These techniques can be employed for fabrication of solar cells on flexible substrates. The continuous deposition is carried out at high coating speed [52]. For deposition of the active HTL and ETL layers, deposition by roll-to-roll printing technique, slot die coating is one the most preferred techniques because of the highly uniform film deposition by this method [42]. Using various roll-to-roll compatible methods, the PSC fabrication is as shown in Fig. 8.

Roll-to-roll deposition is the continuous process of depositing layers on the substrate for making solar cells by using various types of the roll-to-roll compatible techniques [53]. The biggest advantage of this kind of deposition is that very long wafers can be used leading to the bulk production of solar cells, faster technique, highly scalable and cost-effective technique.

2.2.1 Spray Coating

It is one of the most efficient and faster solution-based approaches for producing flexible solar cells. It is not dominating technique, but still it exhibits various advantages over one of the most dominating spin-coating techniques. As already discussed, spin coating is a small-scale thin film deposition technique due to scalability issue [54], while spray coating is a highly scalable thin film deposition

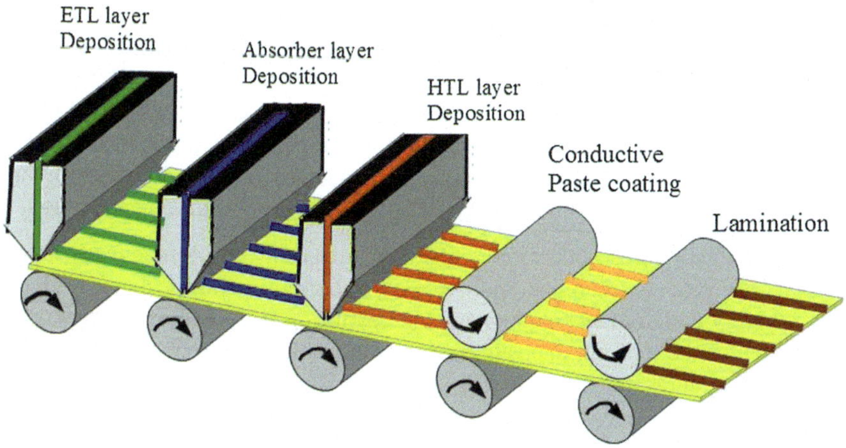

Fig. 8 PSC solar cell production using roll-to-roll printing techniques [42]

technique [54, 55]. There are various other scalable thin film deposition techniques that include ink-jet printing [56], slot die coating [57, 58], doctor blade [51], out of which blade coating is one of the fastest and cheapest thin film deposition techniques, but still PCE achieved using spray-coating technique is about ten times greater than blade-coating technique. Along with this, various other advantages of this technique are rapid film deposition, low-cost processing, and deposition capability for large scale both on flexible and glass-based substrates. Moreover, the spray-coated films exhibit higher thermal stability and also show better optoelectronic properties as compared to spin-coated films due to better charge transfer capability and higher minority carrier lifetime [59].

This technique is compatible for deposition of both planar and inverted perovskite solar cell devices [60]. Recorded efficiency till now using this method is about 11%. Proper control over processing parameters such as carrier gas, substrate temperature, and annealing temperature is necessary [42]. Despite so many advantages, there are certain shortcomings of this technique such as difficulty in achieving fully covered and highly homogeneous layers. These shortcomings arise due to the liquid atomization phenomenon in which random spraying of different size droplets takes place which leaves patches of varying sizes on drying, and hence, the film shows unpredictable characteristics. This leads to an increase in series resistance, affecting the device performance. To overcome this problem, further modified spraying processes such as airbrush pen spray [61], electrostatic spray-coating technique [62], pulsed spray coating [63], ultrasonic spray-coating technique [64] are used.

Ultrasonic spray coating: It is one of the modified forms of spray-coating technique with the additional feature like proper morphology control of perovskite layer in ambiance. The highest PCE achieved so far using this technique is 7.7% [64]. As previously discussed, two types of deposition, i.e., one-step and two-step deposition, processes are used for perovskite deposition [43], out of which two-step deposition is preferred for better morphology control. However, in the case of spin coating (which is highly preferred technique at laboratory scale), it is difficult to control chemical reaction speed. Hence, ultrasonic spray-coating technique is a highly preferred technique for two-step deposition due to proper control over the rate of deposition. Thin films with larger grain size are obtained by this technique which helps in achieving high carrier mobility with low charge recombination.

Moreover, the hysteresis effect in the films obtained using this technique is less than that of spin-coating technique and conventional spray-coating technique. Hence, this is highly preferred technique over spin-coating and conventional spray-coating technique for the uniform deposition of the perovskite layer [65]. The basic schematic for the ultrasonic spray-coating technique is as shown in Fig. 9.

Blow drying: This is one of the complementary processes in manufacturing perovskite solar cells. This process can be clubbed with any solution-processed film deposition techniques such as spraying. Blow drying is a roll-to-roll compatible process and does not require any additional anti-solvent or the sacrificial component. The films produced by this method are very smooth and of excellent quality with the scalable solution processing sequence [66]. This method does not require sophisticated environmental conditions and complicated equipment. The whole process is carried out by dispensing perovskite solution onto the substrate and then absorbing of the excess solution by tissue paper followed by blow drying in the compressed air which leads to faster crystallization and good coverage [67]. The films prepared by blow-drying technique are reported to be darker which indicates higher absorbance of blow-dried films. Blow-dried films are smoother than the spin-coated films on the glass substrate (Fig. 10).

In blow-drying technique, airflow speed greatly affects the film uniformity and the airflow speed depends on the distance between the compressed air outlet and substrate and is measured by wind speedometer. As per the published articles, at

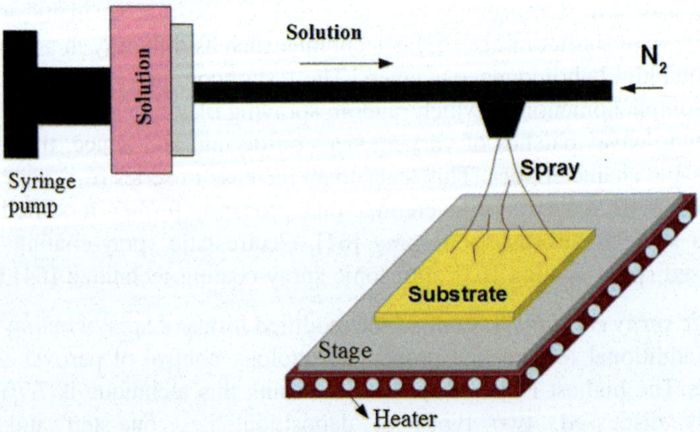

Fig. 9 Schematic diagram of ultrasonic spray-coating process [65]

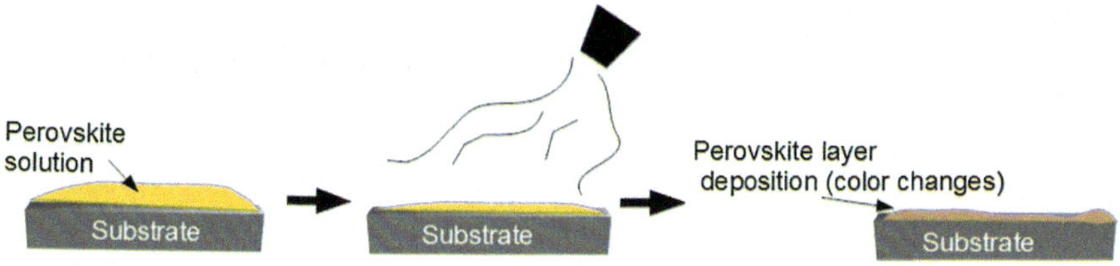

Fig. 10 Basic schematic of blow-drying process [67]

low speed perovskite films exhibit branch like morphology which is commonly found in spin-coated films. As the airflow speed increases, the grain size reduces, and with further increase in speed (>25 ± 2 m/s), highly smooth film is obtained. So the advantages of blow drying over other solution-based techniques are that it does not require any time window for compressed air blowing, is of low cost, and has greater control on the film uniformity. Several other advantages of this technique include highly smooth and good quality films; no additional sacrificial component is required; highly scalable technique; easy component specification [68].

2.2.2 Blade Coating

It is one of the simple, environment-friendly, vacuum free, and low-cost solution-processed methods for the fabricating PSC at industrial scale using blade coater applicator [32]. Blade coater applicator is a simple coating system which consists of a screw of micrometer size used to adjust the height of the blade with respect to the substrate surface [42]. In this technique, film uniformity and quality can be controlled by controlling evaporation rate of inks by adjusting airflow over the substrate or by heating up the substrate to the boiling point of the solvent. The basic arrangement is as shown (Fig. 11).

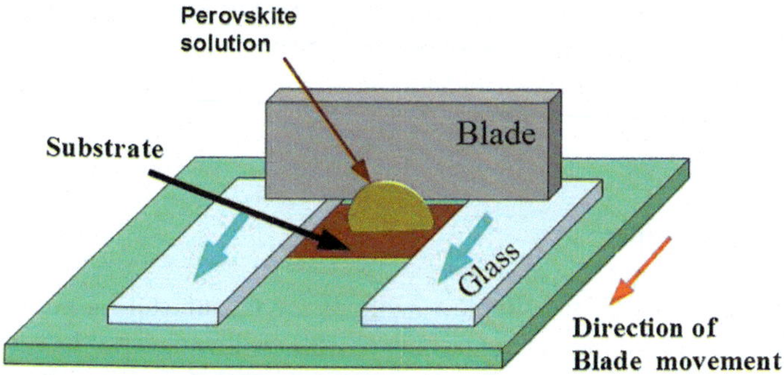

Fig. 11 Perovskite solar cell fabrication by blade-coating technique at laboratory scale [51]

This process delivers better film morphology control as compared to spray-coating and other roll-to-roll compatible techniques. This technique is used for the fabrication of planar heterojunction PSCs (Fig. 3c). Generally, most of the perovskite deposition techniques result in pinhole formation in the perovskite layers due to non-uniform deposition and the presence of moisture and air which often leads to performance degradation of photovoltaic devices. So as to overcome these challenges, blade coating is preferred [69]. It is one of the highly recommended processes for the growth of self-assembled perovskite crystalline domains with the features like uniform film coverage and improved device stability. Furthermore, this technique has advantages of uniform, high-throughput deposition and easy application on flexible substrates [51].

Yang et al. [70] have reported slow crystallization rate in case of blade-coating technique which often leads to large size agglomeration forming air protection patches and thus restricts the moisture permeability of the perovskite layer. Hence, the films obtained by this technique are highly stable with better morphology.

According to Kim et al. [71], due to the feature of depositing electron-transporting layer (ETL) and hole transporting-material (HTM) layer along with perovskite layer, this technique can be implemented at large scale. However, the major loophole behind the implementation of this method is precise control of ink used, and for overcoming that problem, the use of slot die coating is made, which is having same operating principle as blade-coating technique.

2.2.3 Slot Die Coating

This is a solution-processed, roll-to-roll compatible, highly scalable process and is almost similar to blade-coating technique with the advantage of less or no material wastage which is seen in the case of blade-coating technique. Among other techniques, this technique has achieved large success in the production of PVs. This process produces uniform and controlled thickness film in the material and is called as premetered coating process [57]. By controlling the amount of the material feeding, thickness of the film can be controlled.

It is a laboratory to fab transition technique, i.e., having high potential to be used at industrial scale for PSC manufacturing [72]. This process is advantageous over the spin-coating technique in terms of scalability over a large-scale and less material wastage. The basic schematic of the slot die coater using mini roll coater unit is as shown in Fig. 12.

2.2.4 Brush Painting

Brush painting is one of the simplest methods for manufacturing perovskite solar cells. It is a high-speed fabrication technique with the low material loss. It is a faster technique as no annealing process is involved in this process. It is a highly

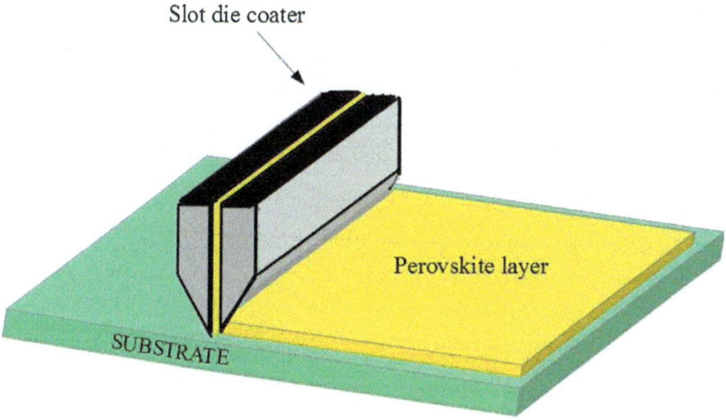

Fig. 12 Perovskite layer deposition using slot die coating technique [73]

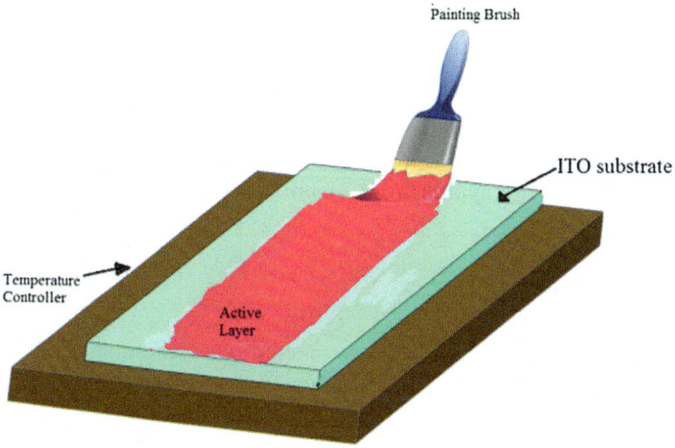

Fig. 13 Perovskite layer deposition using brush painting technique [76]

cost-efficient technique and can be used in large-area flexible solar cells. With proper optimization of the concentration of the solvent and the perovskite material, a high-efficiency solar cell can be attained. Moreover, while using this process, there is no need to control the experimental condition or any other time-consuming procedures [74, 75]. An illustration of PSC fabrication using brush painting technique is shown in Fig. 13.

Advantages of brush painting include the absence of vacuum and fast processing which makes it very cost-effective technique. Brush-painted devices exhibit improvement in efficiency as compared to spin-coating technique as no thermal annealing is required. The downside of brush painting may be controlling the uniform thickness.

2.2.5 Electrodeposition

Electrodeposition is versatile, and roll-to-roll compatible technique is used for producing PSCs. Its cost-effectiveness, rapidness, and a high degree of uniformity make it a desirable technique for perovskite layer deposition [77]. The basic features of this technique are as follows:

- A versatile technique for producing surface coatings,
- Low-temperature variation technique,
- The rapid deposition technique,
- Large-scale productivity,
- Low cost,
- Well-controlled film thickness.

Unlike spin-coating technique, no heating of the substrate takes place in this procedure since heating results in the rough film. When heat is applied, then the film breakdown and island formation take place randomly on the surface of the substrate [78]. The schematic illustration of the PSC fabrication using electrodeposition technique is shown in Fig. 14.

Thin films produced by electrodeposition are highly uniform with large-area coverage with no sheer forces [79, 80]. Perovskite layer deposition on complex shaped substrates, which is not possible using other techniques discussed so far, makes it a very attractive approach for manufacturing at large scale [77].

2.2.6 Ink-Jet Printing

This is a versatile, non-contact, additive patterning technique used for the production of the solar cells, optoelectronic devices, and field-effect transistors. Less

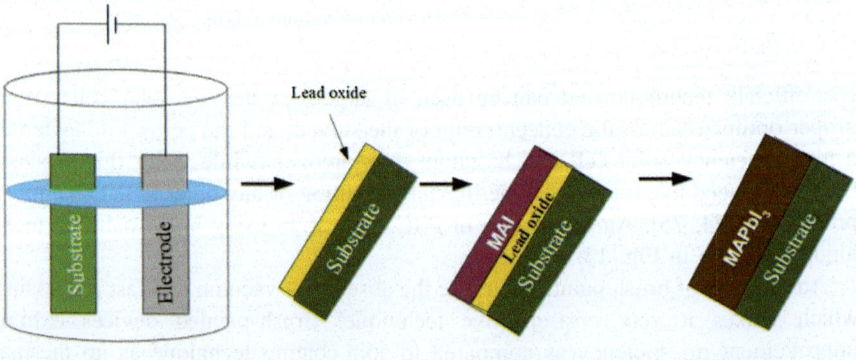

Fig. 14 PSC fabrication using electrochemical deposition [78]

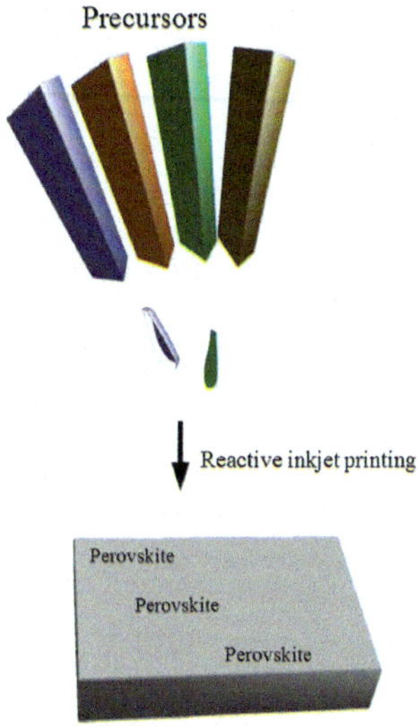

Fig. 15 Schematic illustration of the fabrication of perovskite material using reactive ink-jet printing technique [82]

material wastage, high scalability, and fast deposition of the material on large area make it an attractive approach for large-scale production of the perovskite solar cells [81, 56]. It is used for the highly specific in situ synthesis of the material on the large substrate and thus simplifies the fabrication process. The basic schematic of perovskite solar cell fabrication using ink-jet printing technique is shown in Fig. 15.

This process involves the selective ejection of the ink from the chamber through the nozzle on the substrate. Due to the external bias application, the chambers filled with liquid get contracted, which leads to a shock wave in the liquid and causes ejection of a liquid droplet from the nozzle. According to Jiang et al. [82], the main issue with this technique is blocking of the nozzle which is due to poor solubility of material solvents. Various observations made during ink-jet printing technique are as follows:

a. More viscous ink requires more applied bias and thus results in the formation of the small droplets.
b. If intermediate bias is applied in two different inks, then it results in the jetting of the inks.
c. Higher voltage results in the never recombined small size droplets.
d. Pinch-off point of different liquids matters a lot during printing.

2.3 *Vapor-Based Techniques*

Solution- and vapor-based techniques are the most dominating techniques for fabricating perovskite absorber layers. Out of both, vapor-assisted approach has better film uniformity. Vapor-based techniques are further classified into two main

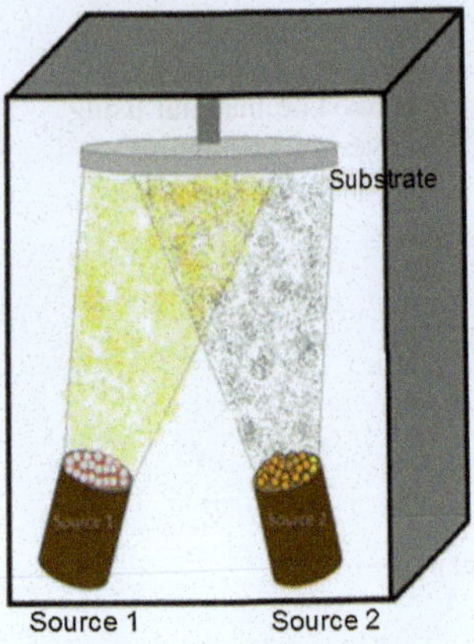

Fig. 16 Vapor deposition of perovskite layers [31]

categories, i.e., physical- and chemical-based techniques. Vapor deposition process results in highly crystalline and uniform films with thickness in nanometers, as compared to micrometer thick films by solution-processed techniques. Moreover, the vapor-deposited thin films are uniform, while the solution-processed layered thin films are of much larger crystal grain sizes than the field of view. As per the reported work, vapor-deposited PSCs are more efficient than the solution deposited. Till now, the reported highest PSC efficiency fabricated by vapor-based technique has reached an efficiency of 15.4% [31] (Fig. 16).

The thickness of perovskite layer significantly affects the optical (absorption) efficiency of the solar cell. If the film is very thin, then the solar cell will not be able to absorb much sunlight, whereas with thick films, the time needed for the electron and hole to reach the contacts increases, resulting in more recombination. If the film is missing in certain areas, then it results in direct contact between ETM and HTM and thus results in the shunting path which leads to lower fill factor (FF) and open-circuit voltage (V_{oc}). The main advantage of the vapor deposition process over the solution processing technique is that by using vapor deposition process, multi-stack thin films over large areas can be produced. Vapor technique is most widely used on the industrial scale including glazing industry, liquid crystal display industry, and thin film solar industry. Moreover, with vapor deposition technique, charge collection at interfaces can be easily tuned. Thus, the vapor deposition is one of the preferred routes for manufacturing uniform thickness solar cell layers [64] (Fig. 17).

The shortcoming of vapor-based techniques is the requirement of vacuum. In vapor-based techniques, a vacuum is employed to increase the mean free path of the vapors for producing highly uniform thin films of very high purity. Various types of the vapor-assisted approach used for deposition of perovskite layers are discussed as follows.

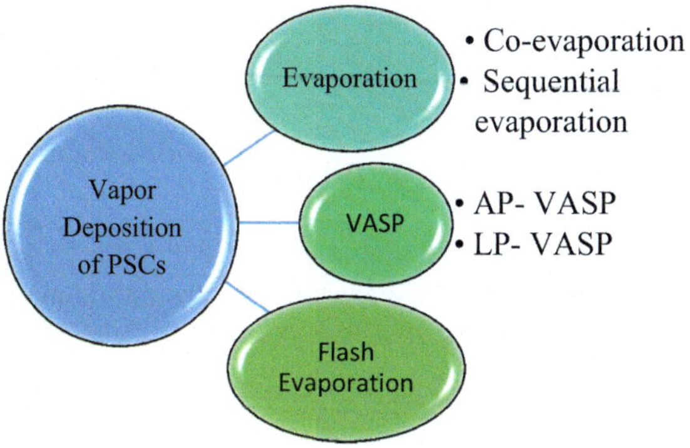

Fig. 17 Classification of various vapor phase deposition techniques for PSCs [64]

2.3.1 CVD

This is one of the promising vapor deposition techniques for large-scale production of highly scalable and uniform pinhole-free perovskite thin films. Unlike other vapor deposition method where challenges were like low material utilization, difficulty in controlling flux deposition, CVD technique was employed for fabricating perovskite films [83–85]. Merits of CVD technique including high material yield ratio and scalability makes it a desirable technique for depositing perovskite layers [86, 87]. The basic schematic of the perovskite layer deposition using a round tube by CVD technique is shown in Fig. 18.

Perovskite layers are deposited by co-evaporation of two different precursors which are heated, mixed, and then transferred to a preheated substrate using carrier gas to form highly uniform, pinhole-free films with larger grain size and long carrier

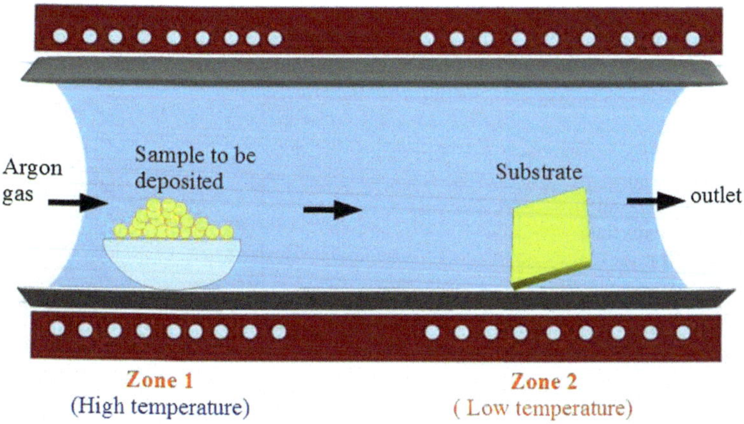

Fig. 18 Schematic of perovskite film fabrication using chemical vapor deposition technique [83]

lifetime. However, this technique requires very high vacuum and uniform co-evaporation of material, which is very challenging for mass production.

2.3.2 Physical Vapor Deposition

This is a very simple and non-reactive thin film deposition process which allows the complete surface coverage of the substrate along with the great stability of the solution against the moisture. Various advantages of this technique are as follows:

- full surface coverage,
- well-defined grain structure,
- high crystallization and reproducibility,
- the tetragonal phase purity,
- more control on film quality, thickness, and morphology (better than the solution process technique).

This method is different from other techniques such as one-step coating technique, two-step sequential vapor-assisted solution process, and dual source vapor deposition method [88]. This method avoids the problems such as high reaction rate, impurities, improper heat treatment.

This approach results in the production of the uniform, smooth, and non-porous films with uniform surface coverage with good crystallization. Hence, we can say that this process is one of the attractive methods for fabrication of high-quality perovskite films. As per the reported data, about 15.4% of efficiency has been attained. The basic schematic diagram of SSPVD technique is shown in Fig. 19.

Very dense, uniform, and compact films can be obtained by using SSPVD technique which indicates that this method can be used at large scale for bulk production of PSCs [88]. The formation of uniform, smooth, and complete surface covering film indicates that no heat treatment is required in this process. PCE about 10.9% was obtained using this process. Hence, it is one of the most promising techniques for the fabrication of the PSCs.

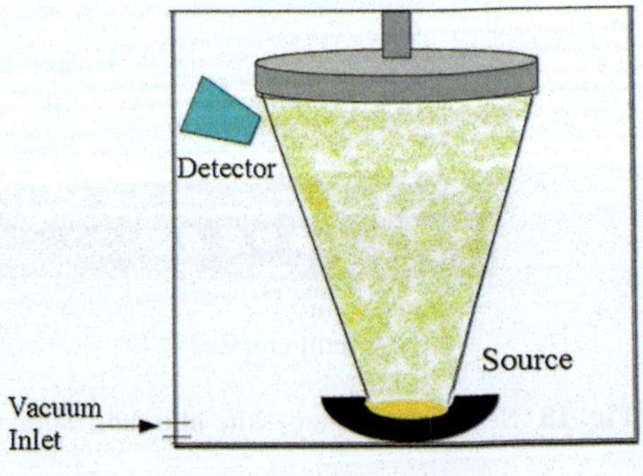

Fig. 19 Single-source physical vapor deposition process of the perovskite $MAPbI_3$ thin film [88]

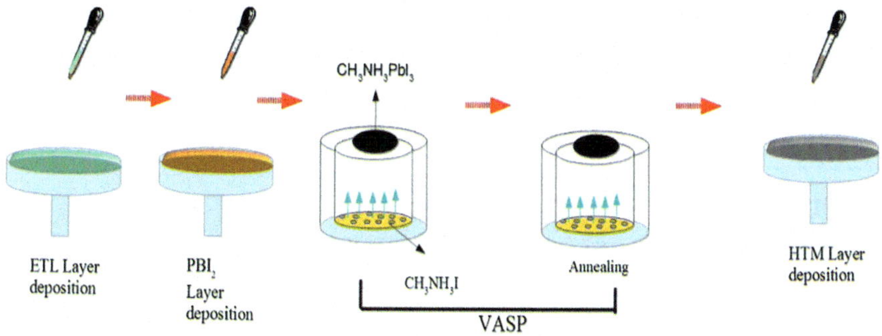

Fig. 20 PSC deposition including VASP process for perovskite deposition [94]

2.3.3 Vapor-Assisted Solution Process

This technique is used for deposition of pinhole-free perovskite layer. Salvation and dehydration problems are encountered in vapor deposition process. In VASP technique, CH_3NH_3I vapors react with PBI_2 forming larger grain size perovskite films. The film morphology obtained by solution-based approach is different from the VASP-based technique [89–93] (Fig. 20).

VASP is a two-step sequential deposition technique where the first step is film deposition in solution-based deposition process, while the second step is vapor deposition process. In the case of solution-assisted techniques, the films get degraded very rapidly. Hence, gas-assisted VASP technique is used to overcome that problem [95–97]. The problem in the deposition of the mixed halides mainly restricts the application of this technique at industrial scale [98].

3 Conclusion

PSCs are thin film-based solar cells with tremendous increase in efficiency up to 22.1% in quite few years, makes it to be the best alternative of the current wafer-based solar cells. Low-cost, thin film-based technology, less bulkiness, etc., make it the next choice of the industry. Manufacturing techniques of perovskite solar cells, i.e., liquid- and vapor-assisted techniques, were discussed, and it was found that both the techniques have their own dos and dont's which directly affect their manufacturing at large scale. Solution-based techniques are very fast and cost-effective, but they have uniformity and stability issues. On the other hand, vapor-based techniques carried out in a vacuum-based environment which helps in achieving uniform films of high purity but makes the deposition costlier and slow too. Depending on all these facts, the major concern is on roll-to-roll compatible techniques to produce large size, flexible PSCs. Apart from the scalable process development, perovskite research is focused on improving the long-term stability of these solar cells and development of new cost-effective materials.

References

1. Asghar MI (2012) Stability issues of dye solar cells, vol 53
2. Nema P, Nema RK, Rangnekar S (2009) A current and future state of art development of hybrid energy system using wind and PV-solar: a review. Renew Sustain Energy Rev 13 (8):2096–2103
3. Lund H, Kempton W (2008) Integration of renewable energy into the transport and electricity sectors through V2G. Energy Policy 36(9):3578–3587
4. Panwar NL, Kaushik SC, Kothari S (2011) Role of renewable energy sources in environmental protection: a review. Renew Sustain Energy Rev 15(3):1513–1524
5. Østergaard PA (2012) Comparing electricity, heat and biogas storages' impacts on renewable energy integration. Energy 37(1):255–262
6. Lewis NS (2007) Solar energy use. Sol Energy 315:798–801
7. Prasanthkumar S, Giribabu L (2016) Recent advances in perovskite-based solar cells. Curr Sci 111(7):1173–1181
8. Rand BP, Genoe J, Heremans P, Poortmans J (2015) Solar cells utilizing small molecular weight organic semiconductors. Prog Photovolt Res Appl 15:659–676
9. T.I.R.E.A. (Irena) (2013) Solar photovoltaics technology brief. The International Renewable Energy Agency, pp 1–28
10. Hibberd CJ, Chassaing E, Liu W, Mitzi DB, Lincot D, Tiwari AN (2010) Non-vacuum methods for formation of Cu(In, Ga)(Se, S)2 thin film photovoltaic absorbers. Prog Photovolt Res Appl 18(6):434–452
11. O'Regan B, Grätzel M (1991) A low-cost, high-efficiency solar cell based on dye-sensitized colloidal TiO2 films. Nature 353(6346):737–740
12. Pandey C (2015) Application of printing techniques in hybrid photovoltaic technologies
13. Cong J, Yang X, Kloo L, Sun L (2012) Iodine/iodide-free redox shuttles for liquid electrolyte-based dye-sensitized solar cells. Energy Environ Sci 5(11):9180
14. Ari. © 1991 Nature Publishing Group
15. Kakiage K, Aoyama Y, Yano T, Oya K, Kyomen T, Hanaya M (2015) Fabrication of a high-performance dye-sensitized solar cell with 12.8% conversion efficiency using organic silyl-anchor dyes. Chem Commun 51(29):6315–6317
16. Ubani CA, Ibrahim MA, Teridi MAM (2017) Moving into the domain of perovskite sensitized solar cell. Renew Sustain Energy Rev 72:907–915
17. Green MA, Emery K, Hishikawa Y, Warta W, Dunlop ED (2016) Solar cell efficiency tables (version 48), version 48, pp 905–913
18. Bai Y et al (2016) Enhancing stability and efficiency of perovskite solar cells with crosslinkable silane-functionalized and doped fullerene. Nat Commun 7:12806
19. Yang S, Fu W, Zhang Z, Chen H, Li C-Z (2017) Recent advances in perovskite solar cells: efficiency, stability and lead-free perovskite. J. Mater Chem A 5:11462–11482
20. Kojima A, Teshima K, Shirai Y, Miyasaka T (2009) Organometal halide perovskites as visible-light sensitizers for photovoltaic cells. J Am Chem Soc 131(17):6050–6051
21. McGehee M (2014) Emerging high-efficiency low-cost solar cell technologies. NREL
22. Sharma S, Jain KK, Sharma A (2015) Solar cells. in research and applications—a review. Mater Sci Appl 6(December):1145–1155
23. Niu G, Guo X, Wang L (2015) Review of recent progress in chemical stability of perovskite solar cells. J Mater Chem A 3(17):8970–8980

24. Xing G, Mathews N, Lim SS, Lam YM, Mhaisalkar S, Sum TC (2013) Reports 10, vol 6960, no 2012, pp 498–500
25. Park NG (2013) Organometal perovskite light absorbers toward a 20% efficiency low-cost solid-state mesoscopic solar cell. J Phys Chem Lett 4(15):2423–2429
26. Grätzel M (2014) The light and shade of perovskite solar cells. Nat Mater 13(9):838–842
27. Koh TM et al (2014) Formamidinium-containing metal-halide: an alternative material for near-IR absorption perovskite solar cells. J Phys Chem C 118(30):16458–16462
28. Karlin KD (2012) Progress in inorganic chemistry, vol 57
29. Asghar MI, Zhang J, Wang H, Lund PD (2017) Device stability of perovskite solar cells—a review. Renew Sustain Energy Rev 77(February):131–146
30. Song Z, Watthage SC, Phillips AB, Heben MJ (2016) Pathways toward high-performance perovskite solar cells: review of recent advances in organo-metal halide perovskites for photovoltaic applications. J Photon Energy 6(2):22001
31. Liu M, Johnston MB, Snaith HJ (2013) Efficient planar heterojunction perovskite solar cells by vapour deposition. Nature 501(7467):395–398
32. Burschka J et al (2013) Sequential deposition as a route to high-performance perovskite-sensitized solar cells. Nature 499(7458):316–319
33. Xiao Z et al (2014) Efficient, high yield perovskite photovoltaic devices grown by interdiffusion of solution-processed precursor stacking layers. Energy Environ Sci 7(8):2619
34. Wang Y, Luo J, Nie R, Deng X (2016) Planar perovskite solar cells using $CH_3NH_3PbI_3$ films: a simple process suitable for large-scale production. Energy Technol 4(4):473–478
35. Sutherland BR et al (2015) Perovskite thin films via atomic layer deposition. Adv Mater 27 (1):53–58
36. Zheng J et al (2017) Spin-coating free fabrication for highly efficient perovskite solar cells. Sol Energy Mater Sol Cells 168(August):165–171
37. Fakharuddin A, Jose R, Brown TM, Fabregat-Santiago F, Bisquert J (2014) A perspective on the production of dye-sensitized solar modules. Energy Environ Sci 7(12):3952–3981
38. Han Y et al (2015) Degradation observations of encapsulated planar $CH_3NH_3PbI_3$ perovskite solar cells at high temperatures and humidity. J Mater Chem A 3(15):8139–8147
39. Li X et al (2015) Outdoor performance and stability under elevated temperatures and long-term light soaking of triple-layer mesoporous perovskite photovoltaics. Energy Technol 3(6):551–555
40. Krebs FC (2009) Polymer solar cell modules prepared using roll-to-roll methods: knife-over-edge coating, slot-die coating and screen printing. Sol Energy Mater Sol Cells 93(4):465–475
41. Galagan Y, Coenen EWC, Verhees WJH, Andriessen R (2016) Towards the scaling up of perovskite solar cells and modules. J Mater Chem A 4(15):5700–5705
42. Razza S, Castro-Hermosa S, Di Carlo A, Brown TM (2016) Research update: large-area deposition, coating, printing, and processing techniques for the upscaling of perovskite solar cell technology. APL Mater 4(9)
43. Im JH, Kim HS, Park NG (2014) Morphology-photovoltaic property correlation in perovskite solar cells: one-step versus two-step deposition of CH3NH3PbI 3. APL Mater 2(8)
44. Jeon NJ, Noh JH, Kim YC, Yang WS, Ryu S, Il Seok S (2014) Solvent engineering for high-performance inorganic–organic hybrid perovskite solar cells. Nat Mater 13(9):897–903
45. Oku T, Matsumoto T, Suzuki A, Suzuki K (2015) Fabrication and characterization of a perovskite-type solar cell with a substrate size of 70 mm. Coatings 5(4):646–655
46. Heo JH, Song DH, Im SH (2014) Planar CH 3 NH 3 PbBr 3 hybrid solar cells with 10.4% power conversion efficiency, fabricated by controlled crystallization in the spin-coating process, pp 8179–8183
47. Yang M, Zhou Y, Zeng Y, Jiang CS, Padture NP, Zhu K (2015) Square-centimeter solution-processed planar CH3NH3PbI3 perovskite solar cells with efficiency exceeding 15%. Adv Mater 27(41):6363–6370
48. Chen W et al (2015) Efficient and stable large-area perovskite solar cells with inorganic charge extraction layers. Science (80) 350(6263):944–948

49. Chang C, Huang Y, Tsao C, Su W (2016) Formation mechanism and control of perovskite films from solution to crystalline phase studied by in situ synchrotron scattering
50. Di Giacomo F, Fakharuddin A, Jose R, Brown TM (2016) Progress, challenges and perspectives in flexible perovskite solar cells. Energy Environ Sci 9(10):3007–3035
51. Deng Y, Peng E, Shao Y, Xiao Z, Dong Q, Huang J (2015) Scalable fabrication of efficient organolead trihalide perovskite solar cells with doctor-bladed active layers. Energy Environ Sci 8(5):1544–1550
52. Krebs FC (2009) Fabrication and processing of polymer solar cells: a review of printing and coating techniques. Sol Energy Mater Sol Cells 93(4):394–412
53. Galagan Y et al (2015) Roll-to-roll slot-die coated organic photovoltaic (OPV) modules with high geometrical fill factors. Energy Technol 3(8):834–842
54. Barrows AT, Pearson AJ, Kwak CK, Dunbar ADF, Buckley AR, Lidzey DG (2014) Efficient planar heterojunction mixed-halide perovskite solar cells deposited via spray-deposition. Energy Environ Sci 7(9):2944
55. Tait JG et al (2016) Rapid composition screening for perovskite photovoltaics via concurrently pumped ultrasonic spray coating. J Mater Chem A 4(10):3792–3797
56. Li S-G et al (2015) Inkjet printing of CH_3 NH_3 PbI_3 on a mesoscopic TiO_2 film for highly efficient perovskite solar cells. J Mater Chem A 3(17):9092–9097
57. Hwang K et al (2015) Toward large scale roll-to-roll production of fully printed perovskite solar cells. Adv Mater 27(7):1241–1247
58. Schmidt TM, Larsen-Olsen TT, Carlé JE, Angmo D, Krebs FC (2015) Upscaling of perovskite solar cells: fully ambient roll processing of flexible perovskite solar cells with printed back electrodes. Adv Energy Mater 5(15):1–9
59. Habibi M, Rahimzadeh A, Bennouna I, Eslamian M (2017) Defect-free large-area (25 cm2) light absorbing perovskite thin films made by spray coating. Coatings 7(3):42
60. Mohamad DK, Griffin J, Bracher C, Barrows AT, Lidzey DG (2016) Spray-cast multilayer organometal perovskite solar cells fabricated in air. Adv Energy Mater 6(22):1–7
61. Ramesh M, Boopathi KM, Huang TY, Huang YC, Tsao CS, Chu CW (2015) Using an airbrush pen for layer-by-layer growth of continuous perovskite thin films for hybrid solar cells. ACS Appl Mater Interfaces 7(4):2359–2366
62. Chandrasekhar PS, Kumar N, Swami SK, Dutta V, Komarala VK (2016) Fabrication of perovskite films using an electrostatic assisted spray technique: the effect of the electric field on morphology, crystallinity and solar cell performance. Nanoscale 8(12):6792–6800
63. Habibi M (2017) Optimization of spray coating for the fabrication of sequentially deposited planar perovskite solar cells, vol 7, no 2
64. Shen P-S, Chiang Y-H, Li M-H, Guo T-F, Chen P (2016) Research update: hybrid organic-inorganic perovskite (HOIP) thin films and solar cells by vapor phase reaction. APL Mater. 4(9):91509
65. Das S et al (2015) High-performance flexible perovskite solar cells by using a combination of ultrasonic spray-coating and low thermal budget photonic curing. ACS Photon 2(6):680–686
66. Zheng J et al (2017) Spin-coating free fabrication for highly efficient perovskite solar cells. Sol Energy Mater Sol Cells 168(February):165–171
67. Zhang M, Yu H, Yun J-H, Lyu M, Wang Q, Wang L (2015) Facile preparation of smooth perovskite films for efficient meso/planar hybrid structured perovskite solar cells. Chem Commun 51(49):10038–10041
68. Zheng J et al (2017) Solar energy materials and solar cells spin-coating free fabrication for highly efficient perovskite solar cells, vol 168, pp 165–171
69. Razza S et al (2015) Perovskite solar cells and large area modules (100 cm2) based on an air flow-assisted PbI2 blade coating deposition process. J Power Sources 277(2015):286–291
70. Yang Z, Chueh CC, Zuo F, Kim JH, Liang PW, Jen AKY (2015) High-performance fully printable perovskite solar cells via blade-coating technique under the ambient condition. Adv Energy Mater 5(13):1–6

71. Kim JH, Williams ST, Cho N, Chueh CC, Jen AKY (2015) Enhanced environmental stability of planar heterojunction perovskite solar cells based on blade-coating. Adv Energy Mater 5 (4):2–7
72. Qin T et al (2017) Amorphous hole-transporting layer in slot-die coated perovskite solar cells. Nano Energy 31:210–217
73. Cai L, Liang L, Wu J, Ding B, Gao L, Fan B (2017) Large area perovskite solar cell module. J Semicond 38(1):14006
74. Lee JW, Na SI, Kim SS (2017) Efficient spin-coating-free planar heterojunction perovskite solar cells fabricated with successive brush-painting. J Power Sources 339:33–40
75. Kim S, Na S, Kang S, Kim D (2010) Solar energy materials & solar cells annealing-free fabrication of P3HT: PCBM solar cells via simple brush painting. Sol Energy Mater Sol Cells 94(2):171–175
76. Kim SS, Na SI, Jo J, Tae G, Kim DY (2007) Efficient polymer solar cells fabricated by simple brush painting. Adv Mater 19(24):4410–4415
77. Chen H, Wei Z, Zheng X, Yang S (2015) A scalable electrodeposition route to the low-cost, versatile and controllable fabrication of perovskite solar cells. Nano Energy 15:216–226
78. Huang J et al (2015) Direct conversion of CH3NH3PbI3 from electrodeposited PbO for highly efficient planar perovskite solar cells. Sci Rep 5(1):15889
79. Su T-S, Hsieh T-Y, Hong C-Y, Wei T-C (2015) Electrodeposited ultrathin TiO2 blocking layers for efficient perovskite solar cells. Sci Rep 5(1):16098
80. Koza JA, Hill JC, Demster AC, Switzer JA (2016) Epitaxial electrodeposition of methylammonium lead iodide perovskites. Chem Mater 28(1):399–405
81. Singh M, Haverinen HM, Dhagat P, Jabbour GE (2010) Inkjet printing-process and its applications. Adv Mater 22(6):673–685
82. Jiang Z, Bag M, Renna L, Jeong SP, Rotello V, Venkataraman D (2016) Aqueous-processed perovskite solar cells based on reactive inkjet printing. Hal, p hal-01386295
83. Tavakoli MM et al (2015) Fabrication of efficient planar perovskite solar cells using a one-step chemical vapor deposition method. Sci Rep 5(1):14083
84. Chen Q et al (2014) Planar heterojunction perovskite solar cells via vapor-assisted solution process. J Am Chem Soc 136(2):622–625
85. Chen CW, Kang HW, Hsiao SY, PF Yang, Chiang KM, Lin HW (2014) Efficient and uniform planar-type perovskite solar cells by simple sequential vacuum deposition. Adv Mater 6647–6652
86. Li M-H, Shen P-S, Chen J-S, Chiang Y-H, Chen P, Guo T-F (2016) Low-pressure hybrid chemical vapor deposition for efficient perovskite solar cells and module. 2016 23rd international workshop on act flatpanel displays devices, pp 256–257
87. Shen PS, Chen JS, Chiang YH, Li MH, Guo TF, Chen P (2016) Low-pressure hybrid chemical vapor growth for efficient perovskite solar cells and large-area module. Adv Mater Interfaces 3(8):1–8
88. Fan P et al (2016) High-performance perovskite CH3NH3PbI3 thin films for solar cells prepared by single-source physical vapour deposition. Sci Rep 6(1):29910
89. Ono LK, Leyden MR, Wang S, Qi Y (2016) Organometal halide perovskite thin films and solar cells by vapor deposition. J Mater Chem A 4(18):6693–6713
90. Luo P et al (2015) Chlorine-conducted defect repairment and seed crystal-mediated vapor growth process for controllable preparation of efficient and stable perovskite solar cells. J Mater Chem A 3(45):22949–22959
91. Luo P, Liu Z, Xia W, Yuan C, Cheng J, Lu Y (2015) A simple in situ tubular chemical vapor deposition processing of large-scale efficient perovskite solar cells and the research on their novel roll-over phenomenon in J-V curves. J Mater Chem A 3(23):12443–12451
92. Luo P, Liu Z, Xia W, Yuan C, Cheng J, Lu Y (2015) Uniform, stable, and efficient planar-heterojunction perovskite solar cells by facile low-pressure chemical vapor deposition under fully open-air conditions. ACS Appl Mater Interfaces 7(4):2708–2714

93. Zhou Z et al (2016) Efficient planar perovskite solar cells prepared via a low-pressure vapor-assisted solution process with fullerene/TiO $_2$ as an electron collection bilayer. RSC Adv 6(82):78585–78594
94. Liu C, Fan J, Zhang X, Shen Y, Yang L, Mai Y (2015) Hysteretic behavior upon light soaking in perovskite solar cells prepared via modified vapor-assisted solution process. ACS Appl Mater Interfaces 7(17):9066–9071
95. Sedighi R, Tajabadi F, Shahbazi S, Gholipour S, Taghavinia N (2016) Mixed-halide CH3NH3PbI3—xXx (X = Cl, Br, I) perovskites: vapor-assisted solution deposition and application as solar cell absorbers. ChemPhysChem 2382–2388
96. Peng Y, Jing G, Cui T (2015) A hybrid physical–chemical deposition process at ultra-low temperatures for high-performance perovskite solar cells. J. Mater Chem A 3(23):12436–12442
97. Sheng R et al (2015) Methylammonium lead bromide perovskite-based solar cells by vapor-assisted deposition. J Phys Chem C 119(7):3545–3549
98. Du T, Wang N, Chen H, Lin H, He H (2015) Comparative study of vapor- and solution-crystallized perovskite for planar heterojunction solar cells. ACS Appl Mater Interfaces 7(5):3382–3388
99. Yang Z et al (2017) Research progress on large-area perovskite thinfilms and solar modules. J Materiomics 1–14

The manufacturer's authorised representative in the EU is Springer
Nature Customer Service Centre GmbH, Europaplatz 3, 69115 Heidelberg,
Germany. If you have any concerns regarding our products, please
contact ProductSafety@springernature.com

Printed and bound by CPI Group (UK) Ltd, Croydon, CR0 4YY
30/09/2026
02225563-0011